12th Conference on Theoretical Nuclear Physics in Italy 2008

Journal of Physics: Conference Series Volume 168

Cortona, Italy
8-10 October 2008

ISBN: 978-1-61738-315-1

Printed from e-media with permission by:

Curran Associates, Inc.
57 Morehouse Lane
Red Hook, NY 12571

Some format issues inherent in the e-media version may also appear in this print version.

Printed by Curran Associates, Inc. (2010)

Institute of Physics
Dirac House, Temple Back
Bristol BS1 6BE UK

Phone: 44 1 17 929 7481
Fax: 44 1 17 920 0979

techtracking@iop.org

Additional copies of this publication are available from:

Curran Associates, Inc.
57 Morehouse Lane
Red Hook, NY 12571 USA
Phone: 845-758-0400
Fax: 845-758-2634
Email: curran@proceedings.com
Web: www.proceedings.com

TABLE OF CONTENTS

FEW-NUCLEON SYSTEMS

NUCLEAR MATTER AND NUCLEAR DYNAMICS

NUCLEAR STRUCTURE

NUCLEAR PHYSICS WITH ELECTROWEAK PROBES

STRUCTURE OF HADRONS AND HADRONIC MATTER

SPECIAL TOPICS

Journal of Physics: Conference Series **168** (2009) 011001 doi:10.1088/1742-6596/168/1/011001

Preface

These Proceedings contain the invited and contributed papers presented at the 12th Conference on Theoretical Nuclear Physics in Italy held in Cortona, Italy, from October 8 to October 10, 2008. As usual, the meeting was held at *il Palazzone*, a 16th century castle owned by the Scuola Normale Superiore di Pisa. The aim of this biennal conference is to bring together Italian theorists working in various fields of Nuclear Physics to discuss their latest results and confront their points of view in a lively and informal way. This offers the opportunity to promote collaborations between different groups.

There were about 50 participants at the conference, coming from 14 Italian Universities (Cagliari, Catania, Ferrara, Firenze, Genova, Lecce, Milano, Napoli, Padova, Pavia, Pisa, Roma, Trento, Trieste). The program of the conference, prepared by the Organizing Committee (Ignazio Bombaci, Aldo Covello, Laura Elisa Marcucci and Sergio Rosati) focused on six main topics: Few-Nucleon Systems, Nuclear Matter and Nuclear Dynamics, Nuclear Astrophysics, Structure of Hadrons and Hadronic Matter, Nuclear Structure, Nuclear Physics with Electroweak Probes. Winfried Leidemann, Maria Colonna, Marcello Lissia, Elena Santopinto, Silvia Lenzi and Omar Benhar took the burden of giving general talks on these topics and reviewing the research activities of the various Italian groups. In addition, 19 contributed papers were presented, most of them by young participants. In the last session of the Conference there were two invited talks related to experimental activities of great current interest. Gianfranco Prete from the Laboratori Nazionali di Legnaro spoke about the Italian radioactive ion beam facility SPES and the status of the European project EURISOL, while Nicola Colonna from the INFN, Bari, gave an overview of the perspectives of development of fourth-generation nuclear reactors.

We would like to thank the authors of the general reports for their hard work in reviewing the main achievements in the various fields as well as our experimental colleagues for having kindly agreed to talk to an audience consisting only of theoretical physicists.

I Bombaci, A Covello, L E Marcucci, S Rosati

Journal of Physics: Conference Series **168** (2009) 011002 doi:10.1088/1742-6596/168/1/011002

List of participants

Armani Paolo (Università di Trento)
Benhar Omar (INFN Roma)
Bombaci Ignazio (Università di Pisa)
Bonanno Luca (Università di Ferrara)
Catara Francesco (Università di Catania)
Cò Giampaolo (Università di Lecce)
Colonna Maria (Laboratori Nazionali del Sud, INFN Catania)
Colonna Nicola (INFN Bari)
Conti Francesco (Università di Pavia)
Coraggio Luigi (INFN Napoli)
Covello Aldo (Università di Napoli)
Cristoforetti Marco (Technische Universität München, Germania)
Cuofano Carmine (Università di Ferrara)
Di Toro Massimo (Università di Catania)
Drago Alessandro (Università di Ferrara)
Faccioli Pietro (Università di Trento)
Farina Nicola (INFN Roma)
Finelli Paolo (Università di Bologna)
Fiorentini Giovanni (Università di Ferrara)
Fortunato Lorenzo (Università di Padova)
Gambacurta Danilo (Università di Catania)
Gandolfi Stefano (Università di Trento)
Gargano Angela (INFN Napoli)
Giannini Mauro (Università di Genova)
Girlanda Luca (INFN Pisa)
Giusti Carlotta (INFN Pavia)
Illarionov Alexei (SISSA Trieste)
Itaco Nunzio (Università di Napoli)
Kievsky Alejandro (INFN Pisa)
Lanza Edoardo (INFN Catania)
Leidemann Winfried (Università di Trento)
Lenzi Silvia (Università di Padova)
Lipparini Enrico (Università di Trento)
Lissia Marcello (Università di Cagliari)
Lo Iudice Nicola (Università di Napoli)
Maieron Chiara (Università di Lecce)
Marcucci Laura Elisa (Università di Pisa)
Matera Francesco (Università di Firenze)
Millo Raffaele (Università di Trento)
Orlandini Giuseppina (Università di Trento)
Pacati Franco (Università di Pavia)
Pastore Alessandro (Univeristy of Jyväskylä, Finlandia)
Pederiva Francesco (Università di Trento)
Pisent Gualtiero (Università di Padova)
Prete Gianfranco (INFN Laboratori Nazionali di Legnaro)
Quarati Piero (Politecnico di Torino)
Rosati Sergio (Università di Pisa)

Journal of Physics: Conference Series **168** (2009) 011002 doi:10.1088/1742-6596/168/1/011002

Salmè Giovanni (INFN Roma)
Santopinto Elena (INFN Genova)
Traini Marco (Università di Trento)
Vigezzi Enrico (INFN Milano)
Vitturi Andrea (Università di Padova)
Viviani Michele (INFN Pisa)

Few-nucleon physics

Winfried Leidemann
Dipartimento di Fisica, Università di Trento and INFN (Gruppo Collegato di Trento), via Sommarive 14, I-38100 Trento, Italy
E-mail: leideman@science.unitn.it

Abstract. Recent progress in few-nucleon physics achieved by Italian research groups is described. Results for relativistic effects in elastic three-nucleon form factors (Rome group), hadronic and electromagnetic observables in three- and four-nucleon systems with special emphasis on three-nucleon force effects (Pisa and Trento groups), and spectra of nuclei with more than 4 nucleons (Padua group) are presented.

1. Introduction

The field of few-nucleon physics constitutes an essential part of nuclear physics. It consists of *ab initio* calculations of hadronic and electromagnetic observables with a given well-defined Hamiltonian of an A-nucleon system, where exact solutions of the quantum mechanical A-body system are requested for bound and/or scattering states. In particular one can consider bound-state spectra and observables from hadronic reactions like scattering lengths, phase shifts, total and differential cross sections. A polarization of beam, target, and/or final particles leads to many other hadronic polarization observables. In addition reactions with electroweak probes (photons, electrons, neutrinos) without and with polarization degrees of freedom give further information on the nuclear dynamics and electroweak structure of few-nucleon systems. For the latter one has to consider a further important aspect, namely the consistency of the electroweak currents with the nuclear Hamiltonian.

Few-nucleon physics has many objectives: development of nuclear force models (nucleon-nucleon (NN) and three-nucleon (3N) forces) and check whether even higher many-body forces are necessary, test of nuclear force models via two- and three-nucleon hadronic observables with possible feedback on force model, further test of nuclear dynamics in electromagnetic reactions with two- and three-nucleon systems; the study of the electroweak structure of few-nucleon systems; the investigation of the importance of relativistic effects; the extension of nuclear force models beyond pion threshold; the determination of neutron electroweak properties in two- and three-nucleon electroweak observables; the determination of observables with astrophysical relevance.

Here follows a very brief summary on the state of the art for some selected aspects of few-nucleon physics. NN potential models have a long history (see e.g. [1]). Today we have the so called modern realistic NN potentials like AV18 [2], various Nijmegen models [3], and CD Bonn [4] as well as chiral models [5, 6, 7]. The potentials [2, 3, 4, 5] all lead to a perfect or almost perfect description of various thousands of NN scattering data for energies up to pion threshold, while the chiral potential [6, 7] is constructed more in the original spirit of effective field theory where open parameters are fitted to low-energy data only. The history of 3N forces started with the inclusion of the Fujita-Miyazawa term (consideration of Δ degrees of freedom). Today, also for 3N forces there are various conventional and chiral models, for example [8, 9, 10, 11, 12].

In the last decade the calculation of bound states with realistic forces has made great progress. Nowadays, such calculations are standard for $A < 4$. For $A = 4$ there exists a benchmark calculation [13], where seven different theory groups have participated by applying seven different theoretical methods. Calculations for p-shell nuclei bound-states with Green's function Monte Carlo (GFMC) and no core shell model (NCSM) techniques have reached $A = 10$ and even beyond (see e.g. [14, 15]). There is hope that with additional methods like coupled cluster [16] and auxiliary field diffusion Monte Carlo [17] even considerably larger systems can be treated.

The calculation of continuum state wave functions has also made quite an improvement, but rigorous solutions with all possible channels exist only up to $A = 3$ (see e.g. [18]). There also exist exact calculations for the $A = 4$ [19, 20, 21] and the $A = 5$ [22] cases, however, for energies below the three-body break-up threshold. On the other hand with the Lorentz integral transform (LIT) approach [23, 24] there exists an alternative method for the calculation of continuum reaction observables, where an explicit calculation of continuum state wave functions is avoided. In fact the LIT approach reduces a scattering state problem to a bound-state like problem, which then can be solved with typical bound-state techniques. Up to now the LIT method has been applied to the calculation of inclusive electromagnetic observables with realistic nuclear forces for three- [25, 26, 27, 28, 29] and four-body systems [30, 31] and with semi-realistic forces for $A = 6, 7$ [32, 33]; for $A = 4$ also exclusive reactions have been considered (see e.g. [34]).

Concerning the electromagnetic interaction with few-nucleon systems it should be pointed out that review articles have been published for the two- [35] and three-nucleon systems [36] quite recently.

Here we will not consider few-nucleon physics beyond pion threshold, but it has to be mentioned that there are two longer recent reports on this topic with authors from Padua [37] and Trento [38].

In the following the work of the Italian few-nucleon physics community of the last two years is reviewed. To this end the work of the Pisa, Padua, Rome and Trento few-nucleon groups is considered. It should be mentioned that the term *group* has to be interpreted in a wider sense, because there are also external collaborators. Since the author of this article is from Trento he would like to mention long-term external collaborators of the Trento group explicitly: Sonia Bacca (TRIUMF), Nir Barnea (Hebrew University), Victor Efros (Kurchatov Institute Moscow) and Edward Tomusiak (University of Victoria).

The paper is organized as follows. In section 2 the basic ingredients of the hyperspherical harmonic (HH) expansion technique, heavily used by Pisa and Trento groups, are briefly described. In section 3 bound and low-energy scattering states of three- and four-body nuclei, calculated with realistic nuclear forces, are discussed. Reactions with electroweak probes are illustrated in section 4, where (i) relativistic effects on the elastic (e, e') form factors of the trinucleons and (ii) LIT calculations for inelastic inclusive reactions with three- and four-body systems are considered. Finally, in section 5, results for weakly bound nuclei with the multi channel scattering approach (MCAS) are presented for $A > 4$.

2. Bound and continuum states with the hyperspherical harmonics expansion

In this section the HH expansion approach is briefly reviewed. More details and further references can be found in two detailed recent papers [19, 24].

The first step of an HH expansion consists of rewriting the $3(A-1)$ internal coordinates of an A-nucleon system, the Jacobi coordinates $\vec{\xi}_i = (\xi_i, \theta_i, \phi_i)$, as follows. One keeps the $2(A-1)$ angles θ_i and ϕ_i and introduces the hyperradius ρ ($\rho^2 = \sum_i \xi_i^2$) and additional $(A-2)$ angles α_i. This leads to a subdivision of the space into a hyperradial ρ and a hyperspherical Ω. As an example the defining equations for the additional angle α_1 in the $A = 3$ case are given: $\xi_1 = \rho\, cos\alpha_1$, $\xi_2 = \rho\, sin\alpha_1$. Next one considers the internal kinetic energy of the A-body system,

$$T = \frac{-1}{2m} \sum_{i=1}^{A-1} \nabla^2_{\xi_i} , \tag{1}$$

where m is a reference mass. With hyperspherical coordinates one has

$$\sum_{i=1}^{A} \nabla^2_{\xi_i} = \frac{\partial^2}{\partial \rho^2} + \frac{3A-4}{\rho}\frac{\partial}{\partial \rho} + \frac{\hat{K}^2}{\rho^2} , \tag{2}$$

where $\hat{K}$ is the generalized angular momentum operator, which has no ρ dependence and is given in terms of the hyperspherical coordinates θ_i, ϕ_i, and α_i. The hyperspherical harmonic functions $\mathcal{Y}_{[K]}$ are eigenfunctions of $\hat{K}^2$ with eigenvalues $K(K + 3A - 5)$ (note that $[K]$ symbolizes a set of quantum numbers). For $A = 2$ one has $\mathcal{Y}_{[K]} \to Y_{lm}$ and $K \to l$, where Y_{lm} are the usual spherical harmonics.

An HH expansion has the following form:

$$\sum_{[K]} \mathcal{Y}_{[K]}(\Omega)\, R_{[K]}(\rho) \tag{3}$$

with the hyperradial part

$$R_{[K]}(\rho) = \sum_j c_{[K],j}\, L_j(\rho)\, exp(-b\rho) , \tag{4}$$

where L_j are the Laguerre polynomials. The expansion is carried out up to a maximal K-value $K_{\rm max}$. The HH functions need to have proper symmetries in order to form together with the nuclear spin and isospin functions an antisymmetric nuclear wave function.

A delicate problem is the convergence of the expansion. Due to the NN short range repulsion one has usually a very slow convergence. The Pisa and Trento groups use various methods to accelerate the convergence: (i) introduction of additional two-nucleon correlation functions $f(r_{ij})$ (HH $\to$ CHH) and multiplication of each HH basis state with $\omega = \prod_{i<j} f(r_{ij})$, the method is also used with a simplified correlation factor for one pair only (PHH, Pisa group [19]) and with spin-isospin dependent correlations $f_{ST}(r_{ij})$ (Trento group [24]); (ii) selection of specific channels (Pisa group); (iii) calculation of an effective interaction by unitary transformation to the $K_{\rm max}$-dependent model space (EIHH [39, 40]), similar as is done in the NCSM approach.

In addition the Pisa group performs HH expansions in momentum space [41]. This facilitates the treatment of non-local NN potentials like [4]. In this case one has a hypersphere with hypermomentum $Q = \sum_i \vec{k}_i^2$, where the $\vec{k}_i$ are the conjugate momenta of the Jacobi vectors $\vec{\xi}_i$.

As mentioned above the Trento group uses the LIT method and does not calculate continuum states explicitly. The Pisa group, however, also performs HH expansions of two-fragment scattering states relevant for N+Y $\to$ N+Y reactions, where N is a nucleon and Y a two- or three-body nucleus. The N-Y scattering state with quantum numbers $[J] = (L, S, J, J_z)$ is given by

$$\Psi^{[J]} = \Psi_C^{[J]} + \Psi_A^{[J]} , \tag{5}$$

where $\Psi_C^{[J]}$ describes the region where N-Y are close, whereas $\Psi_A^{[J]}$ takes care of the asymptotic region. For $\Psi_C^{[J]}$ an HH expansion is carried out, whereas $\Psi_A^{[J]}$ is written as

$$\Psi_A^{[LSJJ_z]} = N\Psi_{A,\,{\rm free}}^{[LSJJ_z]} + \sum_{L'S'} R^J_{LS,L'S'} \Psi_{A,\,{\rm irregular}}^{[L'S'JJ_z]} . \tag{6}$$

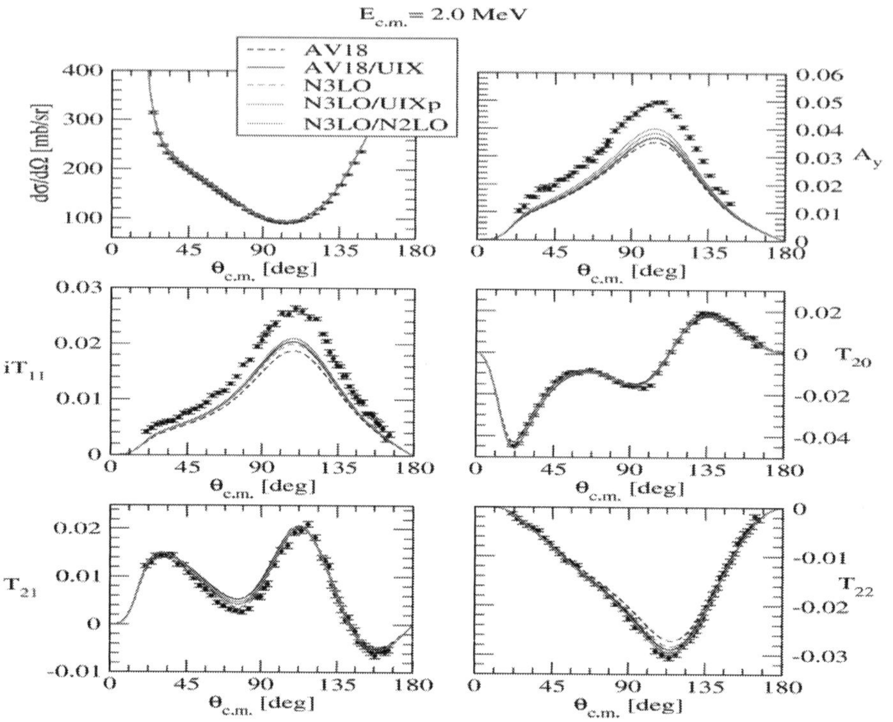

Figure 1. (Color online) Various observables for *pd* elastic scattering at $E_{cm} = 2$ MeV. The curves for the various potential models are defined in the insert. Experimental data from [46].

The matrix R^J is related to the K-matrix which has eigenvalues $tan\delta_{LSJ}$ and where δ_{LSJ} are the phase shifts of N-Y scattering. The matrix elements $R^J_{LS,L'S'}$, the normalization constant N and the expansions coefficients of $\Psi^{[J]}_C$ are determined applying the Kohn variational principle.

3. Three- and four-body bound and low-energy scattering states

In [19] the Pisa group applies the HH expansion technique in order to calculate high-precision results for three- and four-nucleon bound and zero-energy scattering states. There bound-state energies of ^{3}H, ^{3}He, and ^{4}He for semi-realistic and realistic nuclear forces are listed. In case of the zero-energy scattering one finds results for nd/pd doublet and quartet scattering lengths and $n^3\mathrm{H}/p^3\mathrm{He}$ singlet and triplet scattering lengths with various realistic potentials.

Concerning the bound-state energies three realistic NN potentials (AV18 [2], CDBonn [4], N3LO-Idaho [5]) and three 3N forces (UIX[8], TM [10, 19], N2L0 [12]) in the combinations AV18+UIX, CDBonn+TM, and N3LO-Idaho+N2LO have been taken into account. Since the 3N forces are fitted to describe the three-body binding energy it is interesting to see what happens in the four-body case. The results show that for all the considered cases the ^{4}He binding energy is slightly overestimated, namely by 0.06 MeV (N3LO-Idaho+N2LO), 0.16 MeV (AV18+UIX), and 0.6 MeV (CDBonn+TM).

In case of the zero-energy scattering (no results given for CDBonn and CDBonn+TM) the following conclusions are drawn in [19]: (i) *nd* and *pd* quartet scattering lengths are almost model independent, while there is a large effect of the 3N force on the doublet scattering lengths. Once

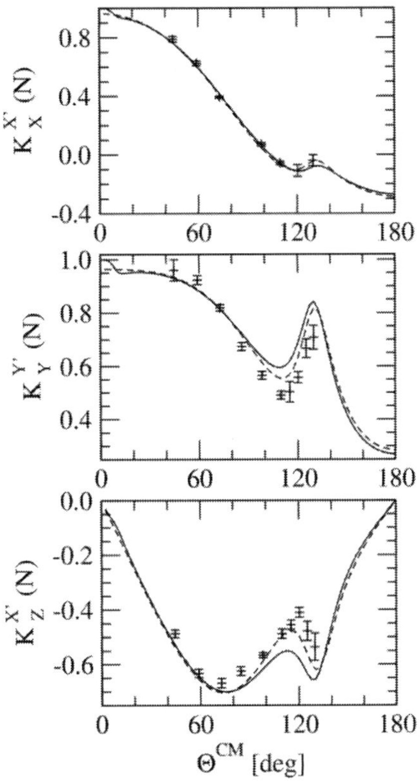

Figure 2. The nucleon to nucleon spin transfer coefficients in Nd elastic scattering at $E_{lab}^N = 22.7$ MeV. The crosses are the pd data from [44, 45]. HH expansion results with AV18 potential without (dashed) and with Coulomb force (solid).

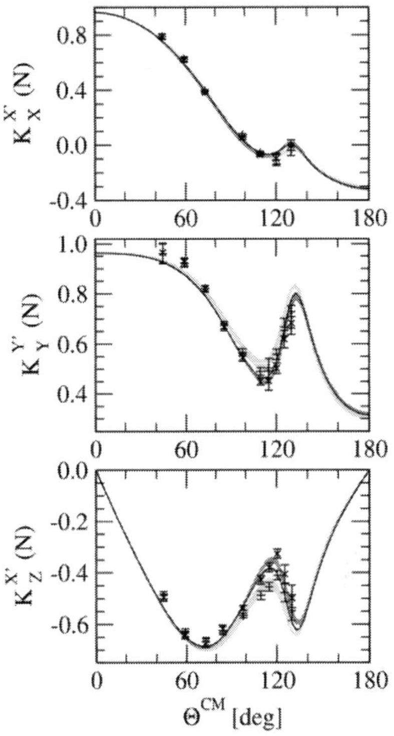

Figure 3. (Color online) As in figure 2, but results from Faddeev calculations (no Coulomb force included) with various potential models: light band (AV18, CDBonn, Nijm I, Nijm II), dark band (additional consideration of TM-3N force), solid line (AV18+UIX). In addition x-es for the pseudo-neutron data (see text).

the 3N force is included also the latter results are rather model independent. Comparison with experiment is made for nd scattering. A very good agreement is found for ${}^4a_{nd}$, while there are small discrepancies in comparison to an experimental high-precision result for ${}^2a_{nd}$. The relatively best agreement with theory is obtained for the N3LO-Idaho+N2LO potential. In case of the 4-body zero-energy scattering the 3N force does not lead to important contributions for the considered isospin T=1 channel. The experimental situation is a little bit less settled than in the three-body case and therefore the comparison with experimental results is not yet very conclusive. Where available, comparisons to results of other calculations for bound-state energies and four-body scattering lengths are made. Quite good and fairly good agreements are obtained for bound-state energies and scattering lengths, respectively.

A more detailed description and interpretation of the results for binding energy and zero-energy scattering can be found in the contribution of A Kievsky in this volume.

In figure 1 results for pd scattering observables at somewhat higher energies ($E_{cm} = 2$ MeV) [42] are shown using as NN and 3N forces AV18 or N3LO-Idaho and UIX or N2LO, respectively. It is evident that the theoretical total cross section shows only negligible potential model effects and that there is an excellent agreement with experiment. Very good agreements are also found for T_{20} and T_{22}. T_{20} is only rather weakly potential model dependent, whereas for T_{22} stronger effects are visible and the 3N force is necessary in order to have a good description of the minimum. Similar potential model effects can be observed for T_{11}, T_{21}, and A_y, but the agreement with experiment is not very good. In particular T_{11} and A_y show rather large discrepancies for the comparison of theory and experiment (well known A_y-puzzle). A more detailed discussion of Nd scattering at low energies is found in the contribution of L E Marcucci in this volume.

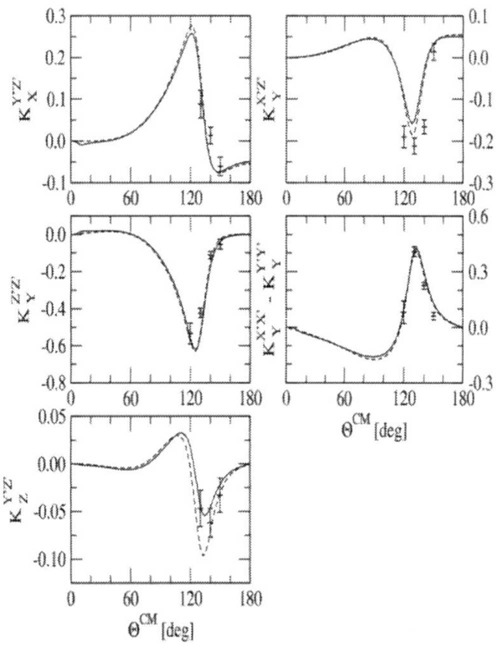

Figure 4. The nucleon to deuteron spin transfer coefficients in Nd elastic scattering at $E_{lab}^{N} = 22.7$ MeV. Notation as in figure 2.

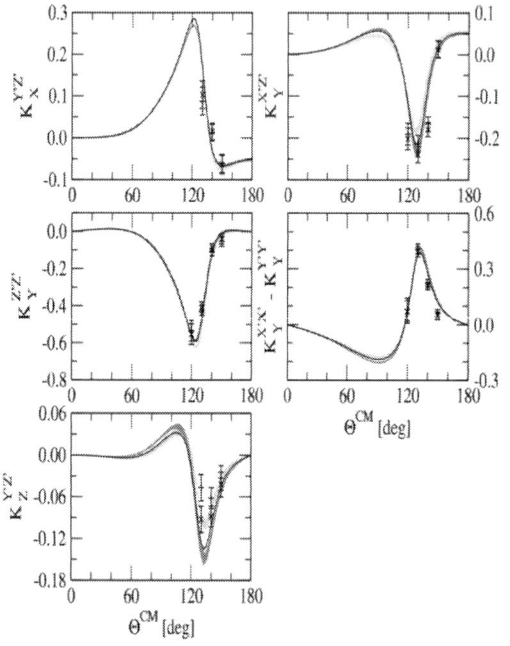

Figure 5. (Color online) The nucleon to deuteron spin transfer coefficients in Nd elastic scattering at $E_{lab}^{N} = 22.7$ MeV. Notation as in figure 3.

In figures 2-5 results for spin transfer coefficients for the reactions $d(\vec{N},\vec{N})d$ and $d(\vec{N},\vec{d})N$ are shown [43] at $E_N^{\rm lab}$ =22.7 MeV. In figures 2 and 4 results from the Pisa group are given for the AV18 potential without (nd scattering) and with (pd scattering) Coulomb force. The Coulomb force effect is quite large for the nucleon to nucleon spin transfer coefficient $K_Z^{X'}$ (figure 2) and for the nucleon to deuteron spin transfer coefficient $K_Z^{Y'Z'}$ (figure 4). The agreement with the experimental data [46] for the $d(\vec{p},\vec{p})d$ and $d(\vec{p},\vec{d})p$ scattering reactions is not sufficiently satisfying even if the Coulomb force is included. This is not surprising since the 3N force has not been considered. In the results of a Faddeev calculation [43] shown in the figures 3 and 5 the 3N force is included, however the Coulomb force is not taken into account. Therefore the proton data of figures 2 and 4 have been corrected for the effect of the Coulomb force resulting in pseudo experimental data for the $d(\vec{n},\vec{n})d$ and $d(\vec{n},\vec{d})n$ reactions. In general this leads to a rather good agreement of theory and pseudo-experiment, however, with the nucleon to deuteron spin transfer coefficient $K_Z^{Y'Z'}$ one finds also one exception. In fact for this case one has already a good agreement with an NN force only, which is then considerably spoiled by the inclusion of a 3N force.

Now a somewhat different argument is discussed, namely the neutron spin rotation in $\vec{n}d$ scattering. Such an effect can only be obtained by parity violating (PV) components in the NN potential. In [47] two models for the PV potential are considered. The so called DHH with π-, ρ and ω exchanges, where weak meson-nucleon couplings are used, and pionless EFT with four-nucleon contact terms. The calculation is performed perturbatively in first order in the PV potentials using the AV18+UIX potential for the n-d wave function. The results show only a rather weak PV model dependence and lead to an effect which is about an order of magnitude larger than in $\vec{n}p$ scattering. Many more details of the calculation are given in the contribution of L Girlanda to this volume.

The references of additional articles on few-nucleon physics of the Pisa group from the last two years, not discussed in this overview, are [48, 49, 50, 51, 52].

4. Reactions with electroweak probes

Perturbation-induced reactions can be divided in inclusive and exclusive processes. In the former case the final state of the particle system undergoing the perturbation is not observed. In the latter case the final state is at least partially observed, like e.g. in an $(e,e'p)$ reaction, where besides the scattered electron also an outgoing proton with energy E_p and scattering angle $\Omega_p = (\theta_p, \Phi_p)$ is detected.

Inclusive cross sections of perturbation-induced reactions have the following general form

$$\frac{d^2\sigma}{d\omega d\Omega_{ext}} = \alpha_{\rm ext} \sum_{i=1}^{M} f_i(\omega, q, \theta_{ext}) R_i(\omega, q)\,, \tag{7}$$

where ω and q and are energy and momentum transfer of the external probe to the particle system, $\Omega_{ext} = (\theta_{ext}, \phi_{ext})$ denotes the scattering angle of the external probe, $\alpha_{\rm ext}$ is a constant characteristic for the external probe, and f_i are kinematic functions. The functions R_i describe the various responses of the particle system to the external probe and thus contain information about the dynamics of the particle system. They are defined as follows

$$R_i(\omega, q) = \sum\!\!\!\!\!\!\int_f |\langle f|\Theta_i|0\rangle|^2 \delta(\omega - (E_f - E_0))\,. \tag{8}$$

Here E_0 and $|0\rangle$ are ground state energy and wave function of the particle system under consideration, E_f and $|f\rangle$ denote final state energy and wave function of the final particle system, and Θ_i is the operator inducing the response function R_i.

In the following only inclusive reactions will be considered and thus no further information about exclusive reactions is given here.

4.1. Trinucleon electromagnetic form factors and the light-front Hamiltonian dynamics

We start the discussion of the electromagnetic probes by considering elastic inclusive responses of the three-nucleon systems, i.e. the trinucleon electromagnetic form factors. With the intention to study relativistic effects they are calculated by the Rome group in continuation of their work on deuteron electromagnetic observables (see e.g. [55]). Their calculation is a manifestly relativistic calculation in the so called light-front (LF) relativistic Hamiltonian dynamics. In this approach one constructs operators under the condition that they fulfill the commutation rules of the Poincaré generators. Two different cases appear: (i) operators not containing the interaction and (ii) operators depending on the mass of the interacting system. It is important to note that the interacting mass operator depends upon intrinsic variables only and has to satisfy

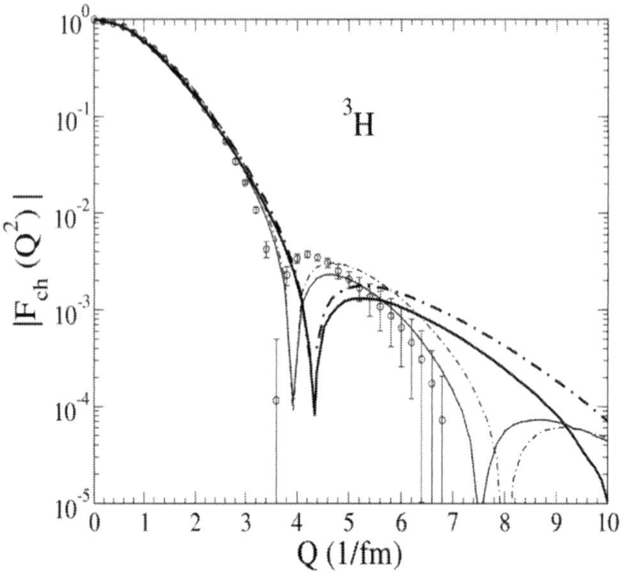

Figure 6. Preliminary results for ^{3}H charge form factor with ^{3}H wave function from Pisa group and with LF nucleon form factors from [53] and with one-body electromagnetic operator only. Thick lines: LF calculations with AV18 NN potential (solid) and additional UIX 3N force (dash-dotted). Thin lines: non relativistic calculation with AV18 (solid) and AV18+UIX (dash-dotted). Data from [54].

the same constraints given by the Galilean group, and thus has the same properties implemented in non relativistic quantum mechanics. Therefore one can exploit realistic wave functions for light nuclei from non relativistic calculations, that depend upon Jacobi coordinates, in order to evaluate matrix elements of a Poincaré covariant current operator [56].

In figures 6-9 preliminary results of the Rome group for the three-nucleon electromagnetic elastic form factors are shown. As had to be expected relativistic effects are small for momentum transfers below 3 fm^{-1}, but become more important beyond the form factor minima. Two-body electromagnetic operators have not yet been considered in the calculation and thus an agreement with experimental data is not yet obtained. Concerning the two-body electromagnetic operators see also G. Salmè's contribution in this volume and [57].

Additional recent publications on theoretical few-nucleon physics of the Rome group are found in [58, 59].

4.2. Inelastic inclusive electromagnetic reactions

As already mentioned above, the Trento group uses the LIT [23] in order to calculate inelastic reactions. The LIT method is motivated by the fact that an explicit calculation of continuum wave functions is avoided. Nonetheless the LIT approach allows the *ab initio* calculation of a reaction cross section, where a many-body continuum is involved in initial and/or final states, but without requiring the knowledge of the generally complicated many-body continuum wave function. In fact, the scattering problem is reduced to a calculation of a localized function with an asymptotic boundary condition similar to a bound-state wave function. Such an approach was already proposed by Efros in 1985, but with the Stieltjes instead of the Lorentz integral transform [60]. However, it has been found that the application of the Stieltjes transform is problematic since it leads to serious inversion problems [61]. Though the LIT can also be applied

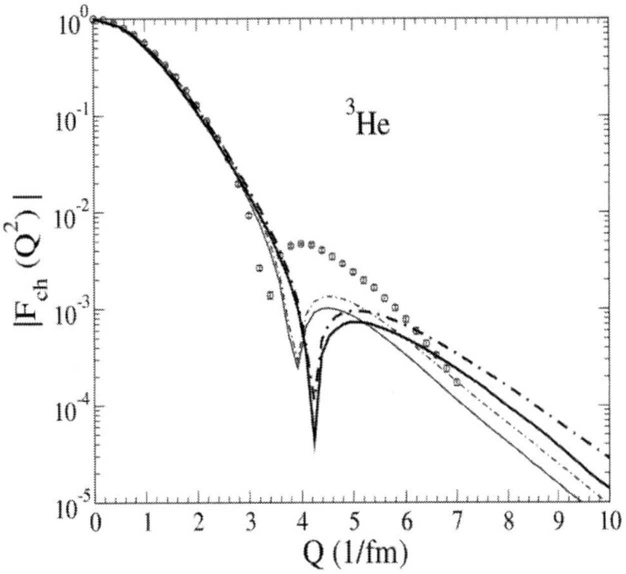

Figure 7. Same as in figure 6 but for ^{3}He charge form factor.

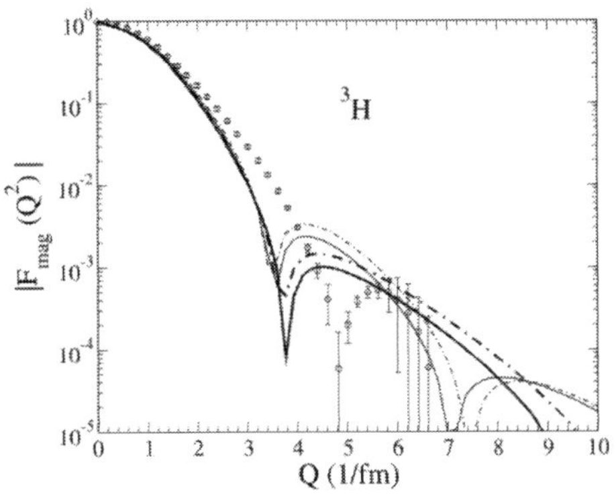

Figure 8. Same as in figure 6 but for ^{3}H magnetic form factor.

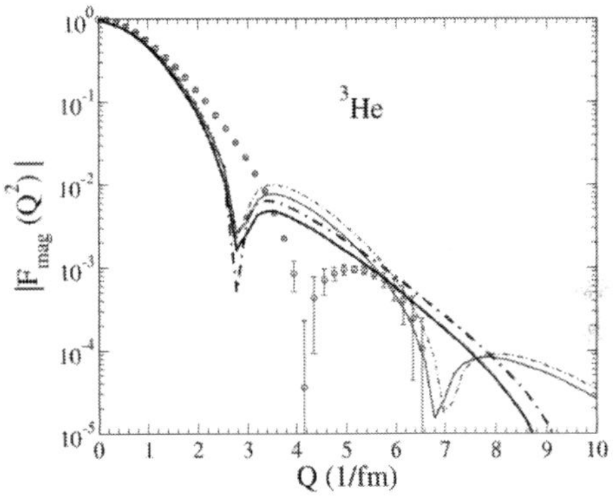

Figure 9. Same as in figure 6 but for ^{3}He magnetic form factor.

to exclusive reactions (see [24]) here only inclusive response functions as defined in Eq. (8) are considered.

For the LIT calculation of R_i of Eq. (8) one proceeds in the following way. First the ground state wave function $|0\rangle$ of the nucleon system under consideration is calculated. Then one solves the equation

$$(H - E_0 - \sigma_R - i\sigma_I)|\tilde{\Psi}_i\rangle = \Theta_i|0\rangle\,, \tag{9}$$

where H is the Hamiltonian of the particle system and $\sigma_{R/I}$ are parameters, whose meaning is explained below. Since the eigenvalues of H have to be real the homogeneous version of Eq. (9) has only the trivial solution $\tilde{\Psi}_i = 0$ and thus Eq. (9) has a unique solution. In addition, due to the asymptotically vanishing ground state wave function $|0\rangle$ also the right-hand-side of Eq. (9) vanishes asymptotically. Therefore, and because of the complex energy, $E_0 - \sigma_R - i\sigma_I$, $\tilde{\Psi}_i$ has a similar asymptotic boundary condition as a bound state. It means that $\tilde{\Psi}_i$ is a so called localized function, i.e. square-integrable with a norm $\langle\tilde{\Psi}_i|\tilde{\Psi}_i\rangle$. This has very important

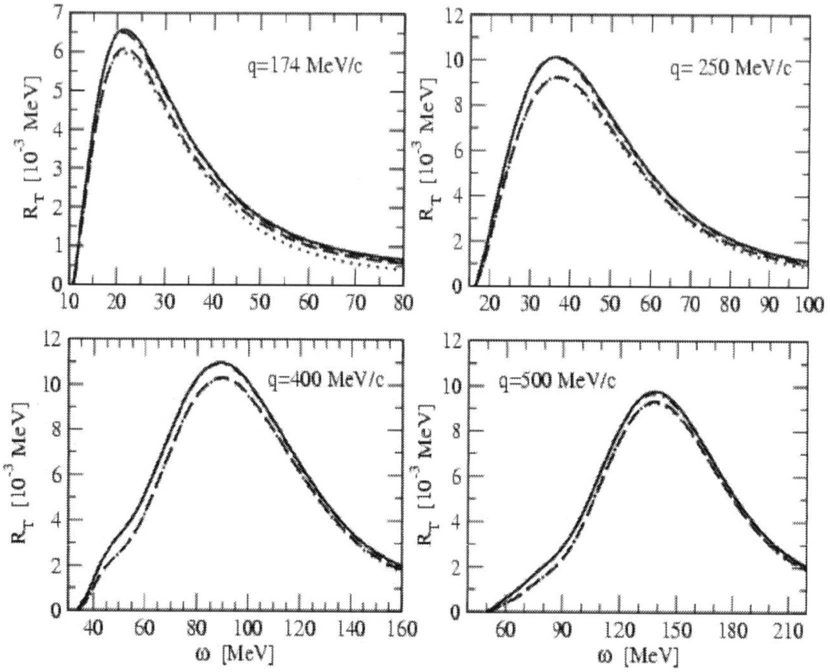

Figure 10. Effects of the various contributions on $R_T(^3\text{He})$ in the quasi-elastic region at q=174 (upper left), 250 (upper right), 400 (lower left), and 500 MeV/c (lower right): one-body (dotted), one-body + implicit MEC via Siegert operator (dashed), one-body + π-MEC + ρ-MEC (dashed dotted), one-body + π-MEC + ρ-MEC + additional MEC via Siegert operator (solid).

consequences: even if one aims at a calculation of a reaction cross section in the continuum, one is not confronted with a scattering state problem any more, in fact one needs to apply only bound-state techniques for the solution of Eq. (9).

The key point of the LIT method consists of the fact that the Lorentz integral transform L_i of the response function R_i,

$$\mathrm{L}_i(\sigma_R, \sigma_I, q) = \int R_i(\omega, q)\, \mathcal{L}(\omega, \sigma_R, \sigma_I)\, d\omega\,, \tag{10}$$

is related to the norm $\langle \tilde{\Psi}_i | \tilde{\Psi}_i \rangle$, which can be obtained from the solution of Eq. (9). In fact one has

$$\mathrm{L}_i(\sigma_R, \sigma_I, q) = \langle \tilde{\Psi}_i(\sigma_R, \sigma_I, q) | \tilde{\Psi}_i(\sigma_R, \sigma_I, q) \rangle \tag{11}$$

(q dependence of L_i and $\tilde{\Psi}_i$ will be dropped in the following). In Eq. (10) $\mathcal{L}$ is a Lorentzian centered at σ_R with a width $\Gamma = 2\sigma_I$:

$$\mathcal{L}(\omega, \sigma_R, \sigma_I) = \frac{1}{(\omega - \sigma_R)^2 + \sigma_I^2}\,. \tag{12}$$

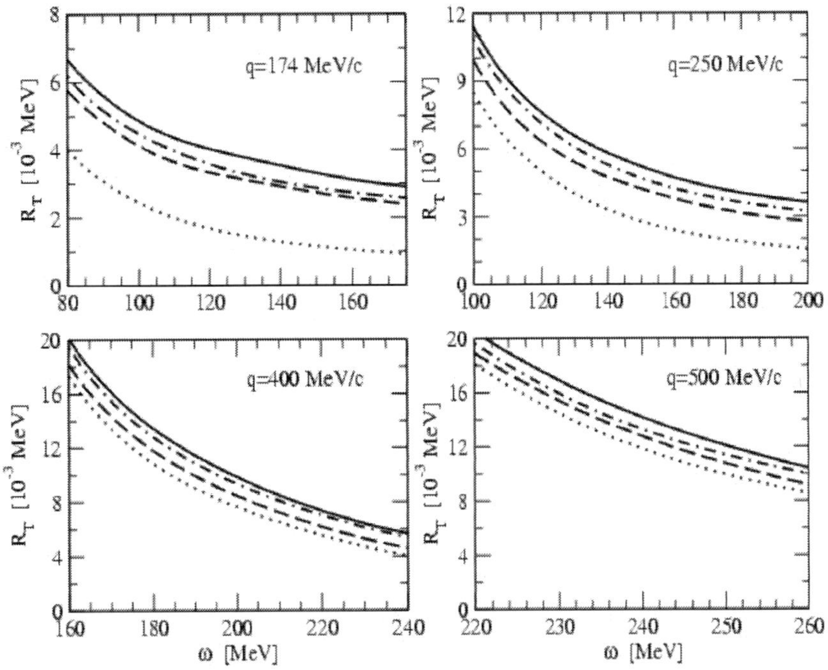

Figure 11. As figure 10 but for the high-energy region.

Now also the meaning of the parameters $\sigma_{R/I}$ becomes evident: σ_I represents a kind of energy resolution, while with σ_R a given energy range can be scanned.

With the above equations the principle idea of the LIT method can be explained: one solves the LIT equation (9) for many values of σ_R and a fixed σ_I and calculates $\mathrm{L}_i(\sigma_R, \sigma_I = \mathrm{const})$, and then one inverts the transform in order to determine $R_i(\omega, q)$. The inversion of the LIT has to be made with some care, since, in principle, it constitutes an ill-posed problem. In fact the LIT method has to be understood as an approach with a controlled resolution. The resolution is governed by σ_I, i.e. structures with a width considerably smaller than σ_I cannot be easily resolved. The LIT inversion is ill-posed, since one can always add an oscillating response with a wave length much smaller than σ_I, which will then lead to an almost identical result for the transform. Therefore the LIT inversion is made with a regularization scheme (for details see [62, 63]). In case that a single resonance with a width much smaller than σ_I is present this will be realized in the inversion process even if it cannot be completely resolved (see discussion of Coulomb monopole resonance in [64, 65]). If one wants to resolve such structures precisely one has to further reduce σ_I (see [67]). Note that one has to make more and more accurate calculations for a smaller and smaller σ_I in order to obtain a sufficiently converged LIT result. Thus, in principle, one can enhance the LIT resolution continuously, but one has to be aware that there can always be structures with an even smaller width that cannot be resolved. This is the cause of the ill-posed problem and also the reason why the LIT approach is a method with a controlled resolution. That the LIT approach works in an excellent way

Journal of Physics: Conference Series **168** (2009) 012001 doi:10.1088/1742-6596/168/1/012001

has been shown in various benchmark calculations for two- and three-nucleon electromagnetic reactions [23, 24, 26, 27, 66, 67].

The LIT results illustrated in the following are calculated using CHH and EIHH expansions (see section 2) for three- and four-nucleon systems, respectively.

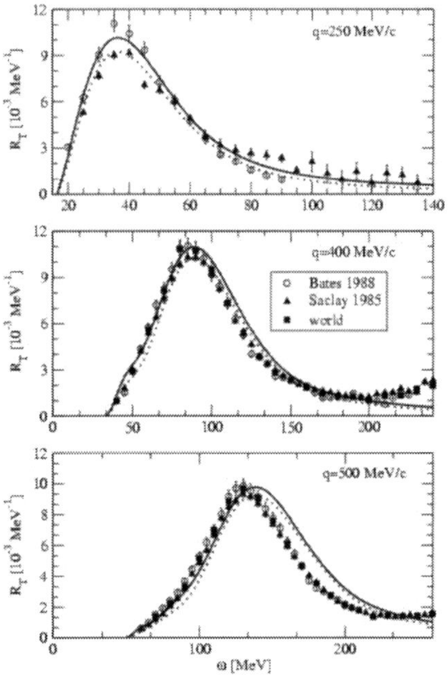

Figure 12. $R_T(^3\mathrm{He})$ at q=250, 400, 500 MeV/c in upper, middle, lower panel. Theoretical R_T with one-body (dotted) and one-body + π-MEC + ρ-MEC + additional MEC via Siegert operator (solid). Data from [68] (triangles), [69] (circles), and [70] (squares).

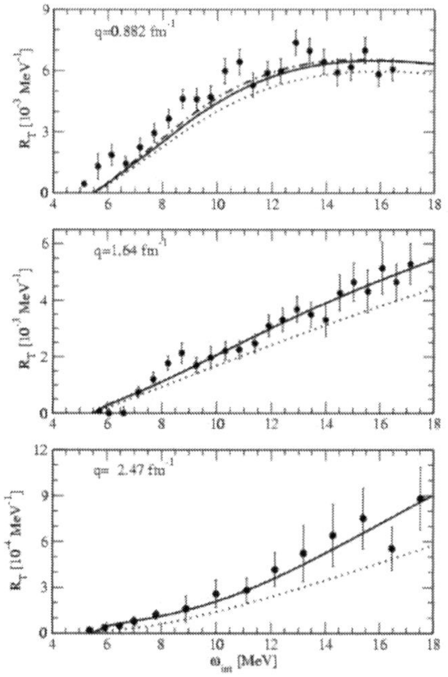

Figure 13. $R_T(^3\mathrm{He})$ at q=0.882, 1.64, 2.47 fm^{-1} in upper, middle, lower panel. Notation as in figure 12, but additional curve in upper panel for total result without Coulomb force in final state interaction (dash-dotted). Experimental data from [71].

As first LIT calculation the transverse response function $R_T(\omega, q)$ of ^{3}He is discussed [29]. As NN potential the BonnRA model has been used and in addition also the TM 3N force has been taken into account. As current operators non relativistic one-body as well as π and ρ meson exchange currents (MEC) are taken. The MEC are uniquely defined by the meson theoretical BonnRA potential. Besides the explicit MEC calculation also an implicit one has been carried out by using Siegert operators. The latter lead to the dominant part of the electric multipole contributions only, while there are no magnetic multipole contributions at all via the Siegert operator. Besides the isovector MEC in the BonnRA additional MEC of isoscalar nature are present. They are in general much less important than the isovector π and ρ MEC. It is important to note that a part of the isoscalar MEC is taken into account by the Siegert operator and thus it can be checked if they become important for a specific kinematics. To this end one has to compare the result of a direct calculation of one- and two-body current operators with a result obtained in a calculation where the Siegert operator is used and where in addition the

IOP Publishing
doi:10.1088/1742-6596/168/1/012001

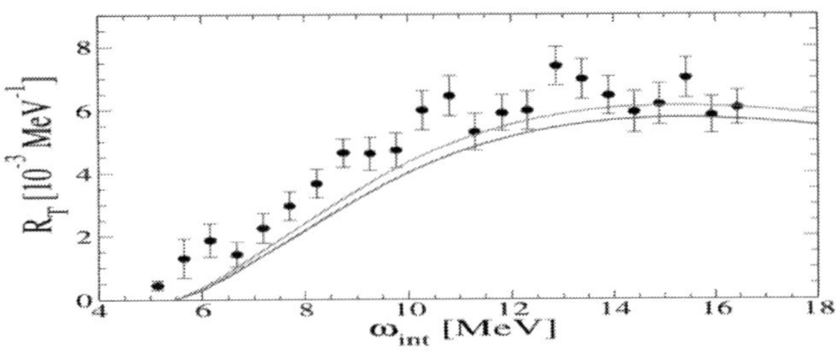

Figure 14. (Color online) As in upper panel of figure 13, but theoretical results with AV18 (lower curve) and AV18+UIX (upper curve).

non-Siegert contributions of the explicitly used current operators are added. A difference in the two calculations is then due to isoscalar exchange currents (for a more detailed description see [29]).

In figures 10 and 11 results for the $R_T(\omega, q)$ of ^{3}He are depicted at various momentum transfers. It is readily seen that there are rather strong MEC effects: at 15-30 MeV above threshold MEC enhance R_T by more than 30% for the two higher q-values (very close to threshold even by up to 200%, see figure 13); they increase the quasi-elastic peak height up to about 10%. Effects due to the Siegert operator remain quite small in and below the quasi-elastic peak. The picture changes at higher energies, where Siegert contributions become more important, particularly at lower q. In the high-energy tail one also finds a small increase of R_T due to residual exchange currents, which are taken into account only in the Siegert and not in the direct MEC calculation. In figure 12 the R_T results are shown in comparison to experimental data. For q= 250 and 400 MeV/c one finds a good agreement. However, data are not precise enough to allow a definite conclusion about the MEC contribution. As opposed to the lower q cases one finds at q=500 MeV/c a difference between the theoretical and experimental peak positions. The shift amounts to about 5-10 MeV. Relativistic effects, in particular those arising from corrections to the kinetic energy, might be responsible for this difference. In fact in [28] it was shown for the longitudinal response function $R_L(q, \omega)$ that such effects lead at q=500 MeV/c to a shift of the peak position by 6 MeV. In figure 13 various R_T theoretical and experimental low-energy results are illustrated at q=0.882, 1.64 and 2.47 fm^{-1} corresponding to about 174, 324 and 487 MeV/c, respectively. One sees that the MEC contribution can be very important, e.g. at q=487 MeV/c one finds an increase of about 200% close to threshold. Contrary to the cases shown in figure 12 one can make a definite conclusion about the MEC contribution. It is evident that they lead to a considerably improved agreement between theory and experiment. For the two higher q-values theoretical and experimental results agree very well, whereas for q=174 MeV/c the theoretical result underestimates experimental data somewhat below 10 MeV. In the upper panel of figure 13 the effect of the Coulomb force is shown in addition. Within 2 MeV above threshold there is effect of more than 10%, while at 5 MeV above threshold the Coulomb effect still amounts to 4%.

In figure 14 preliminary R_T results are shown for a calculation with AV18+UIX, where π and ρ MEC have been taken into account consistently with the AV18 potential applying the Arenhövel-Schwamb technique [76]. One sees that the results are very similar to those of BonnRA+TM potential shown in the lower panel of figure 13.

Now we turn to the ^{4}He electromagnetic responses. Already a few years ago the total ^{4}He photoabsorption cross section has been calculated with the LIT method using realistic nuclear forces [30, 73]. In the meantime also the ^{4}He longitudinal and transverse (e, e') response functions have been considered with the LIT approach. In figure 15 R_L results are shown. They

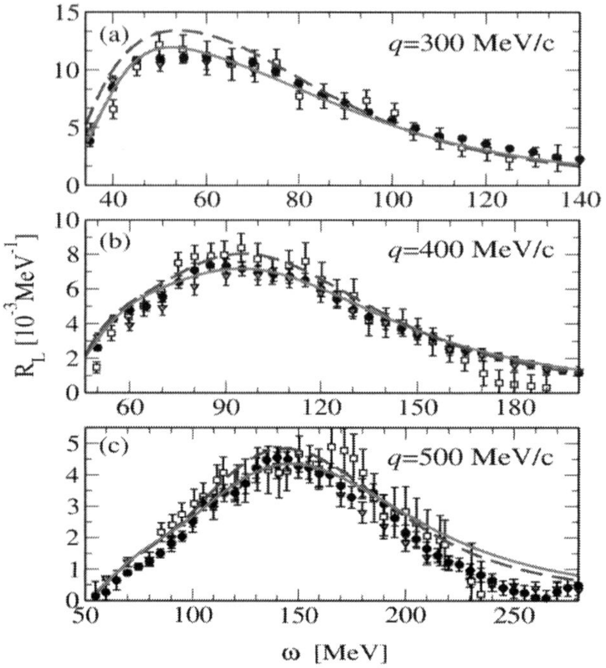

Figure 15. (Color online) $R_L(\omega, q)$ of ^{4}He at various q. Dashed lines: AV18; solid lines: AV18+UIX. Data from Bates [74] (squares), Saclay [75] (circles) and world data-set from [70] (triangles).

are obtained with AV18+UIX [65]. One sees that the 3N force reduces the quasi-elastic peak strength by about 10%. Experimental data are described quite well by our full result, but they are not precise enough to resolve the 3N force effect. In figure 16 R_L is illustrated at lower q. One readily notes a very strong reduction at lower energies due to the 3N force, which reaches up to about 40%. The reduction cannot be attributed to a simple binding effect as becomes evident from the also shown R_L results with a semi-realistic NN force (MT potential [77]). In fact, ^{4}He binding energies are 24.3, 28.4, and 30.6 MeV for AV18, AV18+UIX, and MT potentials, respectively. Even though the MT energy is closer to that of AV18+UIX, the MT R_L is more similar to the R_L of AV18 than to the AV18+UIX R_L. At lower q there is only one data set at 200 MeV/c [78], which is not sufficiently precise to draw concrete conclusions.

Also for the transverse response function R_T there is a LIT calculation, however, presently only for the semi-realistic MT potential [64]. On the other hand in the calculation a consistent current for the MT potential has been taken into account consisting of: one-body spin current, total one-body operator, i.e. spin + convection current (IA), and a consistent MEC for the MT potential, however, the center of mass dependence of the MEC is neglected. In figure 17 the results are shown at various q. At higher q the spin current dominates, but for $q = 100$ MeV/c the convection current presents a stronger effect. By comparing the IA with the full calculation, one can conclude that quite a strong MEC effect is found at low q. For $q = 100$ MeV and $\omega = 100$ MeV MEC lead to an enhancement of the strength by a factor of four with respect to the IA. At this q, the introduced approximation for the MEC is safely reliable. As q increases the MEC effect is reduced dramatically. Unfortunately, there are no experimental data in the quasi-elastic region for the two lower q. For $q = 300$ MeV/c where experimental data from Bates [74] and Saclay [75] exist, one observes a tiny effect of the MEC with respect to IA result. The

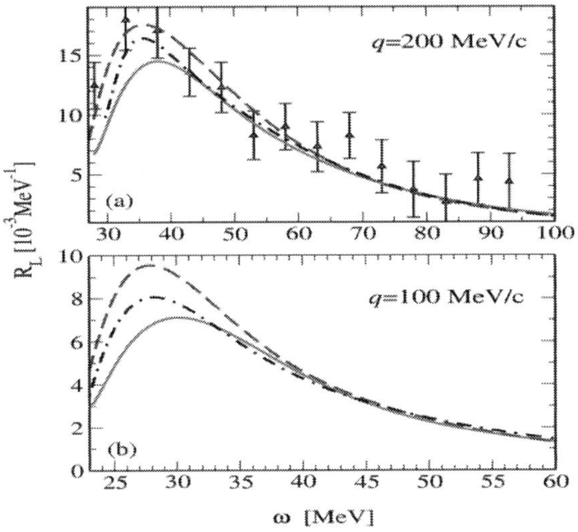

Figure 16. (Color online) $R_L(\omega, q)$ of ^{4}He at various q: AV18 (dashed), AV18+UIX (solid), MT (dash-dotted). Data in (a) from [78].

addition of the MEC leads to an increase of 4% and almost 10% at $\omega = 100$ and 150 MeV, respectively. The full calculation agrees rather well with the data of Bates in the energy range $55MeV \leq \omega \leq 115$ MeV, but is lower than the data of Saclay in the quasi-elastic peak region and above. Here it should be mentioned that the approximation introduced in the MEC is estimated in [64] to be of about 10% to 20% in the peak and negligible in the tail for $q = 200$ to 300 MeV/c. Thus it does not affect the conclusion, since the overall MEC effect is small.

There are other recent publications of the Trento group not discussed here, they are found in [67, 79, 80, 81].

5. Weakly bound light nuclei with the multi channel algebraic scattering approach

Now we turn to a somewhat different subject, namely to calculations with the multi channel algebraic scattering (MCAS) method. Different from all the topics discussed above no realistic NN and 3N forces are used. In fact the Hamiltonian is purely phenomenological. The aim is the treatment of bound and low-energy scattering states of stable nuclei as well as of weakly bound nuclear systems. The MCAS starting point consists of a finite matrix representation of Heisenberg's scattering matrix using the Sturmian theory as expansion method (see e.g. [82, 83]). Sturmians represent an efficient method for determining S-matrix, bound states, resonances, and scattering wave functions. Non-local and Coulomb interactions as well as the scattering process in coupled channel (CC) dynamics can be treated consistently. This is important, for example, when strong couplings to low-lying excitations of the target nucleus play a role.

It follows a brief introduction to the MCAS theory, which is taken from [85]. In case of the Schrödinger equation,

$$(E - H_0)\Psi_E = V\Psi_E \,, \tag{13}$$

one has a spectral variable E and an eigenstate Ψ_E. However, for Sturmians one considers the following equation

$$(E - H_0)\Phi_i(E) = \frac{V\Phi_i(E)}{\eta_i(E)} \,, \tag{14}$$

where E is a parameter and the eigenvalue $\eta_i(E)$ is the potential scale. As boundary conditions one has (i) bound-state like and normalizable for $E < 0$ and (ii) for $E > 0$ purely

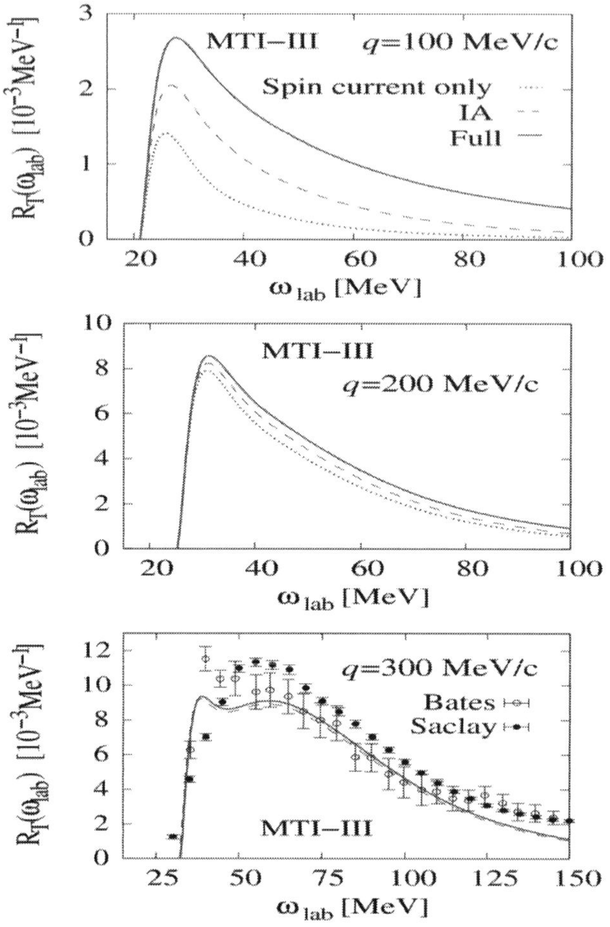

Figure 17. (Color online) $R_T(\omega, q)$ at various q with MT potential: spin current (dotted), IA (dashed), IA+MEC (solid). Data in lower panel from [74].

outgoing/incoming waves, which are not normalizable. The Sturmian spectrum consist of all the potential rescalings that give solution to Eq. (14) for given energy E and with well-defined boundary conditions.

The CC S-matrix can be written as

$$S_{cc'} = \delta_{cc'} - i\pi\sqrt{k_c k_{c'}} \sum_i \hat{\chi}_{ci}(E^+; k_c) \frac{1}{1 - \eta_i(E^+)} \hat{\chi}_{c'i}(E^+; k_{c'}) \,, \tag{15}$$

where in momentum space one has $\hat{\chi}_{ci}(E^+; k_c) = \langle k, c|V|\Phi_I(E)\rangle$. This form stands for a scattering process starting in the asymptotic channel c, then expanded into Sturmians, which freely propagate in the interaction region, and finally decay into the outgoing channels.

The nucleon-nucleus interaction is given as a potential matrix, where different operators are considered:

$$V_{cc'} = \sum_{n=C,LS,LL,SI} V_n \langle (LS)jI; J^\pi | \mathcal{O} f_n(r, R, \theta_{r,R}) | (L'S)j'I'; J^\pi \rangle \,. \tag{16}$$

In the calculations of the Padua group the scattering of a nucleon with a light nucleus with a 0^+ ground state is considered. The operators are expanded up to second order in the core

Table 1. Experimental data and theoretical results for ^{7}Li and ^{7}Be states (energies are in MeV, widths are in keV). All energies are defined with thresholds, $p+^6$He = 9.975 MeV with respect to ^{7}Li ground state and $n+^6$Be = 10.676 MeV with respect to ^{7}Be ground state (^{a} for these states spin and parity are unknown, ^{b} spin-parity of this state has been assigned as $\frac{1}{2}^-$).

J^π	^{7}Li		^{7}Be	
	Exp.	Theory	Exp.	Theory
$\frac{3}{2}^-$	−9.975	−9.975	−10.676	−11.046
$\frac{1}{2}^-$	−9.497	−9.497	−10.246	−10.680
$\frac{7}{2}^-$	−5.323 [69]	−5.323	−6.106 [175]	−6.409
$\frac{5}{2}^-$	−3.371 [918]	−3.371	−3.946 [1200]	−4.497
$\frac{5}{2}^-$	−2.251 [80]	−0.321	−3.466 [400]	−1.597
$\frac{3}{2}^-$	−1.225 [4712]	−2.244	−−	−−
$\frac{1}{2}^-$	−0.885 [2752]	−0.885		−2.116
$\frac{7}{2}^-$	−0.405 [437]	−0.405	−1.406 [?]	−1.704
$\frac{3}{2}^-$	−−	−−	−0.776 [1800]	−3.346
$\frac{3}{2}^-$	1.265 (260)	0.704 (56)	0.334 (320)	−0.539
$\frac{1}{2}^-$		1.796 (1570)		0.727 (699)
$\frac{3}{2}^-$	3.7 (800) ?^{a}	2.981 (990)		1.995 (231)
$\frac{5}{2}^-$	4.7 (700) ?^{a}	3.046 (750)		2.009 (203)
$\frac{5}{2}^-$		5.964 (230)		4.904 (150)
$\frac{7}{2}^-$		6.76 (2240)	6.5 (6500) ?^{b}	5.78 (1650)

deformation parameter. For instance for a single quadrupole deformation, $R = R_0(1+\beta_2 P_2(\theta))$, it is used

$$f_n(r,R,\theta) = f_n^{(0)} - \beta_2 R_0 P_2(\theta)\frac{df_n^{(0)}(r)}{dr} + \frac{\beta_2^2 R_0^2}{2\sqrt{\pi}}\left(P_0 - \frac{2\sqrt{5}}{7}P_2(\theta) + \frac{2}{7}P_4(\theta)\right)\frac{d^2 f_n^{(0)}(r)}{dr^2}. \quad (17)$$

The functions $f_n^{(0)}$ have a standard Woods-Saxon form WS(r) for $n = C, LL, SI$ and for $n = LS$ an additional derivation: $\frac{1}{r}\frac{d}{dr}WS(r)$.

Calculations with the CC Hamiltonian above can easily lead to deeply bound spurious states in the bound spectrum. They originate form a violation of the Pauli principle. The phase space corresponding to target nucleons in fully occupied shells has to be forbidden to the incoming nucleon, however, in the original CC model such a condition is missing. As described in [86] one can use the technique of orthogonalizing pseudo potentials (OPP) [87], which eliminates the spurious states by the addition of a new term in the nuclear potential.

Before describing some selected results obtained by the Padua group another recent publication should be mentioned, [88], where nonlocalities in nucleon-nucleus potentials are considered in greater detail.

There are various calculations of the Padua group with the MCAS theory: $n+^{12}$C system [83, 84]; $n+^{14}$C and $p+^{14}$O [89]; $N+^6$He and $N+^6$Be [90]. Here we discuss only the lightest of these systems, i.e. the $A = 7$ systems. In tables 1 and 2 results for the spectra are shown. A single CC potential has been used for all four $A = 7$ systems, however with a different OPP term for Li/He and Be/B. For the former only the $0s_{1/2}$ shells are blocked, while for the latter

Table 2. Experimental data and theoretical results for ^{7}He and ^{7}B states. All energies are in MeV and relate to thresholds of −0.445 MeV for n+^{6}He and of −2.21 MeV for p+^{6}Be ([a] observed very recently and interpreted as a $\frac{1}{2}^-$ state, [b] spin-parity of this state is unknown).

J^π	^{7}He		^{7}B	
	Exp.	Theory	Exp.	Theory
$\frac{3}{2}^-$	0.445 (150)	0.43 (100)	2.21 (1400)	2.10 (190)
$\frac{7}{2}^-$	– –	1.70 (30)		3.01 (110)
$\frac{1}{2}^-$	1.0 (750) ?[a]	2.79 (4100)		5.40 (7200)
$\frac{5}{2}^-$	3.35 (1990)	3.55 (200)		5.35 (340)
$\frac{3}{2}^-$	6.24 (4000) ?[b]	6.24 (1900)		

a more complex OPP scenario is considered (for details see [90]). It should be mentioned that in [90] also ^{3}H-^{4}He and ^{3}He-^{4}He scattering has been calculated in terms of a di-cluster model and a rather good description of the scattering data has been obtained.

A very recent application of the MCAS approach is the calculation of the radiative capture ^{3}He$(\alpha,\gamma)^7$Be [91]. To this end the MCAS method for nucleon-nucleus scattering has been extended to a treatment of a two-cluster system, where finite sizes and internal excitations of the cluster are taken into account. The initial and final state wave functions were calculated using the nuclear potential of [90], which had been fitted to reproduce best the states and resonance widths of ^{7}Li. In figures 18 and 19 the results of the radiative capture calculation are shown. One sees that the MCAS S-factor reproduces the energy dependence of the data

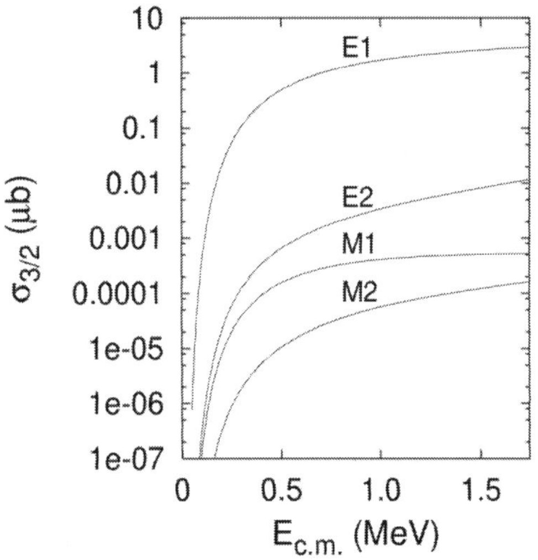

Figure 18. Contribution of the various multipoles to the ^{3}He$(\alpha,\gamma)^7$Be cross section.

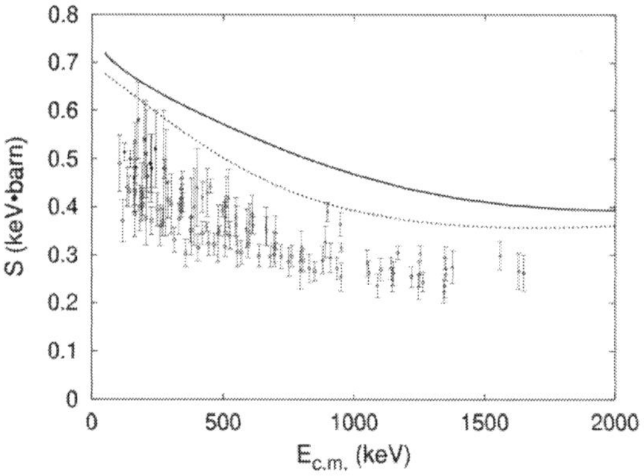

Figure 19. Astrophysical S-factor for the reaction ^{3}He$(\alpha,\gamma)^7$Be with various experimental data (see [91]). Solid line: MCAS result. Dotted line: initial-state wave function for a repulsive hard-core model. Various data from different sources (see [91]).

quite well, but overestimates the experimental S-factor somewhat. A change of the initial-state scattering state wave function by use of a rather simple repulsive hard-core model leads to a decrease of the S-factor showing that the result is rather sensitive to the wave function. However, as the authors point out, for a consistent calculation one would need to construct a better MCAS potential. For a discussion of results from other calculations we refer to [91].

References

[1] Machleidt R 1989 *Adv. Nucl. Phys.* **19** 189
[2] Wiringa R B, Stoks V G J and Schiavilla R 1995 *Phys. Rev.* C **51** 38
[3] Stoks V G J, Klomp R A M, Terheggen C P F and de Swart J J 1994 *Phys. Rev.* C **49** 2950
[4] Machleidt R 2001 *Phys. Rev.* C **63** 024001
[5] Entem D R and Machleidt R 2003 *Phys. Rev.* C **68** 041001
[6] Epelbaum E, Glöckle W and Meissner U G 1998 *Nucl. Phys.* A **637** 107
[7] Epelbaum E 2006 *Prog. Part. Nucl. Phys.* **57** 654
[8] Pudliner B S, Pandharipande V R, Carlson J, Pieper S C and Wiringa R B 1997 *Phys. Rev.* C **56**, 1720
[9] Coelho H T, Das T K and Robilotta M R 1983 *Phys. Rev.* C **28** 1812; Robilotta M R and Coelho H T 1986 *Nucl. Phys.* A **460** 645
[10] Coon S A, Scadron M D, McNamee P C, Barrett B R, Blatt D W E and McKellar B H J 1979 *Nucl. Phys.* A **317** 242; Coon S A and Glöckle W *Phys. Rev.* C **23** 1790 (1981); Coon S A and Han H K 2001 *Few-Body Syst.* **30** 131
[11] Pieper S C, Pandharipande V R, Wiringa R B and Carlson J 2001 *Phys. Rev.* C **64** 014001
[12] Epelbaum E, Nogga A, Glöckle W, Kamada H, Meissner U G and Witala H 2002 *Phys. Rev.* C **66** 064001
[13] Kamada H *et al* 2001 *Phys. Rev.* C **64** 044001
[14] Pieper S C, Varga K and Wiringa R B 2002 *Phys. Rev.* C **66** 044310
[15] Forssén C, Navrátil P, Ormand W E and Caurier E 2005 *Phys. Rev.* C **71** 044312
[16] Hagen G, Dean D J, Hjorth-Jensen M, Papenbrock T and Schwenk A 2007 *Phys. Rev.* C **76** 044305
[17] Gandolfi S, Pederiva F, Fantoni S and Schmidt K E 2007 *Phys. Rev. Lett.* **99** 022507
[18] Witala H and Glöckle W 2008 *Eur. Phys. J* A **37** 87, and references therein.
[19] Kievsky A, Rosati S, Viviani M, Marcucci L E and Girlanda L 2008 *J. Phys. G: Nuclear and Particle Physics* **35** 063101
[20] Lazauskas R and Carbonell J 2004 *Phys. Rev.* C **70** 044002
[21] Deltuva A and Fonseca A C 2007 *Phys. Rev.* C **75** 014005
[22] Nollett K M, Pieper S C, Wiringa R B, Carlson J and Hale G M 2007 *Phys. Rev. Lett.* **99** 022502
[23] Efros V D, Leidemann W and Orlandini G 1994 *Phys. Lett.* B **338** 130
[24] Efros V D, Leidemann W, Orlandini G and Barnea N 2007 *J. Phys. G: Nuclear and Particle Physics* **34** R459
[25] Efros V D, Leidemann W, Orlandini G and Tomusiak E L 2000 *Phys. Lett.* B **484** 223
[26] Golak J, Skibiński R, Glöckle W, Kamada H, Nogga A, Witala H, Efros V D, Leidemann W, Orlandini W and Tomusiak E L 2002 *Nucl. Phys.* A **707** 365
[27] Efros V D, Leidemann W, Orlandini G and Tomusiak E L 2004 *Phys. Rev.* C **69** 044001
[28] Efros V D, Leidemann W, Orlandini G and Tomusiak E L 2005 *Phys. Rev.* C **72** 011002(R)
[29] Della Monaca S, Efros V D, Kughaev A, Leidemann W, Orlandini G, Tomusiak E L and Yuan L P 2008 *Phys. Rev.* C **77** 044007
[30] Gazit D, Bacca S, Barnea N, Leidemann W and Orlandini G 2006 *Phys. Rev. Lett.* **96** 112301
[31] Gazit D and Barnea N 2007 *Phys. Rev. Lett.* **98** 192501
[32] Bacca S, Marchisio M, Barnea N, Leidemann W and Orlandini G 2002 *Phys. Rev. Lett.* **89** 052502; Bacca S, Barnea N, Leidemann W and Orlandini G 2004 *Phys. Rev.* **69** 057001
[33] Bacca S, Arenhövel H, Barnea N, Leidemann W and Orlandini G 2004 *Phys. Lett.* B **603** 159
[34] Quaglioni S, Leidemann W, Orlandini G, Barnea N and Efros V D 2004 *Phys. Rev.* C **69** 044002
[35] Arenhövel H, Leidemann W and Tomusiak E L 2005 *Eur. Phys. J.* A **23** 147
[36] Golak J, Skibiński R, Witala H, Glöckle W, Nogga A and Kamada H 2005 *Phys. Rept.* **415** 89
[37] Korda V Yu, Canton L and Shebeko A V 2007 *An. Phys.* **322** 736
[38] Schwamb M, submitted to *Phys. Rept.*
[39] Barnea N, Leidemann W and Orlandini G 2000 *Phys. Rev.* C **61** 054001; 2001 *Nucl. Phys.* A **693** 565
[40] Barnea N, Efros V D, Leidemann W and Orlandini G 2004 *Few-Body Syst.* **35** 155
[41] Viviani M, Marcucci L E, Rosati S, Kievsky A and Girlanda L 2006 *Few-Body Syst.* **39** 159
[42] Marcucci L E, paper in preparation
[43] Witala H, Golak J, Skibinski R, Glöckle W, Nogga A, Epelbaum E, Kamada H, Kievsky A and Viviani M 2006 *Phys. Rev.* C **73** 044004
[44] Glombik A *et al* 1994 *AIP Conf. Proc.* **334** 486
[45] Kretschmer W *et al* 1995 *AIP Conf. Proc.* **339** 335

[46] Shimizu S, Sagara K, Nakamura H, Maeda K, Miwa T, Nishinori N, Ueno S, Nakashima T and Morinobu S 1995 *Phys. Rev.* C **52** 1193
[47] Schiavilla R, Viviani M, Girlanda L, Kievsky A and Marcucci L E 2008 *Phys. Rev.* C **78** 014002
[48] Girlanda L, Klink W H and Viviani M 2007 *Phys. Rev.* C **76** 044002
[49] Girlanda L 2008 *Phys. Rev.* C **77** 067001
[50] Viviani M, Schiavilla R, Kubis B, Lewis R, Girlanda L, Kievsky A, Marcucci L E and Rosati S 2007 *Phys. Rev. Lett.* **99** 112002
[51] Barletta P and Kievsky A 2009 *Few-Body Syst.* in print
[52] Marcucci L E, Pervin M, Pieper S C, Schiavilla R and Wiringa R B 2008 *Phys. Rev.* C **78** 065501
[53] de Melo J P B C, Frederico T, Pace E, Pisano S and Salmè G 2008 *arXiv*:0804.1511
[54] Sick I 2001 *Prog. Part, Nucl. Phys.* **47** 245
[55] Baroncini F, Kievsky A, Pace E and Salmè G 2008 *AIP Conf. Proc.* **1056** 272
[56] Lev F M, Pace E and Salmè G 1998 *Nucl. Phys.* A **641**
[57] Marinho J A O, Frederico T, Pace E, Salmè G and Sauer P 2008 *Phys. Rev.* D **77** 116010
[58] Pace E, Salmè G, Frederico T, Pisano S, de Melo J P B C 2006 *Nucl. Phys.* A **790** 606
[59] Baroncini F A, Pace E and Salmè G 2008 *Few-Body Syst.* **43** 173
[60] Efros V D 1985 *Yad. Fiz.* **41** 1498 [*Sov. J. Nucl. Phys.* **41** 949]
[61] Efros V D, Leidemann W and Orlandini G 1993 *Few-Body Syst.* **14** 151
[62] Efros V D, Leidemann W and Orlandini G 1999 *Few-Body Syst.* **26** 251
[63] Andreasi D, Leidemann W, Reiss Ch and Schamb M 2005 *Eur. Phys. J.* A **24** 361
[64] Bacca S, Arenhövel H, Barnea N, Leidemann W and Orlandini G 2007 Phys. Rev. C **76** 014003
[65] Bacca S, Barnea N, Leidemann W and Orlandini 2008 *arXiv*:0811.4624, *Phys. Rev. Lett.* in print
[66] La Piana A and Leidemann W 2000 *Nucl. Phys.* A **677** 423
[67] Leidemann W 2008 Few-Body Syst. **42** 139
[68] Marchand C *et al* 1985 *Phys. Lett.* B **153** 29
[69] Dow C *et al* 1988 *Phys. Rev. Lett.* **61** 1706
[70] Carlson J, Jourdan J, Schiavilla R and Sick I 2002 *Phys. Rev.* C **65** 024002
[71] Retzlaff G A *et al* 1994 *Phys. Rev.* **49** 1263
[72] Arenhövel H and Sanzone M 1991 *Few-Body Syst. Suppl.* **3** 1
[73] Quaglioni S and Navràtil P 2007 *Phys. Lett.* B **652** 370
[74] Dytman S A *et al* 1988 *Phys. Rev.* C **38** 800
[75] Zghiche A *et al* 1994 *Nucl. Phys.* A **572** 513
[76] Arenhövel H and Schwamb M 2001 *Eur. Phys. J.* A **12** 207
[77] Malfliet R A and Tjon J 1969 *Nucl. Phys.* A **127** 161
[78] Buki A Yu, Timchenko I S, Shevchenko N G and Nenko I A 2006 *Phys. Lett.* B **641** 156
[79] Barnea N, Leidemann W and Orlandini G 2006 *Phys. Rev.* C **74** 034003
[80] Gazit D, Bacca S, Barnea N, Leidemann W and Orlandini G 2006 *Phys. Rev.* C **74** 061001(R)
[81] Stetcu I, Quaglioni S, Bacca S, Barrett B R, Johnson C W, Navrátil P, Barnea N, Leidemann W and Orlandini G 2007 *Nucl. Phys.* A **785** 307
[82] Canton L and Rawitscher G H 1991 *Journ. of Phys.* G **17** 429
[83] Amos K, Canton L, Pisent G, Svenne J P and van der Knijf D 2003 *Nucl. Phys.* A **728** 65
[84] Svenne J P, Amos K, Karataglidis S, van der Knijf D, Canton L and Pisent G 2006 *Phys. Rev.* C **73** 027601
[85] Canton L 2007 *Few-Body Systems in Nuclear and Hadronic Systems* (Padova: Impremenda - Università degli Studi di Padova) p 79
[86] Canton L, Pisent G, Svenne J P, van der Knijf D, Amos K and Karataglidis S 2005 *Phys. Rev. Lett.* **94** 122503
[87] Kukulin V and Pomerantsev V 1978 *Annals of Phys.* **111** 330
[88] Fraser P, Amos K, Karataglidis S, Canton L, Pisent G and Svenne J P 2008 *Eur. Phys. J.* A **35** 69
[89] Canton L, Pisent G, Svenne J P, Amos K and Karataglidis S 2006 *Phys. Rev. Lett.* **96** 072502
[90] Canton L, Pisent G, Amos K, Karataglidis S, Svenne J P and van der Knijf D 2008 *Phys. Rev.* C **74** 064605
[91] Canton L and Levchuk L G 2008 *Nucl. Phys.* A **808** 192

Journal of Physics: Conference Series **168** (2009) 012002 doi:10.1088/1742-6596/168/1/012002

Light-front projections of the Bethe-Salpeter amplitude and the 4D electromagnetic current for an interacting two-fermion system

T Frederico[1], **J A O Marinho** [1,2], **E Pace**[3] **and G Salmè**[4]

[1] Dep. de Física, Instituto Tecnológico de Aeronáutica, 12.228-900 São José dos Campos, São Paulo, Brazil

[2]Centro de Física Computacional, Department of Physics, University of Coimbra, Portugal

[3] Dipartimento di Fisica, Università di Roma "Tor Vergata" and Istituto Nazionale di Fisica Nucleare, Sezione Tor Vergata, Via della Ricerca Scientifica 1, I-00133 Roma, Italy

[4]Istituto Nazionale di Fisica Nucleare, Sezione Roma I, P.le A. Moro 2, I-00185 Roma, Italy

E-mail: salmeg@roma1.infn.it

Abstract. A recent approach for constructing an exact 3D reduction of the 4D matrix elements of the electromagnetic current for an interacting two-fermion system, is briefly reviewed. The properties of the obtained 3D current, like the fulfillment of the Ward-Takashi Identity and its Fock decomposition, will be illustrated in relation with future applications to few-nucleon systems.

1. Introduction

A detailed investigation of the electromagnetic (em) properties of hadrons and nuclei requests a careful treatment of the current operator for interacting fermion systems. In particular, besides general properties like the Poincaré covariance, the gauge symmetry of the em interaction plays the well-known essential role, that leads to the Ward-Takahashi Identity (WTI) for the em vertex function. In this short presentation, some insights will be given on the main ingredients of our approach [1] aimed to construct a 3D em current for an interacting two-fermion system, that i) fulfills WTI and therefore allows the current conservation (CC) and ii) preserves at large extent the Poincaré covariance, emphasizing the relevant steps for obtaining workable expressions to be used in actual calculations.

Field theoretical approaches, based on the Bethe-Salpeter equation, (see, e.g. [2, 3, 4]) allow one to evaluate matrix elements of the 4D em current for an interacting two-body system, viz

$$\langle P_f, J_f, \sigma_f | j^\mu | P_i, J_i, \sigma_i \rangle = \mathcal{C} \sum_{\tau_i, ..} \int \frac{d^4 k_1}{(2\pi)^4} \bar{\Psi}^{J_f}_{\sigma_f}(k_1, k'_2, P_f) \mathcal{J}^\mu(k_1, k'_2, k_2) \Psi^{J_i}_{\sigma_i}(k_1, k_2, P_i) \quad (1)$$

where, in the present case, $\Psi^{J\sigma}(k_1, k_2, P)$ is the Bethe-Salpeter (BS) amplitude of a two-fermion systems (with $k_1 + k'_2 = P_f$ and $k_1 + k_2 = P_i$ for the final and initial state, respectively), $\mathcal{C}$ contains the normalization factors, and $\mathcal{J}^\mu$ is the 4D fermion-photon vertex (or kernel) that contains one + two +...+ many-body contributions, and must fulfill WTI. The sum in Eq. (1) is performed on isospin, spinorial, etc. indexes.

To solve the BS equation in Minkowski space for obtaining $\Psi^{J\sigma}(k_1, k_2, P)$, in realistic cases, is still a prohibitive task, but new and very effective approaches are coming into play (see, e.g. [5]). On the other hand, 3D reductions are well known tools (see, e.g. [6]) for attempting to overcome the above mentioned difficulties, but still saving at large extent the general properties that the em current has to fulfill.

Aim of our investigation [1] is i) to achieve an exact 3D reduction of the 4D matrix elements in Eq. (1), i.e. without loosing any physics content, and ii) to elaborate a systematic analysis of the obtained expression in order to develop effective approximation. The main ingredients for our approach, that were already applied to the case of two-boson interacting system in [4], are in order i) the Quasi-Potential (QP) approach for the Transition matrix (see [7]), and ii) the projection of the 4D physical quantities onto the 3D Light-front (LF) hyperplane (i.e. $x^+ = x^0 + x^3 = 0$) (see, e.g. [2, 3]). As a first result, one can establish a formally exact correspondence between 4D BS amplitude and the 3D LF "valence" wave function. Then it is possible to express, without approximation, the matrix elements of the 4D em current, where the BS amplitudes appear, in terms of matrix elements of a 3D LF current, that fulfills the WTI. It should be pointed out that in these last matrix elements the valence states are present.

Within the Quasi-Potential approach for the T-matrix, where the so-called *effective interaction* appears, one can develop a meaningful approximation scheme for evaluating the 3D matrix elements, based on the Fock decomposition of the effective interaction. In particular, one expects a decreasing influence of the Fock states that contain an increasing number of intermediate particles. This machinery allows one to construct an iterated approximation of the effective interaction and eventually to evaluate matrix elements of the "truncated" LF em current, that still fulfills at any order the LF WTI. Finally, by using a Yukawa model in ladder approximation, as a pedagogical example, explicit expressions for many-body contributions to the em LF current have been obtained [1].

2. BS amplitudes and EM current

The quantities present in Eq. (1) are the BS amplitude and the em vertex. In particular, the BS amplitude $|\Psi\rangle$, that depends upon *internal variables*, satisfies the following equation (the $\lim_{\varepsilon \to 0}$ is understood)

$$G^{-1}(K)\,|\Psi\rangle = 0 \tag{2}$$

with appropriate boundary conditions for bound and scattering states. In Eq. (2), K is the 4-momentum of the center of mass and the inverse of the Green function is given by

$$G^{-1}(K) = G_0^{-1}(K) - V(K) \tag{3}$$

where $V(K)$ is the 4D interaction, as obtained by the Lagrangian, $G_0(K) = S(k_1)\ S(K - k_1)$, with $S(k)$ the Dirac propagator.

The em vertex, $\mathcal{J}^\mu$, corresponding to the interaction $V(K)$ has a free term, $\mathcal{J}_0^\mu$, and another one, $\mathcal{J}_I^\mu$, which depends on the interaction, as requested by the commutation rules with the generators of the Poincarè group. It also satisfies the following Ward-Takahashi identity

$$Q_\mu \mathcal{J}^\mu(Q) = G^{-1}(K_f)\hat{e}\ - \hat{e}\ G^{-1}(K_i) \tag{4}$$

where Q^μ is the 4-momentum transfer and $\hat{e} = \hat{e}_1 + \hat{e}_2$ is the pointlike charge operator with matrix elements given by $\langle k_j|\hat{e}_j|p_j\rangle = e_j \delta^4\,(k_j - p_j - Q)$

Combining Eqs. (3) and (4) with the decomposition of $\mathcal{J}^\mu$ in terms of the dependence upon $V(K)$, it turns out that each contribution fulfills a suitable constraint, viz

$$Q_\mu \mathcal{J}_0^\mu(Q) = G_0^{-1}(K_f)\hat{e}\ - \hat{e}\ G_0^{-1}(K_i)$$

and

$$Q_\mu \mathcal{J}_I^\mu(Q) = \hat{e}\, V(K_i) - V(K_f)\hat{e} \ .$$

Finally, CC can be explicitly retrieved by using solutions of the BS equation, Eq. (2), i.e.,

$$Q_\mu \langle \Psi_f | \mathcal{J}^\mu(Q) | \Psi_i \rangle = \langle \Psi_f | \left[G^{-1}(K_f)\hat{e} \ - \hat{e}\, G^{-1}(K_i) \right] | \Psi_i \rangle = 0 \quad . \tag{5}$$

After introducing the general equations fulfilled by BS amplitudes and em current, let us illustrate our procedure. As well known, the T-matrix is given in terms of the interaction $V(K)$ by an integral equation: $T(K) = V(K) + V(K)\ G_0(K)\ T(K)$.

In the Quasi-Potential framework [7], $T(K)$ can be rewritten in terms of two coupled equations

$$T(K) = W(K) + W(K)\ G_{aux}(K)\ T(K) \tag{6}$$

where the effective interaction $W(K)$, in turn, is a solution of

$$W(K) = V(K) + V(K)\ \Delta_0(K)\ W(K) \tag{7}$$

with $\Delta_0(K) := G_0(K) - G_{aux}(K)$, and $G_{aux}(K)$ is a smart choice to speed up the convergence of the iterative solutions.

3. Projecting onto the Light-Front hyperplane

In Eq. (1) the 4D integration contains all the difficulty given by the very complicated analytic structure of both BS amplitudes and em vertex in Minkowski space. The adopted strategy for decreasing the degree of difficulty, is based on disentangling the "dynamical" variable, i.e. the time, from the remaining three kinematical variables, that describe the "initial" state. In order to accomplish such a task, the LF kinematics appears a very appealing tool (see, e.g. [8] for an extensive review) since the LF combination of the Poincaré generators has the property to produce the maximal numbers of generators without interaction (7 out of 10). In turn, such a property produces many other relevant features, like the kinematical nature of the LF boosts (to be not confused with the standard boosts, a part the longitudinal one). In particular, the LF components of the 4-momentum and the scalar product are given by

$$k^\pm = k^0 \pm k_z, \qquad \mathbf{k}_\perp \equiv \{k_x, k_y\}, \qquad k \cdot x = \frac{(x^- k^+ + x^+ k^-)}{2} - \mathbf{x}_\perp \cdot \mathbf{k}_\perp \quad . \tag{8}$$

Integrating a given physical quantity, $\phi(k)$, over the minus component of the 4-momentum, k^-, amounts to restrict its dependence upon only 3 variables in the coordinate space. One has

$$\begin{aligned} \int dk^-\ \phi(k) &= \int dk^- \ \frac{1}{2(2\pi)^4} \int dx^+\ dx^-\ d\mathbf{x}_\perp\ e^{ik\cdot x}\ \tilde{\phi}(x) = \\ &= \frac{1}{(2\pi)^3} \int dx^+\ dx^-\ \delta(x^+) d\mathbf{x}_\perp\ e^{i(x^- k^+/2 - \mathbf{x}_\perp \cdot \mathbf{k}_\perp)}\ \tilde{\phi}(x) \quad . \end{aligned} \tag{9}$$

Therefore, the "LF-time", x^+, that labels the dynamical evolution (as the standard "t" in the instant form of the relativistic Hamiltonian), is constrained to its initial value, i.e. $x^+ = 0$. Therefore, once the physical quantities are integrated over k^-, they are constrained to live onto the LF hyperplane defined by $x^+ = 0$.

A crucial step for the LF projection is the separation of the fermion propagator in an on-shell term and in an instantaneous one, i.e

$$i\ S(k) = \frac{\not{k} + m}{k^2 - m^2 + i\varepsilon} = \frac{\not{k}_{on} + m}{k^+(k^- - k^-_{on} + \frac{i\varepsilon}{k^+})} + \frac{\gamma^+}{2k^+} = i\ S_{on}(k) + \frac{\gamma^+}{2k^+} \tag{10}$$

where $k_{on}^- = (\mathbf{k}_\perp^2 + m^2)/k^+$ is the minus-component of k_{on}^μ, such that $k_{on} \cdot k_{on} = m^2$, and the second term leads to an instantaneous (in LF time!) free propagation, since the formal Fourier transform generates $\delta(x^+)$ (see Eq. (9)). It should be pointed out that such a term makes the treatment of a fermionic system basically different from the treatment of a bosonic one. On the other hand, the decomposition in Eq. (10) suggests a strategy for achieving our goal of an exact 3D reduction. As a matter of fact, let us define the 4D on-shell two-body Green's function [1, 4]

$$G_0^{on}(K) := S_{on}(k_1)\, S_{on}(K - k_1) \tag{11}$$

that plays a key role in our approach, since allows us to separate the *trivial* 4D propagation associated to such a contribution.

The 3D LF counterpart of $G_0^{on}(K)$ is introduced by integrating over k_1^- and $k_1'^-$, i.e. by projecting onto the 3D LF hyperplane, the following matrix elements (note that the other three components of the momentum remain operators)

$$g_0(K) = \downarrow G_0^{on}(K) \downarrow \;\equiv \Omega^+ \int dk_1'^- dk_1^- \left\langle k_1'^- | G_0^{on}(K) | k_1^- \right\rangle \tag{12}$$

where Ω^+ is a suitable kinematical factor ensuring the kinematical Poincaré covariance (e.g. $\Omega^+ = k^+$ or $\Omega^+ = \sqrt{k^+(K^+ - k^+)}$), and the symbol $\downarrow$, from now on, indicates the integration over the minus component of the 4-momentum that appears on the left or on the right side, depending upon the position of the symbol itself.

In the case without interaction, it is easy to find the relation between the 3D LF free state, $|\phi_0\rangle$, solution of $g_0^{-1}(K)\, |\phi_0\rangle = 0$ (note that instantaneous terms do not contribute) and the free BS amplitude, solution of $G_0^{-1}(K)\, |\Psi_0\rangle = 0$ (note that the particles are on their own mass-shell in the free case). It turns out [1] that

$$|\Psi_0\rangle = G_0^{on}(K) \downarrow g_0^{-1}(K)\, |\phi_0\rangle = \Pi_0(K)\, |\phi_0\rangle \tag{13}$$

where $\Pi_0(K)$ is the *free reverse LF projector*, that allows to jump from a 3D description to a 4D world. The operator $\Pi_0(K)$ has a simple explicit expression [1, 4] in terms of the positive energy Dirac projector. Let us remind that, by construction $\downarrow |\Psi_0\rangle = |\phi_0\rangle$: this completes the reconstruction pattern in the simple case of a free two-fermion system.

In order to apply the projection technique to the QP approach, namely for an interacting case, one needs an intermediate step, given by a suitable choice for the auxiliary Green's function to be used in the 4D operator Δ_0, see Eq. (7). The chosen 4D auxiliary Green's function is defined as follows in terms of the 3D $g_0(K)$

$$G_{aux}(k) = \Pi_0(K) g_0(K) \overline{\Pi}_0(K) \tag{14}$$

where $\overline{\Pi}_0(K) = g_0^{-1}(K) \downarrow G_0^{on}(K)$. Then, the T-matrix and the effective interaction $W(K)$ can be rewritten as follows

$$\begin{aligned} T(K) &= W(K) + W(K)\ \left[\Pi_0(K) g_0(K) \overline{\Pi}_0(K)\right]\ T(K) \\ W(K) &= V(K) + V(K)\ \Delta_0(K)\ W(K) = \\ &= V(K) + V(K)\ \left[G_0(K) - \left(\Pi_0(K) g_0(K) \overline{\Pi}_0(K)\right)\right]\ W(K) \quad . \end{aligned} \tag{15}$$

The equation for $W(K)$ can be solved by iteration, i.e.

$$W(K) = V(K) \sum_{i=1}^{\infty} [\Delta_0(K) V(K)]^{i-1} \quad . \tag{16}$$

The rate of convergence of this series is related to $\Delta_0(K)$, that describes the difference between the intermediate propagation of two non-interacting fermions and the 4D counterpart of the LF propagator $g_0(K)$. Depending upon the powers of $\Delta_0(K)$, we will have intermediate states with more and more bosons and fermions (once we introduce time-ordered diagrams).

The Rosetta Stone relating 4D and 3D quantities, is readily constructed. The 3D T-matrix and effective interaction are respectively

$$t(K) = \overline{\Pi}_0(K)\ T(K)\ \Pi_0(K) \qquad\qquad w(K) = \overline{\Pi}_0(K)\ W(K)\ \Pi_0(K) \quad . \tag{17}$$

Once $t(K)$ is reexpressed in terms of $g_0(K)$ and $w(K)$, one can introduce the 3D interacting Green's function and its inverse as follows

$$g(K) = g_0(K) + g_0(K)t(K)g_0(K) \qquad\qquad g^{-1}(K) = \overline{\Pi}_0(K)\ G^{-1}(K)\ \Pi(K) \tag{18}$$

where the *interacting LF reverse projection operator* is

$$\Pi(K) = \ [1 + \Delta_0(K)\ W(K)\ ]\ \Pi_0(K) \quad . \tag{19}$$

The solution $|\phi\rangle$ of the following 3D equation,

$$g^{-1}(K)\,|\phi\rangle = \left[g_0^{-1}(K) - w(K)\right]\ |\phi\rangle = 0 \tag{20}$$

with appropriate boundary conditions for bound and scattering states, is the valence component of the LF wave function for two interacting fermions, as can be deduced by the relation with the BS amplitude, shown in what follows. The exact relations between the 3D valence component and the 4D BS amplitude can be readily obtained, and they are given by

$$|\Psi\rangle = \Pi(K)\ |\phi\rangle \qquad\qquad \downarrow G_0^{on}(K)\ G_0^{-1}(K)\ |\Psi\rangle = \ |\phi\rangle \quad . \tag{21}$$

The rightmost equation illustrates the motivation for calling $|\phi\rangle$ valence component, given the presence of the integration on the minus component in lhs (see, e.g [5] and references quoted therein). Such an integration makes only the first component of the Fock decomposition of $|\Psi\rangle$ to survive.

A possible approximation scheme is suggested by the QP expansion of the effective interaction, Eq. (16). Then, $w(K)$, $t(K)$ and the 3D LF Green's function $g(K)$ (all containing $W(K)$ in a more or less explicit way) can be evaluated by truncating the expansion at a given order. It should be pointed out that the full complexity of the Fock-space affects $g(K)$ through the effective interaction $w(K)$, as can be seen through a time-ordered analysis of the diagrams associated to each contribution $[\Delta_0(K)V(K)]^{i-1}$, and reminding that $\Delta_0(K)$ describes propagation. Indeed, the truncation of the QP expansion limits the number of Fock states involved in the construction of the effective interaction. As a final remark, one could argue that the convergence rate of the QP expansion is related to the small probability of the higher Fock-components.

4. The 3D EM current and the LF Ward-Takahashi Identity

Once we have obtained the relation between the 4D BS amplitude and the 3D LF valence wave function, one can establish a direct link between the matrix element of the 4D current and the matrix elements of the 3D LF em current operator.

For both scattering and bound states one has [1, 4, 2, 3]

$$\langle\Psi_f|\ \mathcal{J}^\mu(Q)\,|\Psi_i\rangle = \langle\phi_f|j^\mu(K_f, K_i)|\phi_i\rangle \tag{22}$$

where the 3D LF current operator, acting on the valence wave functions, is defined as follows

$$j^\mu(K_f, K_i) = \overline{\Pi}(K_f)\mathcal{J}^\mu(Q)\Pi(K_i) = \\ = \overline{\Pi}_0(K_f)\left[1 + W(K_f)\Delta_0(K_f)\right]\mathcal{J}^\mu(Q)\left[1 + \Delta_0(K_i)W(K_i)\right]\Pi_0(K_i) \quad . \tag{23}$$

Since the 4D current must fulfill the WTI, Eq. (4), the 3D LF current fulfills the following LF WTI [1]

$$Q_\mu j^\mu(K_f, K_i) = \overline{\Pi}(K_f)\left[G^{-1}(K_f)\hat{e} - \hat{e}G^{-1}(K_i)\right]\Pi(K_i) = \\ = g^{-1}(K_f)\ \widehat{\mathcal{Q}}^L_{LF} - \widehat{\mathcal{Q}}^R_{LF}\ g^{-1}(K_i) \tag{24}$$

where $\mathcal{Q}^{L(R)}_{LF}$ is called the left (right) LF charge operator (note that it is *interaction free !*). For the sake of concreteness, let us show the explicit expression of the left charge, viz

$$\widehat{\mathcal{Q}}^L_{1LF} = \Lambda_+(\widehat{k}_{1on})\frac{m_1}{k_1^+}\gamma_1^+\ \widehat{e}_{1LF}\Lambda_+(\widehat{k}_{1on})\Lambda_+(\widehat{k}_{2on}) \tag{25}$$

where Λ_+ is the positive energy Dirac projector, and $\widehat{e}_{1LF}$ the 3D pointlike charge operator, given by

$$\langle k_1'^+, \vec{k}'_{1\perp}|\widehat{e}_{1LF}|k_1^+, \vec{k}_{1j\perp}\rangle := e_1\delta\left(k_1'^+ - k_1^+ - Q^+\right)\ \delta^2\left(\vec{k}'_{1\perp} - \vec{k}_{1\perp} - \vec{Q}_\perp\right) \quad . \tag{26}$$

The current conservation straightforwardly follows by taking the matrix elements of Eq. (24) between 3D interacting states $|\phi_{i(f)}\rangle$, solutions of $g^{-1}(K)\ |\phi\rangle = 0$

For the actual calculations, one needs to truncate the iterated expression of the effective interaction $W(K)$, Eq. (16), and therefore one ends up with a truncated 3D em current. The truncation scheme for the LF current cannot be carried out in a naive way, if one has to fulfill a WTI. Indeed, at each truncation order it will correspond a truncated effective interaction, $w^{(n)}(K)$, a truncated LF Green's function $g^{(n)}(K)$, a truncated valence wave function, $|\phi^{(n)}\rangle$, a properly truncated 3D LF current, and finally a truncated WTI. Let us see in details this chain.

Once we truncate the QP expansion of the effective interaction at the order n,

$$W^{(n)}(K) = V(K)\sum_{i=1}^{n}\left[\Delta_0(K)V(K)\right]^{i-1} \tag{27}$$

we can immediately generate an approximate 3D LF Green's function, $g^{(n)}(K)$, leading to an approximate valence wave functions, $|\phi^{(n)}\rangle$, solutions of the following equation

$$[g^{(n)}(K)]^{-1}\ |\phi^{(n)}\rangle = \left[g_0^{-1}(K) - w^{(n)}(K)\right]\ |\phi^{(n)}\rangle = 0 \quad . \tag{28}$$

The corresponding em current, that fulfills CC, if one uses $|\phi^{(n)}\rangle$ for evaluating the matrix elements, cannot be directly constructed inserting in the definition of the 3D current, Eq. (23), the truncated $W^{(n)}(K)$, i.e.

$$Q \cdot j^{(n)}(K_f, K_i) \neq [g^{(n)}(K_f)]^{-1}\ \widehat{\mathcal{Q}}^L_{LF} - \widehat{\mathcal{Q}}^R_{LF}\ [g^{(n)}(K_i)]^{-1} \tag{29}$$

with

$$j^{(n)\mu}(K_f, K_i) = \overline{\Pi}_0(K_f)\left[1 + W^{(n)}(K_f)\Delta_0(K_f)\right]\mathcal{J}^\mu(Q)\ \times \\ \left[1 + \Delta_0(K_i)W^{(n)}(K_i)\right]\Pi_0(K_i) \quad . \tag{30}$$

The fulfillment of the LF WTI excludes such a naive approximation scheme for the truncated current, but, at the same time, gives us the correct hint. One should implement a power counting of the effective interaction that appears in the definition of the LF current (the interaction $V(K)$ is present both in the projectors $\Pi(K)$ and in the 4D current $\mathcal{J}$) and in the rhs of the LF WTI, where only $[g^{(n)}(K_f)]^{-1}$ and $[g^{(n)}(K_i)]^{-1}$ are affected by the interaction, as in the exact case, i.e. Eq. (24).

For the truncation at the n-th order, a truncated current that satisfies a truncated LF WTI is given by [1]

$$j^{c(n)\mu} = j^{c(n-1)\mu} + \overline{\Pi}_0 \left[\sum_{i=0}^{n} W_i \Delta_0 \mathcal{J}_0^\mu \Delta_0 W_{n-i} + \sum_{i=0}^{n-1} W_i \Delta_0 \mathcal{J}_I^\mu \Delta_0 W_{n-1-i} \right] \Pi_0 \quad (31)$$

where $W_i = V\,[\Delta_0 V]^{i-1}$ and it has been formally defined $W_0\Delta_0 = \Delta_0 W_0 = 1$. It should be reminded that $\mathcal{J}_0^\mu$ is $O(V^0)$ and $\mathcal{J}_I^\mu$ is $O(V^1)$.

Then $j^{c(n)\mu}$ satisfies a LF WTI given by

$$Q^\mu j_\mu^{c(n)} = [g^{(n)}]^{-1}(K_f)\; \hat{\mathcal{Q}}^L_{LF} - \hat{\mathcal{Q}}^R_{LF}\; [g^{(n)}]^{-1}(K_i) \quad . \quad (32)$$

For obtaining CC, the matrix elements should be taken between solutions of $[g^{(n)}]^{-1}|\phi_n\rangle = 0$.

5. Two-body current in the LF approach: a pedagogical example

A Yukawa model in ladder approximation has been adopted in order to elaborate an application of our general approach. Within such a model, a couple of fermions interacts by the exchange of a boson, in the present case chargeless, for simplicity. From the Lagrangian, one can deduce immediately the 4D interaction to be used in the definition of the T-matrix and of the 4D current. In ladder approximation, the 4D current is nothing else but the free 4D current that fulfills the suitable WTI (see, e.g. [9]). Let us stress that, though in Minkowski space the current of the model under investigation is the free one, its projection onto a 3D hyperplane becomes interaction dependent. The matrix elements of the first-order LF current, in a 3D Fourier space, are given by

$$\langle k_1'^+ \vec{k}'_{1\perp} | j^{c(1)\mu} | k_1^+ \vec{k}_{1\perp} \rangle = \langle k_1'^+ \vec{k}'_{1\perp} | \left[j^{c(0)\mu} + \overline{\Pi}_0 \left[V\; \Delta_0 \mathcal{J}_0^\mu + \mathcal{J}_0^\mu \Delta_0\; V \right] \Pi_0 \right] | k_1^+ \vec{k}_{1\perp} \rangle \quad (33)$$

with $j^{c(0)\mu} = \overline{\Pi}_0 \mathcal{J}_0^\mu \Pi_0$ (remind that Π_0 is the free-case projector). In figure 1, the diagrammatic analysis of the first-order LF current is shown. It should be pointed out that i) the diagrams are LF-time ordered, ii) the LF-time flows from the right to the left and iii) the external legs are particles on their-own mass shell (therefore only three components can independently change). Such a diagrammatic analysis allows one to appreciate the physical content inside the operator $j_\mu^{c(1)}$. The peculiar feature of the LF approach is the presence of instantaneous contributions, generated by the term proportional to γ^+ in Eq. (10). This contributions are labeled by horizontal dashes. Another interesting contribution is the Z-diagram, or pair production diagram, generated by the time-ordering of the 4D Dirac propagator. In this respect, we should emphasize that the full Dirac structure, i.e. the spinorial structure, is exactly taken into account, since the LF projection only works on the minus components of the four-momenta.

This first-order current will be applied for evaluating the Deuteron electromagnetic properties, in order to extend the analysis at the zero-order already performed [10]. But, one can anticipate that i) the pair term affects all the three electromagnetic form factors, while the instantaneous terms contribute only to the magnetic one (since the property $\gamma^+\;\gamma^+ = 0$ is acting in the charge form factor), ii) the pair term vanishes for $q^+ \to 0$, for the conservation law of the plus components, while the instantaneous ones survive, iii) the pair term should be maximal at $q^+ \sim m_N$ [11], iv) the remaining terms affect all the Deuteron form factors, in the whole range of q^+.

Journal of Physics: Conference Series **168** (2009) 012002 doi:10.1088/1742-6596/168/1/012002

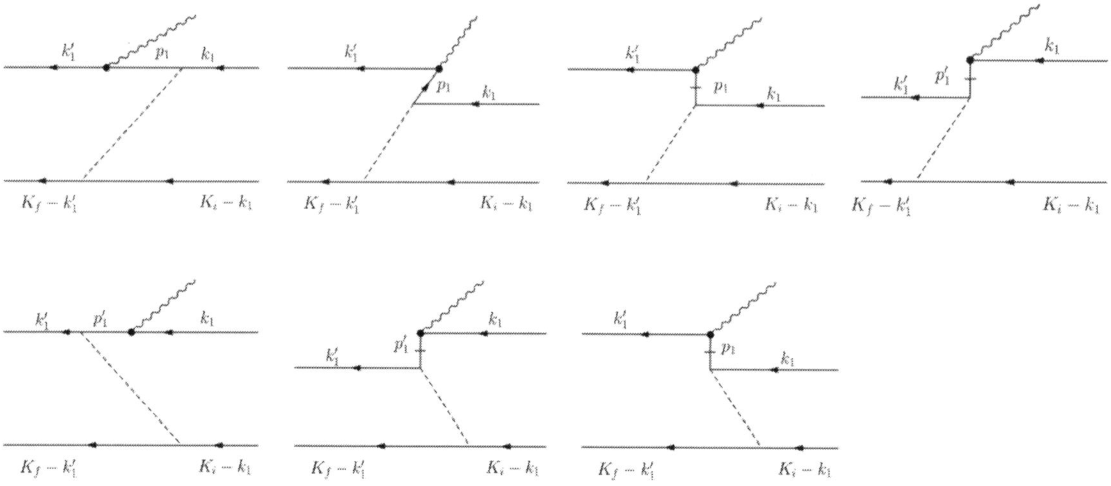

Figure 1. LF-time ordered diagrams for the first-order LF current for a Yukawa model in the ladder approximation, see Eq. (33) and text. Horizontal dashes indicate instantaneous propagations. (Adapted from [1])

6. Conclusions

For two interacting fermions, an exact correspondence between 4D BS amplitudes and 3D LF valence wave functions has been obtained [1], by using the projection onto the LF hyperplane. The approach fulfills the covariance with respect to the kinematical Poincaré subgroup (7 generators out of 10).

The obtained relation allows one to express the matrix elements of the 4D em current in terms of matrix elements of the 3D LF em current between valence wave functions. For such a current operator, the corresponding LF WTI can be constructed by introducing interaction free LF charge operators, then current conservation can be trivially retrieved.

Finally, by using a suitable auxiliary Green's function, within the Quasi-Potential approach, that allows an ordering in terms of intermediate Fock states, an approximation scheme has been developed, for both dynamical equation and the em current operator. In particular, at any order of the effective interaction the em current operator fulfills a proper LF WTI, leading to CC.

A systematic analysis of LF two-body currents, obtained within a Yukawa model in ladder approximation, is in progress for the Deuteron case.

References

[1] Marinho J A O, Frederico T, Pace E, Salmè G and Sauer PU, 2008 *Phys. Rev.* D **77** 116010
[2] Sales J H O, Frederico T, Carlson B V and Sauer P U, 2000 *Phys. Rev.* C **61** 044003
[3] Sales J H O, Frederico ., Carlson B V and Sauer P U, 2001 *Phys. Rev.* C **63** 064003
[4] Marinho J A O, Frederico T and Sauer P U, 2007 *Phys. Rev.* D **76** 096001, and references quoted therein
[5] Karmanov V A and Carbonell J, 2006 *Eur. Phys. J.* A **27** 1; 2006 *Eur. Phys. J.* A **27** 11
[6] Gross F, "Relativistic Quantum Mechanics and Field Theory" (John Wiley & Sons, New York 1993)
[7] Woloshyn R M and Jackson A D, 1973 *Nucl. Phys.* B **64** 269
[8] Brodsky S J, Pauli H C and Pinsky S S, 1998 *Phys. Rep.* **301** 299
[9] Gross F and Riska D O, 1987 *Phys. Rev.* C **36** 1928
[10] Lev F M, Pace E and Salmè G, 2000 *Phys. Rev.* C **62** 0640004
[11] De Melo J P B C, Frederico T, Pace E and Salmè G, 2002 *Nucl. Phys.* A **707** 399

Hadronic parity violation in few-body systems

L Girlanda[1,2], R Schiavilla[3,4], M Viviani[2,1], A Kievsky[2,1] and L E Marcucci[1,2]

[1] Department of Physics, University of Pisa, I-56127 Pisa, Italy
[2] INFN, Sezione di Pisa, I-56127 Pisa, Italy
[3] Jefferson Lab, Newport News, Virginia 23606, USA
[4] Department of Physics, Old Dominion University, Norfolk, Virginia 23529, USA

E-mail: girlanda@pi.infn.it

Abstract. Recent interest, both from experimental and theoretical point of view, on hadronic parity violation is reviewed, with particular emphasis on an effective theory description. After discussing the minimal form of the parity-violating NN contact effective Lagrangian, we concentrate on the calculation of the neutron spin rotation in $\vec{n}-d$ scattering at zero energy. We find that this observable is sensitive to the long-range component of the parity-violating NN pontential due to the pion exchange, and that it is expected to be one order of magnitude larger than in $\vec{n}-p$ scattering.

1. Introduction

The understanding of weak interaction, while satisfactory at the level of quarks and leptons, is complicated in the hadronic sector by the non-perturbative character of QCD in this regime. At the nuclear level, weak interactions can only be studied by analyzing small effects induced by characteristic properties of this interaction, like flavour changing or parity violation.

There remain indeed many puzzles in the domain of hadronic weak interactions (HWI) [1], which are yet to be explained: among the most prominent ones are the origin of the $\Delta I = 1/2$ rule in the non-leptonic decays of kaons and hyperons, largely dominated by transition amplitudes with $\Delta I = 1/2$ [2], the difficulties in describing simultaneously the S- and P-waves of hyperon non-leptonic decays [3], and the anomalously large parity-violating asymmetries in the radiative hyperon decays [4]. All these problems involve the strange quark and might be ascribed to a lack of convergence of the SU(3) chiral series, for several possible reasons (see e.g. Ref. [5]). In order to isolate the genuine properties of HWI, without being much affected by the presence of dynamical strange quarks, it is therefore interesting to focus on the $\Delta S = 0$ component of HWI, by examining the hadronic parity violation (HPV). In this respect, few-nucleon systems provide a clean theoretical laboratory, due to the possibility of performing accurate *ab initio* calculations. Improved experimental techniques allow now to measure on such systems the tiny effects of HPV, which are of the order of $G_F m_\pi^2 \sim O(10^{-7})$.

The main observables under experimental consideration are: i) the longitudinal analyzing power in $\vec{p}p$ scattering; ii) the asymmetry A_γ^d of the emitted photon in the radiative capture $\vec{n}p \to d\gamma$ with respect to the initial neutron polarization, with a sensitivity of 10^{-8} expected at the NPDGamma experiment [6]; iii) the asymmetry A_γ^t in $\vec{n}d \to t\gamma$, proposed as a follow up

Journal of Physics: Conference Series **168** (2009) 012003 doi:10.1088/1742-6596/168/1/012003

of the latter experiment at the Spallation Neutron Source (SNS), Oak Ridge; *iv*) neutron spin rotation in $\vec{n}p$, $\vec{n}d$ and $\vec{n}\alpha$, also planned at the SNS.

In the following, we will concentrate on the calculation of the neutron spin rotation in $\vec{n}-d$ scattering at zero energy [7]. In section 2 we discuss the two models for the PV NN potential that we have used, with particular emphasis on the leading-order effective Lagrangian, exhibited in its minimal form [8] (see also Ref. [9]); the neutron spin rotation observable is briefly described in section 3, while section 4 contains our results.

2. Models of parity-violating NN interaction

The traditional framework for the analysis of HPV has been, for almost three decades, the meson-exchange parity-violating NN potential of Desplanques, Donoghue and Holstein [10], which incorporates π, ρ and ω exchange. It is parametrized by 7 parity-violating meson-nucleon couplings, to be determined from experiment. These couplings have rather broad allowed ranges, and moreover, the consistency between different experiments is not entirely satisfactory [11]. A more modern point of view has been advocated recently in Ref. [12], by recasting HPV in the language of effective field theory (EFT). This has the advantage of being a model-independent approach, based on the separation of the physical scales involved in the problem: subleading corrections can in principle be evaluated systematically in a $p/\Lambda_{\rm H}$ expansion, p being the typical momentum and $\Lambda_{\rm H}$ the hadronic scale, characteristic of hadrons whose mass is not protected by chiral symmetry. It is then ideally suited for a situation in which experimental data are scarce. The effective theory is formulated in terms of nucleons and pions only, transforming non-linearly under the broken chiral symmetry of QCD [13]. In addition to the (strong) parity-conserving couplings of chiral perturbation theory, one has to consider the induced PV couplings from weak interactions. Virtual W and Z exchanges induce (at the quark level) four-quark, current-current operators in the effective hamiltonian, which have definite transformation rules under the broken chiral symmetry of strong interactions. Hadronic operators, built in terms of pions and nucleons, are therefore introduced in the effective theory, with the same transformation properties [14], yielding a PV effective Lagrangian, whose infinite number of terms is ordered according to the power counting of chiral perturbation theory. The effect of heavier particles is encoded in the value of the low-energy coupling constants. At the leading order, the relevant Lagrangian is the sum of two pieces,

$$\mathcal{L}^{\rm PV} = \mathcal{L}^{\rm PV}_{\pi N} + \mathcal{L}^{\rm PV}_{NN}, \tag{1}$$

the first one containing the interaction of pions with a single nucleon, the second one containing 2-nucleon contact vertices. At energies much smaller than the pion mass, even pions can be integrated out, and the pion mass be taken as the ultraviolet cutoff Λ. The "pionless effective field theory" ($\not\pi$EFT) is then expressed in terms of nucleons only, with interactions described by the effective Lagrangian $\mathcal{L}^{\rm PV}_{NN}$. In this case however, the value of the low-energy constants will be changed, compared to Eq. (1), since in $\not\pi$EFT they include the effects of the integrated-out pions.

In the following we will focus our attention on the contact Lagrangian $\mathcal{L}^{\rm PV}_{NN}$. It has been written in Ref. [12] as consisting of 12 operators,

$$\mathcal{L}^{\rm PV}_{NN} = \sum_{i=1}^{6}(C_i O_i + \tilde{C}_i \tilde{O}_i), \tag{2}$$

Table 1. Components of the PV NN potential corresponding to the DDH model and to the $\not\pi$EFT.

n	$c_n^{\rm DDH}$	$f_n^{\rm DDH}(r)$	$c_n^{\rm EFT}$	$f_n^{\rm EFT}(r)$	$O_{ij}^{(n)}$
1	$+\frac{g_\pi h_\pi^1}{2\sqrt{2}\,m}$	$f_\pi(r)$	$\frac{2\mu^2}{\Lambda_\chi^3}\,C_6$	$f_\mu(r)$	$(\boldsymbol{\tau}_i \times \boldsymbol{\tau}_j)_z\,(\boldsymbol{\sigma}_i + \boldsymbol{\sigma}_j)\cdot \mathbf{X}_{ij,-}^{(1)}$
2	$-\frac{g_\rho h_\rho^0}{m}$	$f_\rho(r)$	0	0	$\boldsymbol{\tau}_i \cdot \boldsymbol{\tau}_j\,(\boldsymbol{\sigma}_i - \boldsymbol{\sigma}_j)\cdot \mathbf{X}_{ij,+}^{(2)}$
3	$-\frac{g_\rho h_\rho^0(1+\kappa_\rho)}{m}$	$f_\rho(r)$	0	0	$\boldsymbol{\tau}_i \cdot \boldsymbol{\tau}_j\,(\boldsymbol{\sigma}_i \times \boldsymbol{\sigma}_j)\cdot \mathbf{X}_{ij,-}^{(3)}$
4	$-\frac{g_\rho h_\rho^1}{2m}$	$f_\rho(r)$	$\frac{\mu^2}{\Lambda_\chi^3}\,(C_2+C_4)$	$f_\mu(r)$	$(\boldsymbol{\tau}_i + \boldsymbol{\tau}_j)_z\,(\boldsymbol{\sigma}_i - \boldsymbol{\sigma}_j)\cdot \mathbf{X}_{ij,+}^{(4)}$
5	$-\frac{g_\rho h_\rho^1(1+\kappa_\rho)}{2m}$	$f_\rho(r)$	0	0	$(\boldsymbol{\tau}_i + \boldsymbol{\tau}_j)_z\,(\boldsymbol{\sigma}_i \times \boldsymbol{\sigma}_j)\cdot \mathbf{X}_{ij,-}^{(5)}$
6	$-\frac{g_\rho h_\rho^2}{2\sqrt{6}\,m}$	$f_\rho(r)$	$-\frac{2\mu^2}{\Lambda_\chi^3}\,C_5$	$f_\mu(r)$	$(3\,\tau_{i,z}\tau_{j,z} - \boldsymbol{\tau}_i \cdot \boldsymbol{\tau}_j)\,(\boldsymbol{\sigma}_i - \boldsymbol{\sigma}_j)\cdot \mathbf{X}_{ij,+}^{(6)}$
7	$-\frac{g_\rho h_\rho^2(1+\kappa_\rho)}{2\sqrt{6}\,m}$	$f_\rho(r)$	0	0	$(3\,\tau_{i,z}\tau_{j,z} - \boldsymbol{\tau}_i \cdot \boldsymbol{\tau}_j)\,(\boldsymbol{\sigma}_i \times \boldsymbol{\sigma}_j)\cdot \mathbf{X}_{ij,-}^{(7)}$
8	$-\frac{g_\omega h_\omega^0}{m}$	$f_\omega(r)$	$\frac{2\mu^2}{\Lambda_\chi^3}\,C_1$	$f_\mu(r)$	$(\boldsymbol{\sigma}_i - \boldsymbol{\sigma}_j)\cdot \mathbf{X}_{ij,+}^{(8)}$
9	$-\frac{g_\omega h_\omega^0(1+\kappa_\omega)}{m}$	$f_\omega(r)$	$\frac{2\mu^2}{\Lambda_\chi^3}\,\tilde{C}_1$	$f_\mu(r)$	$(\boldsymbol{\sigma}_i \times \boldsymbol{\sigma}_j)\cdot \mathbf{X}_{ij,-}^{(9)}$
10	$-\frac{g_\omega h_\omega^1}{2m}$	$f_\omega(r)$	0	0	$(\boldsymbol{\tau}_i + \boldsymbol{\tau}_j)_z\,(\boldsymbol{\sigma}_i - \boldsymbol{\sigma}_j)\cdot \mathbf{X}_{ij,+}^{(10)}$
11	$-\frac{g_\omega h_\omega^1(1+\kappa_\omega)}{2m}$	$f_\omega(r)$	0	0	$(\boldsymbol{\tau}_i + \boldsymbol{\tau}_j)_z\,(\boldsymbol{\sigma}_i \times \boldsymbol{\sigma}_j)\cdot \mathbf{X}_{ij,-}^{(11)}$
12	$-\frac{g_\omega h_\omega^1 - g_\rho h_\rho^1}{2m}$	$f_\rho(r)$	0	0	$(\boldsymbol{\tau}_i - \boldsymbol{\tau}_j)_z\,(\boldsymbol{\sigma}_i + \boldsymbol{\sigma}_j)\cdot \mathbf{X}_{ij,+}^{(12)}$
13	$-\frac{g_\rho h_\rho^{\prime 1}}{2m}$	$f_\rho(r)$	0	0	$(\boldsymbol{\tau}_i \times \boldsymbol{\tau}_j)_z\,(\boldsymbol{\sigma}_i + \boldsymbol{\sigma}_j)\cdot \mathbf{X}_{ij,-}^{(13)}$

where

$$
\begin{aligned}
O_1 &= \bar\psi\gamma^\mu\psi\bar\psi\gamma_\mu\gamma_5\psi, & \tilde O_1 &= \bar\psi\gamma^\mu\gamma_5\psi\partial^\nu(\bar\psi\sigma_{\mu\nu}\psi),\\
O_2 &= \bar\psi\gamma^\mu\psi\bar\psi\tau_3\gamma_\mu\gamma_5\psi, & \tilde O_2 &= \bar\psi\gamma^\mu\gamma_5\psi\partial^\nu(\bar\psi\tau_3\sigma_{\mu\nu}\psi),\\
O_3 &= \bar\psi\tau^a\gamma^\mu\psi\bar\psi\tau^a\gamma_\mu\gamma_5\psi, & \tilde O_3 &= \bar\psi\tau^a\gamma^\mu\gamma_5\psi\partial^\nu(\bar\psi\tau^a\sigma_{\mu\nu}\psi),\\
O_4 &= \bar\psi\tau_3\gamma^\mu\psi\bar\psi\gamma_\mu\gamma_5\psi, & \tilde O_4 &= \bar\psi\tau_3\gamma^\mu\gamma_5\psi\partial^\nu(\bar\psi\sigma_{\mu\nu}\psi),\\
O_5 &= \mathcal{I}_{ab}\bar\psi\tau^a\gamma^\mu\psi\bar\psi\tau^b\gamma_\mu\gamma_5\psi, & \tilde O_5 &= \mathcal{I}_{ab}\bar\psi\tau^a\gamma^\mu\psi\partial^\nu(\bar\psi\tau^b\sigma_{\mu\nu}\psi),\\
O_6 &= i\epsilon^{ab3}\bar\psi\tau^a\psi\bar\psi\tau^b\gamma_5\psi, & \tilde O_6 &= i\epsilon^{ab3}\bar\psi\tau^a\gamma^\mu\psi\partial^\nu(\bar\psi\tau^b\sigma_{\mu\nu}\gamma_5\psi),
\end{aligned}
\tag{3}
$$

the nucleon field ψ is a doublet in isospin space and $\mathcal{I}$ =diag$(1,1,-2)$. However, this set of operators is redundant: by using fields' equations of motion and Fierz rearrangements, the following relations may be shown to hold [8],

$$
\begin{aligned}
O_1 &= O_3, & O_2 - O_4 &= 2O_6,\\
\tilde O_3 &= 2m(O_1 + O_3), & \tilde O_2 + \tilde O_4 &= m(O_2 + O_4),\\
O_5 &= \tilde O_5, & \tilde O_2 - \tilde O_4 &= -2mO_6 - \tilde O_6,
\end{aligned}
\tag{4}
$$

so that the number of independent operators can be reduced to 6. One can then make use of the fact that the operators O_6 and $\tilde O_6$ give rise, in the leading order of the non-relativistic reduction, to the same structure, as already observed in Ref. [12]. Therefore, the number of independent operators can be further reduced to 5 up to order $O(Q)$, and the minimal PV two-nucleon non-relativistic contact Lagrangian may be taken to assume the form

$$\begin{aligned}\mathcal{L}_{NN}^{PV} \;=\; & \frac{1}{\Lambda_\chi^3}\Big\{C_1(N^\dagger\vec{\sigma}N\cdot N^\dagger i\overleftrightarrow{\nabla}N - N^\dagger N N^\dagger i\overleftrightarrow{\nabla}\cdot\vec{\sigma}N)\\ & -\tilde{C}_1\epsilon_{ijk}N^\dagger\sigma^i N\nabla^j(N^\dagger\sigma^k N)\\ & -(C_2+C_4)\epsilon_{ijk}[N^\dagger\tau_3\sigma^i N\nabla^j(N^\dagger\sigma^k N)+N^\dagger\sigma^i N\nabla^j(N^\dagger\tau_3\sigma^k N)]\\ & -\tilde{C}_5\mathcal{I}_{ab}\epsilon_{ijk}N^\dagger\tau^a\sigma^i N\nabla^j(N^\dagger\tau^b\sigma^k N)\\ & +C_6\epsilon^{ab3}\vec{\nabla}(N^\dagger\tau^a N)\cdot N^\dagger\tau^b\vec{\sigma}N\Big\}, \end{aligned} \tag{5}$$

where the notations for the coupling constants have been chosen so as to conform to Ref. [12]. It is worth noting that the reduction of the number of independent operators down to five has no practical consequences at the present stage of phenomenological analyses since only five combinations of low-energy constants are relevant at low energies. This was already noticed in Refs. [12, 1] on the basis of the observation that, since only S- and P-wave amplitudes are important in this regime, several operators give rise to identical matrix elements (see also the discussion in Ref. [15]). Nevertheless, in view of a description of nuclear parity violation with the chiral effective theory, it is important to use a truly minimal set of operators.

The resulting PV NN potential consists of 5 different components, as opposed to the 13 components of the DDH model. It is convenient to express both potential in a unified notation,

$$v_{ij}^{\alpha} = \sum_{n=1}^{13} c_n^{\alpha} O_{ij}^{(n)}, \quad \alpha = \text{ DDH or EFT.} \tag{6}$$

The parameters $c_n^{\rm EFT}$ are in one-to-one correspondence with the 5 low-energy constants of the effective Lagrangian, while the $c_n^{\rm DDH}$ are expressed in terms of 7 unknown PV meson-nucleon couplings (h_π^1, $h_\rho^{0,1,2}$, $h_\rho^{\prime 1}$ and $h_\omega^{0,1}$) as displayed in table 1 together with the corresponding space-spin-isospin structures. The vector spatial operators are defined as

$$\mathbf{X}_{ij,+}^{(n)} \equiv [\mathbf{p}_{ij}, f_n(r_{ij})]_+, \quad \mathbf{X}_{ij,-}^{(n)} \equiv i[\mathbf{p}_{ij}, f_n(r_{ij})]_-, \tag{7}$$

where $[...]_-$ ($[...]_+$) stands for the commutator (anticommutator) , and $\mathbf{p}_{ij} = (\mathbf{p}_i - \mathbf{p}_j)/2$ for the relative momentum of nucleons i and j. In the DDH model the radial functions $f_x(r)$ (with $x = \pi, \rho, \omega$) are Yukawa functions modified by the inclusion of monopole form factors,

$$f_x(r) = \frac{1}{4\pi r}\left\{ {\rm e}^{-m_x r} - {\rm e}^{-\Lambda_x r}\left[1 + \frac{\Lambda_x r}{2}\left(1 - \frac{m_x^2}{\Lambda_x^2}\right)\right]\right\}, \tag{8}$$

while in the EFT there is a single cutoff function $f_\mu(r)$, taken as a Yukawa function of range $1/\mu$,

$$f_\mu(r) = \frac{1}{4\pi r}{\rm e}^{-\mu r}. \tag{9}$$

3. Neutron spin rotation in $\vec{n}d$ scattering

Due to the presence of PV interactions (as exemplified in figure 3 by a $\boldsymbol{\sigma}\cdot\boldsymbol{p}$ interaction), neutrons polarized in the transverse direction (e.g. along $\mathbf{x}$, with $|x\rangle \propto |+\rangle + |-\rangle$), while traveling along $\mathbf{z}$, accumulate different phases along $|+\rangle$ and $|-\rangle$, and the spin $\boldsymbol{\sigma}$ is then viewed to rotate by an angle ϕ proportional to the forward scattering amplitude with the deuteron,

$$\frac{1}{\rho}\frac{d\phi}{dz} = \frac{1}{3v_{\rm rel}}\sum_{m_n,m_d} \epsilon_{m_n}\,{}_{\rm in}\langle p\hat{\mathbf{z}}; m_n, m_d|v^{\rm PV}|p\hat{\mathbf{z}}; m_n, m_d\rangle_{\rm out}. \tag{10}$$

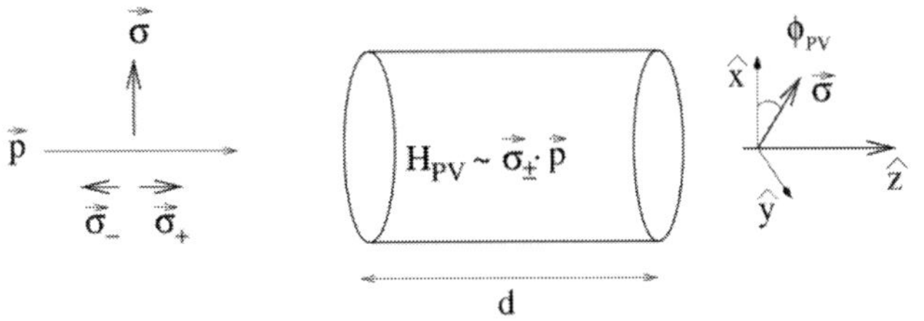

Figure 1. Neutron spin rotation produced by a PV interaction.

Here ρ is the deuteron density, $v^{\rm PV}$ is the PV NN potential, $\mathbf{p} = p\hat{\mathbf{z}}$ is the relative momentum, taken along the spin-quantization axis, $v_{\rm rel} = p/\mu$ is the relative velocity, μ being the $n-d$ reduced mass, and m_d (m_n) the deuteron (neutron) spin projection. The phase factor $\epsilon_{m_n} = \pm 1$ depending on whether $m_n = \pm 1/2$. The scattering states are calculated from the AV18 potential [16], supplemented by the three-nucleon interaction Urbana IX [17], by means of the hyperspherical harmonics method [18]. They are expanded in partial waves as [19]

$$|p\hat{\mathbf{z}}; m_n, m_d\rangle_{\rm in/out} = 4\pi \sum_{LSJ} \langle \frac{1}{2}m_n; 1m_d|SS_z\rangle i^L \langle SS_z; L0|JJ_z\rangle Y^*_{L0}(\hat{\mathbf{z}})|p; LS; J, J_z\rangle_{\rm in/out}. \qquad (11)$$

For very low-energy neutrons, only the S to P transitions are relevant, and the spin rotation observable is written, using the time-reversal transformation properties of the ingoing and outgoing states, as

$$\begin{aligned} \frac{1}{\rho}\frac{{\rm d}\phi}{{\rm d}z} &= \frac{8\pi}{\sqrt{3}v_{\rm rel}} \sum_{m_n m_d} \sum_{SJ} \epsilon_{m_n} \langle \frac{1}{2}\, m_n, 1\, m_d \mid J J_z\rangle \langle \frac{1}{2}\, m_n, 1\, m_d \mid S J_z\rangle \langle S J_z, 1\, 0 \mid J J_z\rangle \\ & \quad {\rm Im}\Big[{}_{\rm in}\langle\, p; 1S; J, J_z \mid v^{\rm PV} \mid p; 0J; J, J_z\,\rangle_{\rm out} \Big]\,. \end{aligned} \qquad (12)$$

The PV potential connects the doublet S-wave state ${}^2S_{1/2}$ to both the doublet ${}^2P_{1/2}$ and quartet ${}^4P_{1/2}$ P-wave states, and similarly for $J = 3/2$.

4. Results

In the calculation of the spin rotation observable, we do not consider any definite value for the coupling constants. Instead, we express the result as a sum

$$\frac{1}{\rho}\frac{d\phi}{dp} = \sum_{n=1}^{13} c_n^\alpha I_n^\alpha \qquad (13)$$

where the contribution from the single components of the PV NN potential are computed separately. Details of the calculation can be found in Ref. [7]. The numerical procedure has also been tested for the case of gaussian-like wave functions, for which the calculations can be done almost analytically. We have selected two sets of ultraviolet cutoffs, as indicated in table 2, for both the DDH model and the $\not\pi$EFT, in order to test the cutoff dependence.

The results for the individual contributions I_n as defined in Eq. (13) are displayed in table 3 for the DDH model with and without cutoff. The effect of the inclusion of the three-nucleon interaction is also shown. The coefficients $I^{\rm DDH}_{n=6,7}$ vanish because of isospin selection rules, while the results for $I^{\rm DDH}_{13}$ are not reported, because the coupling constant h'^{1}_{ρ} that multiplies this contribution has been estimated to be much smaller [20] than the "best" values estimated for

Table 2. Different sets of values, in GeV units, for the short-range cutoffs Λ_x and μ used in the DDH model and in the $\not\pi$EFT. The masses m_π, m_ρ, and m_ω are taken respectively as 0.138, 0.771, and 0.783 in units of GeV. The row with $\Lambda_x = \infty$ corresponds to point-like couplings.

	Λ_π	Λ_ρ	Λ_ω		μ
DDH-I	1.72	1.31	1.50	EFT-I	0.138
DDH-II	∞	∞	∞	EFT-II	1.0

Table 3. The coefficients $I_n^{\rm DDH}$ (in fm) in Eq. (13) corresponding to the DDH PV potential in combination with the AV18 and AV18/UIX strong-interaction potentials for the two different sets of cutoff parameters given in table 2. The statistical errors associated with the Monte Carlo integrations are not shown, but are typically at the 1-2% level.

	DDH-I		DDH-II	
n	AV18	AV18/UIX	AV18	AV18/UIX
1	$0.256E+03$	$0.270E+03$	$0.257E+03$	$0.274E+03$
2	$-0.444E+01$	$-0.691E+01$	$-0.719E+01$	$-0.118E+02$
3	$0.444E+01$	$0.401E+01$	$0.732E+01$	$0.761E+01$
4	$-0.231E+01$	$-0.881E+00$	$-0.332E+01$	$-0.148E+01$
5	$-0.247E+01$	$-0.122E+01$	$-0.387E+01$	$-0.234E+01$
8	$0.420E+01$	$0.362E+01$	$0.543E+01$	$0.516E+01$
9	$0.111E+01$	$-0.117E+00$	$0.136E+01$	$-0.189E+00$
10	$-0.253E+01$	$-0.991E+00$	$-0.314E+01$	$-0.141E+01$
11	$-0.280E+01$	$-0.144E+01$	$-0.369E+01$	$-0.225E+01$
12	$0.316E+01$	$0.339E+01$	$0.455E+01$	$0.546E+01$

the other PV meson-nucleon couplings in the DDH model [10]. As it appears, the matrix element of the pion-range component of the DDH potential is larger by two orders of magnitude than the components of shorter range induced by ρ and ω exchanges. Moreover, the long-range component is quite independent of the cutoff, and it is not much affected by the inclusion of the three-nucleon force. On the contrary, the other individual contributions due to vector meson exchanges exhibit considerable dependence on the strong-interaction input and on the cutoff. This model dependence, though, has little impact on the neutron spin rotation observable. This is also shown in table 4 where the contributions from the long- and short-range components to this observable are indicated separately, for the cutoff version of the DDH model (DDH-I): the long-range component accounts for almost the 90% of the entire contribution. For the sake of comparison, two different set of strong- and weak-interaction coupling constants are used, corresponding to those reported in tables I and II of Ref. [21]: the "best" values are taken from Ref. [10], while the "adjusted" set is the one found in Ref. [22] to reproduce precise measurements of the longitudinal asymmetry in $\vec{p}-p$ scattering. The PV pion-nucleon coupling h_π^1 is the same for both sets. Interestingly enough, the neutron spin rotation in deuterium is predicted to be one order of magnitude larger than in the case of $\vec{n}-p$ [21], which should make the measurement easier, in principle.

In table 5 we show the individual contributions $I_n^{\rm EFT}$ from the PV NN potential derived in $\not\pi$EFT, corresponding to $n = 1$, 4, 8 and 9. The term $n = 6$ is isotensor, and has vanishing matrix

Table 4. Spin rotation for $\vec{n} - d$ scattering in units of 10^{-7} rad/cm at zero energy, obtained for the DDH-I model assuming a liquid deuterium density of $\rho = 0.4 \times 10^{23}$ atoms cm^{-3}. The columns labeled DDH-best (DDH-adj) list the spin rotation predicted using the "best" ("adjusted") values for the weak-interaction coupling constants, as reported in table II (I) of Ref. [21]. The row labeled "π" ("ρ-ω") lists the results obtained by including only the pion (ρ and ω mesons), while that labeled "TOT" gives the total contributions.

	AV18		AV18/UIX	
	DDH − best	DDH − adj	DDH − best	DDH − adj
π	0.457	0.457	0.485	0.485
ρ-ω	0.063	0.092	0.046	0.075
TOT	0.520	0.550	0.531	0.560

Table 5. Same as in table 3 but for the pionless EFT PV potential. Note that there are no potential components with $O_{ij}^{(n)}$ with n=2, 3, 5, 7, 10 and 11, and that the coefficient $I_{n=6}^{\rm EFT}$ vanishes because of isospin selection rules.

	EFT-I		EFT-II	
n	AV18	AV18/UIX	AV18	AV18/UIX
1	$0.257E+03$	$0.274E+03$	$0.392E+01$	$0.706E+01$
4	$-0.154E+03$	$-0.508E+02$	$-0.127E+01$	$-0.682E+00$
8	$0.260E+03$	$0.189E+03$	$0.233E+01$	$0.251E+01$
9	$0.244E+00$	$-0.359E+02$	$0.557E+00$	$-0.955E-01$

elements. Contrary to the DDH model, all contributions are of the same order of magnitude and correspond to radial functions with a same range μ. Of course, they depend significantly on the value of the cutoff, chosen as $\mu = m_\pi$ (EFT-I), as appropriate in $\not\pi$EFT, or as $\mu = 1$ GeV (EFT-II), as appropriate in the pionful version of the EFT. In the latter case, the leading-order component of the PV NN potential has the same form as the pion-exchange term in the DDH model. Also notice that, while the $n = 1$ contribution is weakly dependent on the strong-interaction input, the same is not true for the other coefficients. However, we did not attempt to establish predictions for the spin rotation observable in $\not\pi$EFT, since at the present stage the value of the low-energy constants is essentially unknown.

5. Conclusions

The EFT approach, in conjunction with precise *ab initio* techniques for handling the few-body problem, allows to investigate on a firm theoretical basis the hadronic parity violation. Future and on-going experimental work will hopefully allow to pin down the PV low-energy constants, and shed light on long-standing problems of non-perturbative QCD. We have constructed the relevant leading-order NN contact Lagrangian in its minimal form, which is an essential ingredient in both the pionful and pionless versions of the EFT. We have computed the neutron spin rotation in liquid deuterium at zero energy, and parametrized it in terms of the low-energy constants or the parameters of the DDH model. Within the DDH model, this observable is dominated by the contribution of the long-range part of the PV potential due to pion exchange, and this contribution is almost model-independent. The predicted value, $d\phi/dz \sim 0.5 \times 10^{-7}$ rad/cm^{-1}, is an order of magnitude larger than expected in $\vec{n}-p$ scattering [21]. An experimental investigation of this observable could therefore provide a further constraint, complementary to that coming from measurements of the photon asymmetry in $\vec{n}-p$ radiative capture, on the strength of the long-range component of the hadronic weak interaction.

References

[1] Ramsey-Musolf M J and Page S A 2006 *Ann. Rev. Nucl. Part. Sci.* **56** 1
[2] Donoghue J F, Golowich E and Holstein B R 1992 *Dynamics of the standard model* (Cambridge University Press)
[3] Bijnens J, Sonoda H and Wise M 1985 *Nucl. Phys.* B **261** 185; Jenkins E 1992 *Nucl. Phys.* B **375** 561
[4] Borasoy B 2005 *Nucl. Phys.* A **754** 191, and references therein
[5] Descotes-Genon S, Girlanda L and Stern J 2000 *JHEP* **0001** 041; Descotes-Genon S, Fuchs N H, Girlanda L and Stern J 2003 *Eur. Phys. J.* C **34** 201
[6] Lauss B *et al* 2006 *AIP Conf. Proc.* **842** 790
[7] Schiavilla R, Viviani M, Girlanda L, Kievsky A, Marcucci L E 2008 *Phys. Rev.* C **78** 014002
[8] Girlanda L 2008 *Phys. Rev.* C **77** 067001
[9] Phillips D R, Schindler M R and Springer R P 2009 *Nucl. Phys.* A, in press; *Preprint* arXiv:0812.2073 [nucl-th], doi:10.1016/j.nuclphysa.2009.02.011
[10] Desplanques B, Donoghue J F and Holstein B R 1980 *Annals Phys.* **124** 449
[11] Haxton W C 2008 *Preprint* arXiv:0802.2984 [nucl-th].
[12] Zhu S L, Maekawa C M, Holstein B R, Ramsey-Musolf M J and van Kolck U 2005 *Nucl. Phys.* A **748** 435
[13] Coleman S R, Wess J and Zumino B 1969 *Phys. Rev.* **177** 2239; Callan C G, Coleman S R, Wess J and Zumino B 1969 *Phys. Rev.* **177** 2247 (1969)
[14] Kaplan D B and Savage M J 1993 *Nucl. Phys.* A **556** 653.
[15] Liu C P 2007 *Phys. Rev.* C **75** 065501
[16] Wiringa R B, Stoks V G J and Schiavilla R 1995 *Phys. Rev.* C **51** 38
[17] Pudliner B S, Pandharipande V R, Carlson J and Wiringa R B 1995 *Phys. Rev. Lett.* **74** 4396
[18] Kievsky A, Rosati S and Viviani M 1994 *Nucl. Phys.* A **577** 511; Kievsky A, Viviani M and Marcucci L E 2004 *Phys. Rev.* C **69** 014002; Kievsky A, Rosati S, Viviani M, Marcucci L E and Girlanda L 2008 *J. Phys.* G **35** 063101 (2008)
[19] Viviani M, Schiavilla R and Kievsky A 1996 *Phys. Rev.* C **54** 534
[20] Holstein B R 1981 *Phys. Rev.* D **23** 1618
[21] Schiavilla R, Carlson J and Paris M 2004 *Phys. Rev.* C **70** 044007
[22] Carlson J, Schiavilla R , Brown V R and Gibson B F 2002 *Phys. Rev.* C **65** 035502

Three-nucleon force study in $A = 3, 4$ systems

A Kievsky[1], S Rosati[2,1], M Viviani[1], L E Marcucci[2,1] and L Girlanda[1]

[1] Istituto Nazionale di Fisica Nucleare, Sezione di Pisa, Largo B. Pontecorvo 3, I-56127, Pisa, Italy

[2] Department of Physics "E. Fermi", University of Pisa, Largo B. Pontecorvo 3, I-56127, Pisa, Italy

E-mail: kievsky@pi.infn.it

Abstract. Using modern nucleon-nucleon interactions in the descritpion on $A = 3, 4$ nuclei, it is not possible to reproduce both the three- and four-nucleon binding energies. This is one manifestation of the need to include a three-nucleon force in the nuclear Hamiltonian. In this paper we will analyze the parametrization of some three-nucleon force models and their capability to describe simultaneously the aforementioned binding energies and selected $N-d$ low energy scattering observables.

1. Introduction

Realistic nucleon-nucleon (NN) potentials reproduce the experimental NN scattering data up to energies of 350 MeV with a χ^2 per datum close to 1. However, the use of these potentials in the description of the three- and four-nucleon bound and scattering states gives a χ^2 per datum much larger than 1 (see for example Ref.[1]). In order to improve this situation, different three-nucleon force (TNF) models have been introduced. Widely used in the literature are the Tucson-Melbourne (TM) and the Urbana IX (URIX) models [2, 3]. These models are based on the exchange mechanism of two pions between three nucleons with the intermediate excitation of a Δ resonance. The TM model has been revisited within a chiral symmetry approach [4], and it has been demonstrated that the contact term present in it should be dropped. This new TM potential, known as TM', has been subsequently readjusted [5]. The final operatorial structure coincides with that one given in the TNF of Brazil already derived many years ago [6]. More recently, TNF models have been derived [7] based on chiral effective field theory at next-to-next-to-leading order. A local version of this interaction (hereafter referred as N2LO) can be found in Ref. [8]. All these models contain a certain number of parameters that fix the strength of the different terms. It is a common practice to determine these parameters from the three- and four-nucleon binding energies. In fact, a TNF is in general associated to a NN potential and the sum of the two interactions forms the nuclear potential energy. Therefore, the parametrization of a particular TNF can change when used with different NN potentials since different NN potentials predict different $A = 3, 4$ binding energies. In this paper we will make a critical analysis of the agreement obtained in some observables of the $A = 3, 4$ systems, when the potential energy consists of a NN potential plus a TNF. We will study the corresponding binding energies as well as the predictions for the doublet $n-d$ scattering length ${}^2a_{nd}$ and some polarization observables in $p-d$ scattering. For this purpose we use the hyperspherical harmonic

(HH) method as developed by some of the authors to describe bound and scattering states in $A = 3, 4$ systems [9, 10, 11, 12] (for a recent review see Ref. [13]).

2. Binding energies and scattering lengths for $A = 3, 4$

From the results obtained in Ref. [13], we report in table 1 the triton and ^{4}He binding energies, and the doublet $n - d$ scattering length $^2a_{nd}$. These results were obtained using the AV18 or the N3LO-Idaho two-nucleon potentials together with the AV18+URIX, AV18+TM' and N3LO-Idaho+N2LO TNF models. The results are compared to the experimental values also reported in the table. Worthy of notice is the recent very accurate datum for $^2a_{nd}$ [14].

Table 1. The triton and ^{4}He binding energies B (MeV), and doublet scattering length $^2a_{nd}$ (fm) calculated using the AV18 and the N3LO-Idaho two-nucleon potentials, and the AV18+URIX, AV18+TM' and N3LO-Idaho+N2LO two- and three-nucleon interactions.

Potential	$B(^3\text{H})$	$B(^4\text{He})$	$^2a_{nd}$
AV18	7.624	24.22	1.258
N3LO-Idaho	7.854	25.38	1.100
AV18+TM'	8.440	28.31	0.623
AV18+URIX	8.479	28.48	0.578
N3LO-Idaho+N2LO	8.474	28.37	0.675
Exp.	8.482	28.30	0.645±0.003±0.007

From the table we may observe that only the results obtained using an interaction model that includes a TNF are close to the corresponding experimental values. In the case of the AV18+TM', the strength of the TM' potential has been fixed to reproduce the ^{4}He binding energy and, as can be seen from the table, the triton binding energy is underpredicted. Conversely, the strength of the URIX potential has been fixed to reproduce the triton binding energy giving too much binding for ^{4}He. The parametrization of the N2LO potential has been fixed to reproduce simultaneously the triton and the ^{4}He binding energies. In the three cases the predictions for the doublet scattering length are not in agreement with the experimental value, in particular for the AV18+URIX model.

The TM' potential has three parameters fixing the strengths of the three terms arising from the exchange of two pions bewteen three nucleons. One term is coming from πN S-wave scattering and the corresponding constant is called a. The other two terms, which are the most important, come from πN P-wave scattering and the corresponding constants are called b and d. Following Ref. [15], we have used the values $a = -0.87\ m_\pi$, $b = -2.58\ m_\pi^3$, and $d = -0.753\ m_\pi^3$, with m_π the pion mass. In addition, this model includes a monopole form factor at the diagram vertices

$$f(q, \Lambda) = \left[\frac{\Lambda^2 - m_\pi^2}{\Lambda^2 + q^2}\right]^2 , \qquad (1)$$

with Λ a cutoff in the momenta, that regularizes the behaviour of the potential at short distances. In table 1 the calculations have been done using the value $\Lambda = 4.756\ m_\pi$.

We would like to see if it is possible to reproduce simultaneously the triton binding energy and the doublet $n - d$ scattering length for some values of the parameters. The a-term gives a very small contribution to this quantities, therefore we maintain it fixed at the value given above. The other parameters have been varied as shown in figure 1. In figure 1(a) the parameter d and the cutoff Λ have been varied maintaining the value $b = -2.58\ m_\pi^3$ fixed. In figure 1(b)

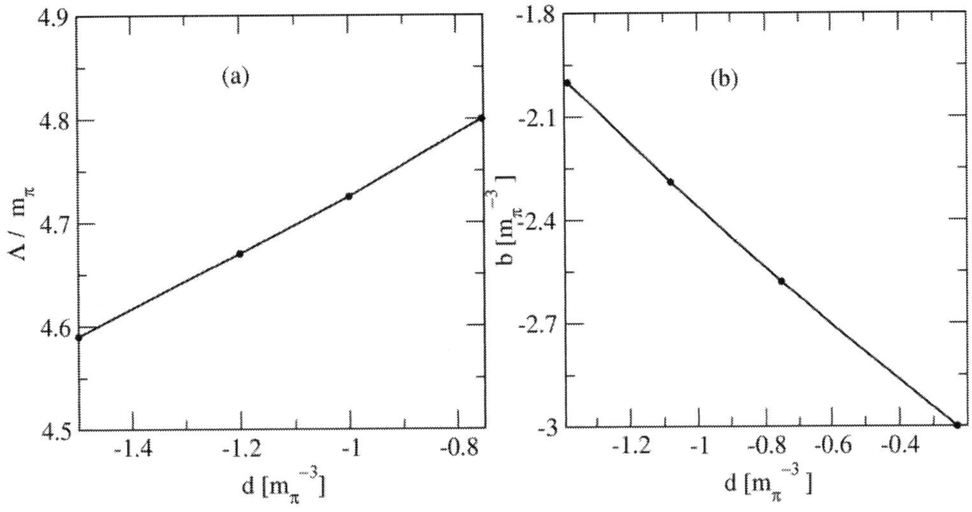

Figure 1. (a) The parameter d as a function of Λ for $b = -2.58\ m_\pi^3$, and (b) the parameter d as a function of b for $\Lambda = 4.8\ m_\pi$. At each point of the two lines the set of parameters describe correctly the triton binding energy.

the parameters d and b have been varied maintaining the value of the cutoff $\Lambda = 4.8\ m_\pi$ fixed. Each point on the two lines of the figure corresponds to a set of parameters that, in connection with the AV18 potential, reproduces the triton binding energy.

The predictions for the doublet scattering length $^2a_{nd}$ are given in table 2 for selected values of the parameters extracted from figure 1 and indicated as points in the figure. The last point of panel (a) and the first point of panel (b) correspond to the same set of parameters. In the first four rows the values are extracted from panel (a) whereas the last four rows are from panel (b) (the fourth row belongs to both panels). We can observe that the doublet scattering length remains almost constant around the value $^2a_{nd} \approx 0.595$ fm. The same occurs using set of parameters corresponding to other points on the two lines of figure 1. We can compare these results to that one given in table 1 for the AV18+TM' in which the triton binding energy is not well reproduced. We observe that when the model is forced to reproduce the triton binding energy, the doublet scattering length goes to a value around 0.595 fm. So, we can extract the first, very important conclusion: with reasonable ranges for the parameters of the AV18+TM' potential it is not possible to describe simultaneously the triton binding energy and the $n-d$ doublet scattering length.

The motivation for this fact is the following. The main contribution of the TM' potential comes from the b- and d- terms which are attractive. Therefore the TM' potential is an attractive potential; it does not include explicitely a repulsive term. Fixing appropriately its strength, it can supply the extra binding needed to correctly describe the three-nucleon binding energy. On the other hand, the scattering length is sensitive to the balance between the attractive part and the repulsive part of the complete interaction. Therefore, it seems that supplying an attraction fixed to reproduce the triton binding energy, this balance results to be incorrect. From this analysis we conclude that the TNF has to include, not only an attractive part, but also a repulsive term, and their strengths could be fixed to reproduce simultaneously the triton binding energy and the $n-d$ doublet scattering length $^2a_{nd}$.

As discussed above the TM' potential is a modification of the original TM potential compatible with chiral symmetry. At the same order (next-to-next-to-leading order) in the

Table 2. The doublet scattering length $^2a_{nd}$ for different choices of the TM' parameters. In all cases the parameters give a correct description of the triton binding energy in connection with the AV18 potential.

a $[m_\pi^{-1}]$	b $[m_\pi^{-3}]$	d $[m_\pi^{-3}]$	Λ $[m_\pi]$	$^2a_{nd}$[fm]
-0.87	-2.58	-1.500	4.590	0.596
-0.87	-2.58	-1.200	4.670	0.596
-0.87	-2.58	-1.000	4.725	0.597
-0.87	-2.58	-0.753	4.800	0.595
-0.87	-2.00	-1.380	4.800	0.591
-0.87	-2.29	-1.080	4.800	0.593
-0.87	-3.00	-0.230	4.800	0.594

chiral effective field theory two extra terms appear (see Ref. [7] and references therein). The configuration space form of these terms is not unique, however a local form is possible with a particular choice of the cutoff [8]. Here we introduce the following additional term to the TM' potential based on a contact term of three nucleons

$$W_{3R} = W_0 \sum_{cyc} Z_0(r_{ij}) Z_0(r_{jk}) \tag{2}$$

with

$$Z_0(r) = \frac{1}{2\pi^2 m_\pi^3} \int_0^\infty dq\, q^2 j_0(qr) \left[\frac{\Lambda^2 - m_\pi^2}{\Lambda^2 + q^2}\right]^2 = \frac{1}{8\pi} \frac{\Lambda^3}{m_\pi^3} \left(1 - \frac{m_\pi^2}{\Lambda^2}\right)^2 \mathrm{e}^{-\Lambda r} \quad . \tag{3}$$

A similar term exists in the URIX potential, though with a different form of the function $Z_0(r)$. For positive values of W_0 the new term is repulsive, and we include it in the following analysis of the TM' potential. Fixing the value of the parameters a and b at the value normally used in the literature, $a = -0.87\ m_\pi$ and $b = -2.58\ m_\pi^3$, we vary the parameter d and the new parameter W_0 in order to reproduce simultaneously the triton binding energy and $^2a_{nd}$. The calculations have been done for several values of the cutoff Λ. The results are shown in figure 2 for Λ/m_π as a function of d. To be noticed that in each point of the curve the value of W_0 is chosen to reproduce the aforementioned quantites. It ranges from a value of 1.228 MeV at $\Lambda/m_\pi = 7.2$ to 0.308 MeV at $\Lambda/m_\pi = 5.0$.

It is interesting to analyze the different contributions of the binding energy produced by the inclusion of the new term in the AV18+TM' model from sets of parameters taken at different points of the curve. In table 3, the mean values of the kinetic energy and potential energy of the two-body AV18 interaction, as well as those ones of the attractive part of the TNF $V_A(3N)$, corresponding to the sum of the a, b and d terms, and the repulsive part $V_R(3N)$ corresponding to the W_0 term, are given for selected values of the parameters (indicated as points in the curve). The last two columns show the triton binding energy and $^2a_{nd}$. As can be seen, the sets of parameters on the curve reproduce these quantities.

In the following we analyze the URIX potential which has two free parameters that can be conveniently fixed. The first one, called $A_{2\pi}^{PW}$, is related to the strength of the term produced by a 2π-exchange with an intermediate Δ excitation. This term is constructed from the sum of two contributions, corresponding to the b-term and d-term of the TM potential. In the URIX potential they are not independent, their relative strength $D_{2\pi}^{PW}$ is taken to be 1/4. The second constant, called A_R, fixes the strength of a purely central repulsive term introduced

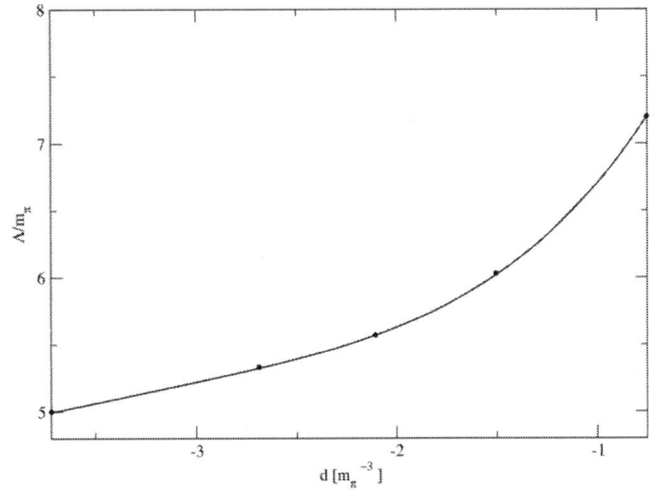

Figure 2. The cutoff Λ as a function of d. In each point of the curve the parameter W_0 is chosen to reproduce simultaneously the triton binding energy and the doublet scattering lenth $^2a_{nd}$.

Table 3. Mean values of the kinetic energy T, the two-body potential $V(2N)$ and the attractive $V_A(3N)$ and repulsive $V_R(3N)$ contributions of the modified TM' potential for the indicated values of the parameters. In the last two columns the triton binding energy and the doublet $n-d$ scattering length are shown.

d $[m_\pi^{-3}]$	W_0 [MeV]	Λ $[m_\pi]$	T [MeV]	$V(2N)$ [MeV]	$V_A(3N)$ [MeV]	$V_R(3N)$ [MeV]	$B(^3\text{H})$ [MeV]	$^2a_{nd}$ [fm]
-0.753	1.228	7.2	51.420	-57.911	-1.996	2.042	8.486	0.644
-1.506	0.512	6.0	51.011	-57.901	-3.180	1.553	8.478	0.644
-2.106	0.402	5.6	50.953	-57.852	-3.000	1.423	8.477	0.645
-2.688	0.333	5.3	50.890	-57.815	-2.863	1.313	8.475	0.644
-3.718	0.308	5.0	50.846	-57.706	-2.942	1.327	8.475	0.644

to compensate the attraction of the previous term, which by itself would produce a large overbinding in infinite nuclear matter. Its form is similar to the one given in eq.(2), with the $Z_0(r) = T^2(m_\pi r)$, and

$$T(x) = \left(1 + \frac{3}{x} + \frac{3}{x^2}\right)\frac{e^{-x}}{x}\xi_T^2(r), \tag{4}$$

where $x = m_\pi r$ and $\xi_T(r) = 1 - e^{-cr^2}$ is a cutoff function with $c = 2.1$ fm^{-2}. The original values of the parameters of the URIX model are $A_{2\pi}^{PW} = -0.0293$ MeV, $A_R = 0.0048$ MeV and $D_{2\pi}^{PW} = 0.25$. The corresponding results for the quantities of interest are given in table 1.

In order to improve the agreement with $^2a_{nd}$ and the ^{4}He binding energy, we have varied the constants $A_{2\pi}^{PW}$, A_R and the relative strength $D_{2\pi}^{PW}$. For a given value of $A_{2\pi}^{PW}$, we have varied A_R and $D_{2\pi}^{PW}$ to reproduce $B(^3\text{H})$ and $^2a_{nd}$. Then we have calculated $B(^4\text{He})$. Surprisingly this last result turned out to be quite close to the experimental value. The results of the analysis are given in table 4 where five sets of values which reproduce the mentioned quantities are reported.

Table 4. Different choices for the parameters of the URIX potential and the corresponding triton and ^{4}He binding energies (in MeV) and scattering length $^2a_{nd}$ (in fm), calculated with the AV18+URIX potential.

$A_{2\pi}^{PW}$(MeV)	$D_{2\pi}^{PW}$	A_R(MeV)	$B(^3\text{H})$	$B(^4\text{He})$	$^2a_{nd}$
-0.0200	1.625	0.0176	8.474	28.33	0.644
-0.0250	1.25	0.0182	8.474	28.34	0.644
-0.0293	1.00	0.0181	8.474	28.33	0.643
-0.0350	0.8125	0.0191	8.474	28.33	0.645
-0.0400	0.6875	0.0198	8.474	28.38	0.645

From the table we observe that the values considered for $D_{2\pi}^{PW}$ and A_R are quite far from the original ones. In particular, the relative strength $D_{2\pi}^{PW}$ differs from the original value of 1/4. To extend the analysis further, the obtained set of values can be used to study $p-d$ scattering at low energy. In figure 3 the $p-d$ analyzing power A_y at $E_{lab} = 3$ MeV is shown in correspondence to calculations performed using the original AV18+URIX model (solid line) and using the first three sets of values given in table 4 and indicated by (a), (b) and (c). The calculations have been done considering the coulomb potential as well as the magnetic moment interaction as explained in Ref. [16]. As can be seen, the results for the models (a), (b) and (c) are very close to each other. The same is true for the last two choices of table 4; however, corresponding results are not shown for the sake of clarity. Besides the usual underprediction of this observable found within the AV18+URIX model, we observe a substantially worse description when the new sets of constants are used. In figure 4 the tensor analyzing power T_{21} is shown corresponding to the same choice of parameters. The original AV18+URIX model overpredicts the minimum close to 90°. Again the curves listed (a), (b) and (c) nearly overlap and there is a substantially worse description of the observable between 40° and 120°.

3. Conclusions

Stimulated by the fact that the commonly used TNF models do not reproduce simultaneously the triton and ^{4}He binding energy and the $n-d$ doublet scattering length $^2a_{nd}$, we have analyzed possible modifications of the AV18+TM' and AV18+URIX potentials. In the case of the AV18+TM' interaction we have shown the impossibility of describing simultaneously the triton binding energy and $^2a_{nd}$. The TM' potential is attractive, accordingly, it provides more binding. Its strength can be fixed in order to reproduce the triton binding energy. However, the doublet scattering length is sensitive to the balance between the attraction and the repulsion provided by the whole potential. From the analysis given here, it has emerged that the TNF, besides providing more attraction, has to provide a certain amount of repulsion. To this end we have included a local repulsive term in the TM' potential. This term is based on a three-nucleon contact term which appears at next-to-next-to-leading order in the chiral effective field theory and is similar to the repulsive term already present in the URIX model. Varying the d parameter and the cutoff Λ, and adjusting the strength of the repulsive term W_0, we found a continuity of sets of parameters that reproduce the triton binding energy and $^2a_{nd}$.

In the case of the AV18+URIX interaction, we have varied the original parameter values of the model so as to improve the description of triton binding energy and $^2a_{nd}$. Five choices of the parameters have been considered and we have shown that it is possible to reproduce simultaneously these quantities. The obtained values of the parameters substantially differ from

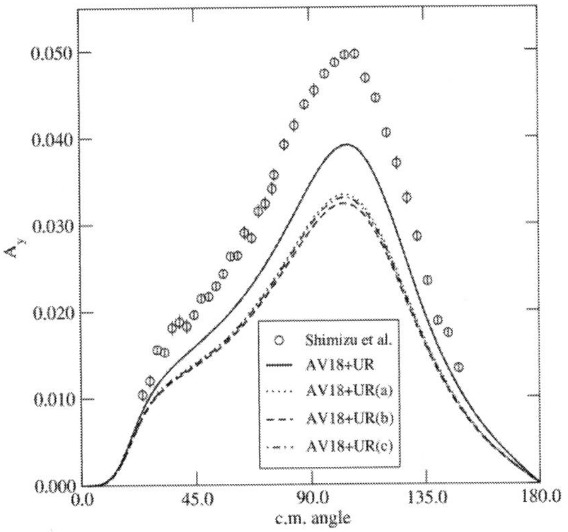

Figure 3. The $p-d$ analyzing power A_y for the models discussed in the text. Experimental data are from Ref. [17].

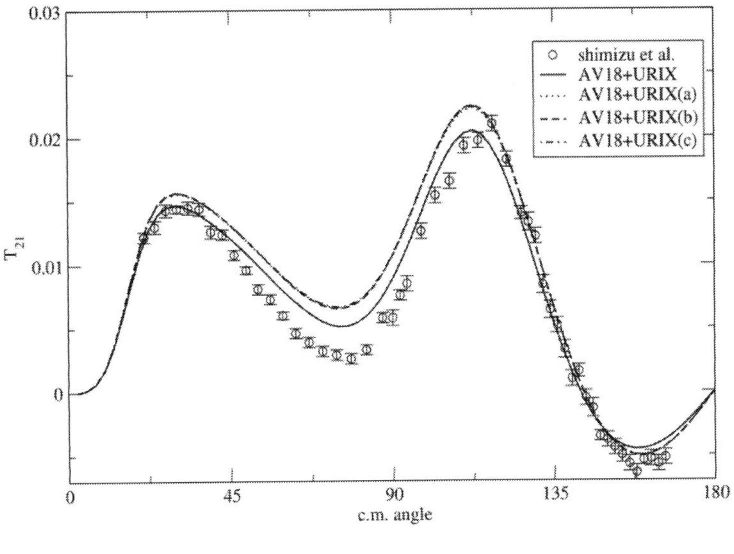

Figure 4. The $d-p$ analyzing power T_{21} for the models discussed in the text. Experimental data are from Ref. [17].

the original ones. Values of the relative strength $D_{2\pi}^{PW}$ ranging from 0.68 to 1.63 have been considered. Its original value of 1/4 produces unnatural values of the other constants $A_{2\pi}^{PW}$ and A_R. In the range considered, the value of A_R results almost constant. We have extended the analysis of the AV18+URIX interaction to the descritpion of $p-d$ scattering. It has emerged that the selected set of parameters worsen the description of some $p-d$ polarization observables at low energies. Further work on this problem is in progress.

References

[1] Kievsky A, Viviani M and Rosati S 2001 *Phys. Rev.* C **64** 024002
[2] Coon S A and Glöckle W. 1981 *Phys. Rev.* C **23** 1790
[3] Pudliner B S *et al* 1995 *Phys. Rev. Lett.* **51** 4396
[4] Friar J L, Hüber D and van Kolck U 1999 *Phys. Rev.* C **59** 53
[5] Coon S A and Han H K 2001 *Few-Body Syst.* **30** 131
[6] Coelho H T, Das T K and Robilotta M R 1983 *Phys. Rev.* C **28** 1812; Robilotta M R and Coelho H T 1986 *Nucl. Phys.* A **460** 645
[7] Epelbaum E *et al* 2002 *Phys. Rev.* C **66** 064001
[8] Navratil P 2007 *Few-Body Syst.* **41** 117
[9] Kievsky A, Viviani M and Rosati S 1993 *Nucl. Phys.* A **551** 241; Kievsky A, Viviani M and Rosati S 1993 *Nucl. Phys.* A **577** 511
[10] Kievsky A 1997 *Nucl. Phys.* A **624** 125
[11] Viviani M, Kievsky A and Rosati S 1995 *Few-Body Syst.* **18** 25; Viviani M, Kievsky A and Rosati S 2005 *Phys. Rev.* C **71** 024006
[12] Viviani M *et al* 2001 *Phys. Rev. Lett.* **86** 3739; Fisher B M *et al* 2006 *Phys. Rev.* C **74** 034001
[13] Kievsky A, Rosati S, Viviani M, Marcucci L E and Girlanda L 2008 *J. Phys. G: Nucl. Part. Phys.* **35**, 063101
[14] Schoen K *et al* 2003 *Phys. Rev.* C **67** 044005
[15] Nogga A, Kamada H, Glöckle W and Barrett B R 2002 *Phys. Rev.* C **65** 054002
[16] Kievsky A, Viviani M and Marcucci L E 2004 *Phys. Rev.* C **69** 014002
[17] Shimizu S *et al* 1995 *Phys. Rev.* C **52** 1193

$N-d$ elastic scattering at low-energies with local and non-local realistic nuclear interactions

L E Marcucci[1,2], L Girlanda[2], A Kievsky[2], S Rosati[1,2] and M Viviani[2]

[1] Department of Physics "E. Fermi", University of Pisa, 56127 Pisa, Italy
[2] Istituto Nazionale di Fisica Nucleare, Sezione di Pisa, 56127 Pisa, Italy

E-mail: laura.marcucci@df.unipi.it

Abstract. The application of the hyperspherical harmonic approach to the case of the $N-d$ scattering problem is described. The nuclear Hamiltonian includes local and non-local two- and three-nucleon interactions. Also, Coulomb repulsion is included in the $p-d$ case. Accurate calculation for $p-d$ and $n-d$ cross section and polarization observables at two center-of-mass energies are performed and the theoretical results are compared with the available experimental data. A satisfactory agreement between theory and experiment is found, except for the vector polarization observables. A quantitative estimate of the agreement or disagreement between theory and experiment is obtained performing a χ^2 analysis.

1. Introduction

One of the main inputs for any study on nuclear systems within a non-relativistic framework is the model used to describe the nuclear interaction, i.e. the nuclear Hamiltonian. Nowadays, it is common practice to use, at least for the few-nucleon systems, Hamiltonian models composed of a two-nucleon (NN) plus, for $A \geq 3$, a three-nucleon interaction (TNI). The modern NN interaction models have a large number of parameters and can reproduce the deuteron properties and the nucleon-nucleon scattering data up to the pion threshold with a χ^2/datum$\simeq$ 1. The available models for the TNI contain, instead, a small number of parameters, usually fixed to reproduce the ^{3}H binding energy and, in some cases, the nuclear matter equilibrium density. A crucial point for this type of studies is to test the model for the nuclear Hamiltonian studying $A \geq 4$ bound states and $A \geq 3$ scattering states. In the present work, we focus our attention to the $A = 3$ scattering problem.

To study the $A = 3$ scattering problem with a realistic Hamiltonian, we have chosen to use the hyperspherical harmonics (HH) method (for a recent review, see Ref. [1]). One of the main advantages of the HH method is the possibility of including the Coulomb interaction in $p-d$ scattering without any additional difficulty compared to the $n-d$ case. On the other hand, the HH method, being first developed in coordinate-space, was limited in the choice of the Hamiltonian models only to the local ones. A first step to implement the method to work in momentum-space has been done in Ref. [2], where the $A = 3, 4$ bound states have been studied using modern non-local interactions. In the present work, we apply the HH method to the $N-d$ scattering problem again using non-local interactions. We limit this first analysis to the case of $N-d$ elastic scattering at center-of-mass energy $E_{cm} = 0, 1.33, 2$ MeV $(0, 1.66, 2$ MeV) for

$n-d$ $(p-d)$. A more complete calculation at several center-of-mass energies will be given elsewhere [3].

The paper is organized as follows: in section 2, the HH method for the bound and the low-energy scattering problem is briefly outlined. In section 3, the results for the zero-energy scattering lengths and some low-energy elastic scattering observables will be presented and discussed. Some concluding remarks are given in section 4.

2. Formalism

In this section we present the HH method for scattering states. The method for bound states has been most recently reviewed in Ref. [1], and we briefly summarize its main characteristics.

The nuclear wave function for the three-body bound state can be written as

$$|\Psi\rangle = \sum_\mu c_\mu |\Psi_\mu\rangle \ , \tag{1}$$

where $|\Psi_\mu\rangle$ is a suitable complete set of states, and μ is an index denoting the set of quantum numbers necessary to completely specify the basis elements. The coefficients of the expansion are determined using the Rayleigh-Ritz variational principle, i.e. $\langle \delta_c\Psi \,|\, H - E \,|\Psi\rangle = 0$, where $\delta_c\Psi$ indicates the variation of Ψ for arbitrary infinitesimal changes of the linear coefficients c_μ. Then, the problem of determining c_μ and the energy E is reduced to a generalized eigenvalue-eigenstate problem. The main difficulty of the method is to compute the matrix elements of the Hamiltonian H with respect to the basis states $|\Psi_\mu\rangle$. Usually H is given as a sum of terms (kinetic energy, NN potential, etc.). The calculation of the matrix elements of some parts of H can be more conveniently performed in coordinate-space, while for other parts it could be easier to work in momentum-space. Therefore, it is important that the basis states $|\Psi_\mu\rangle$ have simple expressions in both spaces. The HH functions indeed have such a property.

Let us first consider the expression of the HH functions in coordinate space. The internal dynamics of a system of three nucleons of identical mass m is conveniently described in terms of the standard Jacobi vectors $\boldsymbol{x}_1, \boldsymbol{x}_2$. It is convenient to replace the moduli of $\boldsymbol{x}_2$ and $\boldsymbol{x}_1$ with the so-called hyperradius $\rho = \sqrt{\boldsymbol{x}_1^2 + \boldsymbol{x}_2^2}$ and hyperangle $\tan\phi = \frac{x_1}{x_2}$ [4]. The complete set of hyperspherical coordinates is then given by $\{\rho, \Omega^{(\rho)}\}$, with $\Omega^{(\rho)} = [\hat{\boldsymbol{x}}_2, \hat{\boldsymbol{x}}_1; \phi]$ and the suffix (ρ) recalls the use of the coordinate space.

The expansion states $|\Psi_\mu\rangle$ of Eq. (1) are then given by

$$|\,\Psi_\mu^{(\rho)}\,\rangle = f_l(\rho)\mathcal{Y}_{\{G\}}(\Omega^{(\rho)}) \ , \tag{2}$$

where $f_l(\rho)$ for $l = 1, \ldots$ is a complete set of hyperradial functions, here chosen to be Laguerre polynomials multiplied by a decreasing exponential function of ρ. The functions $\mathcal{Y}_{\{G\}}(\Omega^{(\rho)})$ contain all the dependence on the angular, spin and isospin quantum numbers of the three nucleons, schematically included in $\{G\}$, and on the angular-hyperangular variables $\Omega^{(\rho)}$. They are known as the spin-isospin HH functions of grand-angular momentum G. More details can be found in Ref. [1].

The expansion states of Eq. (1) in momentum space can be obtained as follows. Let $\hbar\boldsymbol{k}_1, \hbar\boldsymbol{k}_2$ be the standard conjugate Jacobi momenta of the Jacobi vectors. We then define a hypermomentum Q and a set of angular-hyperangular variables as

$$Q = \sqrt{\boldsymbol{k}_1^2 + \boldsymbol{k}_2^2} \ , \qquad\qquad \Omega^{(Q)} = [\hat{\boldsymbol{k}}_2, \hat{\boldsymbol{k}}_1; \varphi] \ , \tag{3}$$

where $\tan\varphi = \frac{k_1}{k_2}$. Then, the momentum-space version of the wave function given in Eq. (2) is

$$|\,\Psi_\mu^{(Q)}\,\rangle = g_{G,l}(Q)\mathcal{Y}_{\{G\}}(\Omega^{(Q)}) \ , \tag{4}$$

Table 1. ^{3}H, ^{3}He and ^{4}He binding energies in MeV calculated with the HH method using different two- and three-nucleon Hamiltonian models.

Interaction	^{3}H	^{3}He	^{4}He
AV18/UIX	8.479	7.750	28.46
CDBonn/TM	8.474	7.720	29.00
N3LO-Idaho/UIXp	8.481	7.730	28.53
N3LO-Idaho/N2LO	8.474	7.733	28.36
Experimental	8.482	7.718	28.30

where $\mathcal{Y}_{\{G\}}(\Omega^{(Q)})$ is the same as $\mathcal{Y}_{\{G\}}(\Omega^{(\rho)})$ with $\boldsymbol{x}_i \to \boldsymbol{k}_i$, and

$$g_{G,l}(Q) = (-i)^G \int_0^\infty d\rho \, \frac{\rho^3}{Q^2} \, J_{G+2}(Q\rho) \, f_l(\rho) \; . \tag{5}$$

Here $J_{G+2}(Q\rho)$ is a Bessel function. With the adopted form of $f_l(\rho)$, the corresponding functions $g_{G,l}(Q)$ can be easily calculated, and they are explicitly given in Ref. [2].

In table 1 we list the binding energies in MeV of ^{3}H, ^{3}He and ^{4}He calculated using different two- and three-nucleon Hamiltonian models. In particular, we have considered the local AV18 [6] NN potential in conjunction with the Urbana IX [7] (UIX) TNI, the non-local CDBonn 2000 (CDBonn) [8] NN potential in conjunction with the Tucson-Merlbourne (TM) [9] TNI, and the non-local N3LO-Idaho [10] NN potential in conjunction with the UIX or the N2LO [11] TNIs. To be noticed that in the case of the N3LO-Idaho plus UIX Hamiltonian model, the parameter in front of the spin-isospin independent part of the UIX TNI has been rescaled by a factor of 0.384 to fit the triton binding energy [14] (UIXp). From inspection of the table, we can conclude that the theoretical predictions are in good agreement with the experimental data, although some discrepancies of the order of 10 keV can be found for the $A = 3$ binding energies, whereas for $A = 4$ the discrepancies can be even of few hundreds of keV, as for the CDBonn/TM potential.

We consider now the extension of the HH technique to describe $N-d$ scattering states below deuteron breakup threshold. Both local and non-local interaction models are considered.

The wave function $\Psi_{N-d}^{LSJJ_z}$ describing the $N-d$ scattering state with incoming orbital angular momentum L and channel spin S, parity $\pi = (-)^L$, and total angular momentum J, J_z, can be written as

$$\Psi_{N-d}^{LSJJ_z} = \Psi_C^{LSJJ_z} + \Psi_A^{LSJJ_z} \; . \tag{6}$$

Here $\Psi_C^{LSJJ_z}$ must describe the system in the region where the particles are close to each other and their mutual interactions are strong, while $\Psi_A^{LSJJ_z}$ must describe the relative motion between the nucleon N and the deuteron d in the asymptotic region, where the $N-d$ nuclear interaction is negligible. The function $\Psi_C^{LSJJ_z}$, which has to vanish in the limit of large $N-d$ separation, can be expanded on the HH basis as it has been done in the case of bound states (see Eq. (1)), and the expansion basis $|\Psi_\mu\rangle$ is given in Eqs. (2) and (4) in coordinate- and momentum-space, respectively. The function $\Psi_A^{LSJJ_z}$ is the appropriate solution of the Schrödinger equation in the case of a two-body scattering problem, and can be casted in the form [1]

$$\Psi_A^{LSJJ_z} \propto F_{LS}(pr_{Nd}) + \sum_{L'S'} \mathcal{R}^J_{LS,L'S'} G_{L'S'}(pr_{Nd}) \; , \tag{7}$$

where p (r_{Nd}) is the $N-d$ relative momentum (distance), and $F_{LS}(pr_{Nd})$ and $G_{L'S'}(pr_{Nd})$ are the regular and irregular Coulomb (Bessel) functions for the $p-d$ ($n-d$) scattering problem. The parameters $\mathcal{R}^J_{LS,L'S'}(q)$ give the relative weight between the regular and irregular components, and are closely related to the reactance matrix ($\mathcal{K}$-matrix) elements. By definition, the $\mathcal{K}$-matrix eigenvalues are proportional to the phase shifts of the scattering problem.

The matrix elements $\mathcal{R}^J_{LS,L'S'}(q)$ and the linear coefficients c_μ occurring in the expansion of $\Psi_C^{LSJJ_z}$ are determined applying the Kohn variational principle [12], which states that the functional

$$[\mathcal{R}^J_{LS,L'S'}(q)] = \mathcal{R}^J_{LS,L'S'}(q) - \left\langle \Psi_{N-d}^{L'S'JJ_z} |H-E| \Psi_{N-d}^{LSJJ_z} \right\rangle , \tag{8}$$

has to be stationary with respect to variations of the trial parameters in $\Psi_{N-d}^{LSJJ_z}$. Here E is the total energy of the system. Performing these variations, the problem reduces to find the solution of a system of linear inhomogeneous equations for c_μ and a set of algebraic equations for $\mathcal{R}^J_{LS,L'S'}(q)$. This can be done with standard numerical techniques. A second order estimate of $\mathcal{R}^J_{LS,L'S'}(q)$ is given by the quantities $[\mathcal{R}^J_{LS,L'S'}(q)]$, obtained by substituting in Eq. (8) the first order results. Such second-order calculation provides a symmetric reactance matrix. This condition is not *a priori* imposed, and therefore it is a useful test of the numerical accuracy.

In the particular case of $q=0$ (zero-energy scattering), the scattering occurs only in the channel $L=0$ and the observables of interest are the scattering lengths. Within the present approach, they can be easily obtained as ${}^{(2J+1)}a_{Nd} = -\lim_{q\to 0}\mathcal{R}^J_{0J,0J}(q)$.

The approach presented so far for the $N-d$ scattering process does not have too many differences compared to the method presented for instance in Ref. [13], and known as pair-correlated hyperspherical harmonics (PHH) method. In the PHH method, a correlation factor is included in the HH expansion to take into account the strong short-range correlations induced by the realistic NN potentials, like the AV18. The presence of correlation functions makes the convergence of the expansion much faster than in the uncorrelated case. However, the PHH method cannot be easily implemented when non-local NN interactions are considered, as in this work. Here, instead, the calculation involving $\Psi_C^{LSJJ_z}$ can be performed in coordinate- or in momentum-space, depending on what is more convenient, as it has been previously explained for the bound state. Some difficulties, however, arise for the calculation of the potential energy matrix elements which involve $\Psi_A^{LSJJ_z}$, like $\langle\Psi_\mu|V|\Psi_A^{LSJJ_z}\rangle$ and $\langle\Psi_A^{LSJJ_z}|V|\Psi_A^{LSJJ_z}\rangle$, when non-local potentials expressed in momentum-space are used. On the contrary to $\Psi_C^{LSJJ_z}$, which can be alternatively expressed in coordinate- or in momentum-space, the asymptotic state $\Psi_A^{LSJJ_z}$ does not have an easy expression in momentum-space, and is more conveniently given and used in coordinate-space. This is especially true when the Coulomb interaction is considered, as for the $p-d$ case. Therefore, we have decided to perform the Fourier transform of the non-local potential $v(\boldsymbol{k},\boldsymbol{k}')$ and to work only in coordinate-space. The integrations over k and k', which run from 0 to ∞, are easily performed when the potential model considered does not have a high-momentum tail, but goes rapidly to zero at rather low values of k and k'. This is true for the N3LO-Idaho potential model, but not for the CDBonn 2000. Since the main goal of the present work is to perform a first test of the applicability of the HH method to the $A=3$ scattering problem using non-local realistic interactions, only the N3LO-Idaho NN potential has been considered.

3. Results

The $N-d$ elastic scattering observables have been studied using the following interaction models: the two-nucleon AV18 and N3LO-Idaho and the two- plus three-nucleon AV18/UIX, N3LO-Idaho/UIXp and N3LO-Idaho/N2LO models. In the case of $p-d$ elastic scattering, only the point Coulomb electromagnetic interaction has been considered, also when the AV18 potential is used.

Table 2. $n-d$ and $p-d$ doublet and quartet scattering lengths in fm calculated with the HH technique using different Hamiltonian models.

Interaction	$^2a_{nd}$	$^4a_{nd}$	$^2a_{pd}$	$^4a_{pd}$
AV18	1.275	6.325	1.185	13.588
AV18/UIX	0.610	6.323	-0.035	13.588
N3LO-Idaho	1.100	6.342	0.876	13.646
N3LO-Idaho/UIXp	0.623	6.343	-0.007	13.647
N3LO-Idaho/N2LO	0.675	6.342	0.072	13.647
Experimental [15]	0.65±0.04	6.35 ±0.02		
Experimental [16]	0.645±0.003±0.007	—		

The results for the $n-d$ and $p-d$ doublet and quartet scattering lengths are given in table 2 and are compared with the available experimental data [15, 16]. The calculation has been performed including also the $T=3/2$ components of the wave function. To make the calculation faster, the PHH expansion has been used in the AV18 and AV18/UIX cases. The dimension of the basis and the number of the integration points have been increased in order to reach an accuracy of 0.002 fm in the calculation of the scattering lengths. Comparing the theoretical and experimental results for $^2a_{nd}$ and $^4a_{nd}$, we can conclude that $^4a_{nd}$ is very little model-dependent (as well as $^4a_{pd}$), and there is a satisfactory agreement between theory and experiment. On the contrary, $^2a_{nd}$ is strongly model-dependent. Only the inclusion of the TNI brings the theoretical value close to the experimental one, but some disagreement is still observed, in particular for the AV18/UIX model. Note that the results presented have been calculated without including the neutron-proton mass difference contribution, which is expected to further decrease the value of $^2a_{nd}$. Finally, the recent measurement of Ref. [16] is not well described by any of the potential models considered. However, the N3LO-Idaho with the UIXp and N2LO gives better results. Further analysis of this aspect can be found in Ref. [17].

The $n-d$ elastic scattering observables, including differential cross section, neutron vector analyzing power A_y, deuteron vector and tensor analyzing powers iT_{11}, T_{20}, T_{21} and T_{22}, at $E_{cm}=1.33$ MeV and 2 MeV are given in figure 1. In figure 2 the same observables are presented but for $p-d$ at $E_{cm}=1.66$ MeV and 2 MeV. The experimental data are from Refs. [18, 19, 20, 21, 22] and Ref. [23] for the $n-d$ and $p-d$ case, respectively. From inspection of the figures, we can observe that there is an overall agreement between experimental data and theoretical prediction except for few remarkable cases: (i) the $n-d$ differential cross section at $E_{cm}=1.33$ MeV for large values of the center-of-mass angle θ_{cm}. Such a discrepancy disappears at $E_{cm}=2$ MeV and is not present in the $p-d$ case. Its origin has still to be clarified [24]. (ii) The $p-d$ tensor analyzing power T_{21} is poorly described in the first minimum region by the Hamiltonian models which include both two- and three-nucleon interactions. However, the non-local potential models seem to do a slightly better job than the local ones. (iii) The $n-d$ and $p-d$ vector analyzing powers A_y and iT_{11} are poorly described by the theory, in particular in the maximum region. Such a discrepancy, since long time known as the "A_y-puzzle", is more remarked with two-nucleon only Hamiltonian models. The inclusion of the TNI slightly improves the situation, but not significantly. In particular, it should be noticed that the differences between the AV18 and N3LO-Idaho results could be explained by the fact that the N3LO-Idaho potential model, on the contrary to the AV18 one, includes the effects of the magnetic moment interaction. On the other hand, the better description of the experimental data given by the N3LO-Idaho/N2LO with respect to the N3LO-Idaho/UIXp Hamiltonian model, could be an

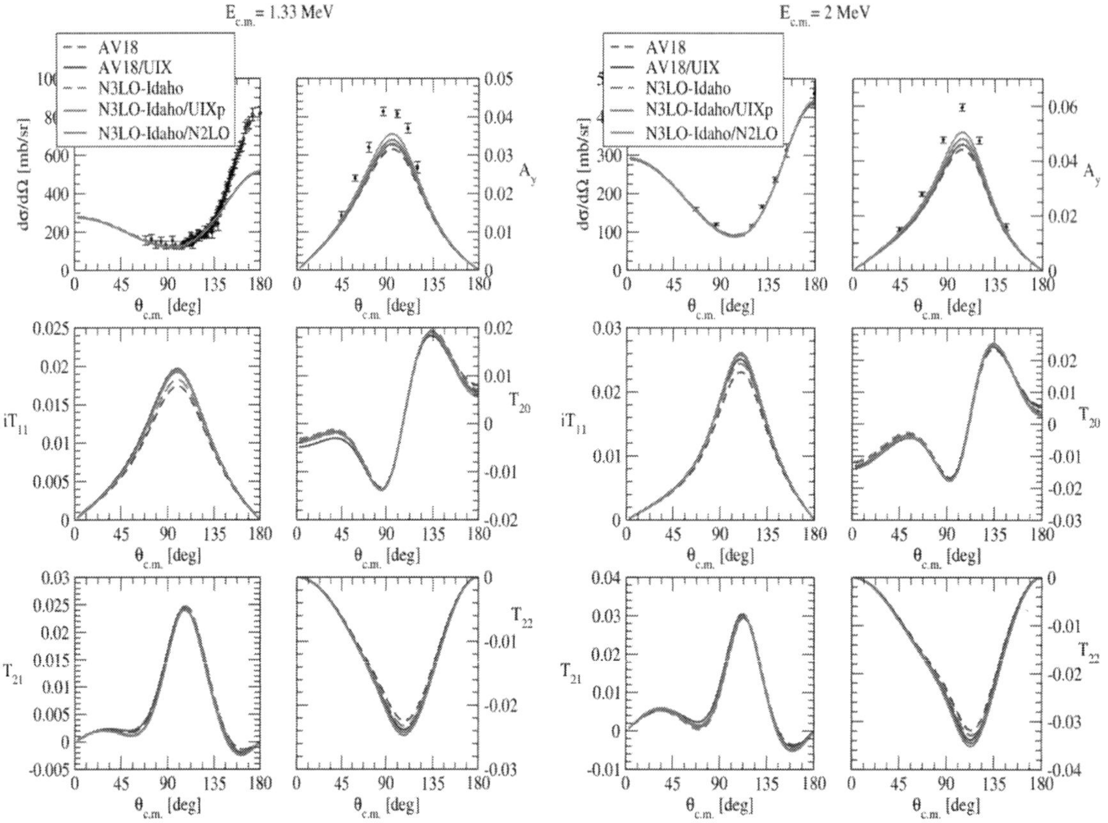

Figure 1. (Color on-line) The theoretical values of the differential cross section, A_y, iT_{11}, T_{20}, T_{21} and T_{22}, at $E_{cm} = 1.33$ MeV and $E_{cm} = 2$ MeV are compared to the experimental data of Refs. [18, 19, 20] for $E_{cm} = 1.33$ MeV and Refs. [21, 22] for $E_{cm} = 2$ MeV. The calculation are done with the AV18 (dashed blue line), the AV18/UIX (solid blue line), the N3LO-Idaho (dashed green line), the N3LO-Idaho/UIXp (solid green line), and the N3LO-Idaho/N2LO (solid red line) potential models. The incident neutron (deuteron) is $E_n = 2$ MeV ($E_d = 4$ MeV) on the left panel and $E_n = 3$ MeV ($E_d = 6$ MeV) on the right panel.

indication that the N2LO TNI model is somehow "better" than the UIX one. Further studies on this aspect are currently underway [3].

Since the comparison among the different potential models done by looking at figures 1 and 2 can be obtained only at a qualitative level, we have performed, in the example case of $p - d$ elastic scattering, a χ^2 analysis. To this aim, following Ref. [25], $\chi^2/\text{datum} = \frac{1}{N}\sum_i \frac{(f_i^{exp} - f_i^{th})^2}{(\Delta f_i)^2}$, where f_i^{exp} is the ith datum at center-of-mass angle θ_i and Δf_i is its experimental error, f_i^{th} is the theoretical value at the same angle, and N is the total number of data points included. The results for the χ^2/datum obtained using the different Hamiltonian models considered in the present study for all the scattering observables are listed in table 3. Also listed are the results for the total χ^2/datum. Note that the main goal of this analysis is to give a more quantitative argument to the statement that the N2LO TNI is somewhat "better" than the UIX. In fact, the χ^2/datum calculated with the N3LO-Idaho/UIXp is comparable with the one calculated with the N3LO-Idaho/N2LO, except for A_y at $E_{cm} = 1.66$ and 2 MeV, where the N3LO-Idaho/UIXp χ^2/datum is higher than the N3LO-Idaho/N2LO one. Therefore, a slight improvement in the

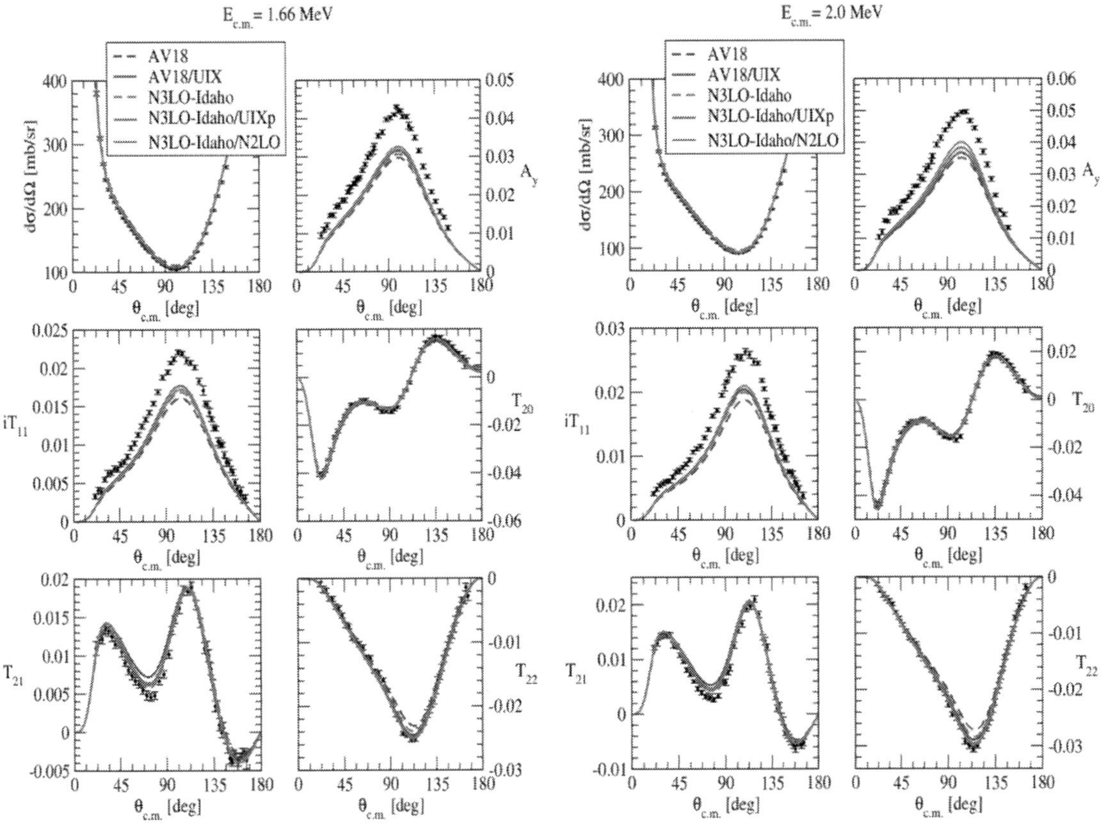

Figure 2. (Color on-line) Same as figure 1 but for $p-d$ scattering at $E_{cm} = 1.66$ MeV and $E_{cm} = 2$ MeV. The experimental data are from Ref. [23] The incident proton (deuteron) is $E_p = 2.5$ MeV ($E_d = 5$ MeV) on the left panel and $E_p = 3$ MeV ($E_d = 6$ MeV) on the right panel.

description of the experimental observables is present for the N3LO-Idaho/N2LO calculation, compared with the N3LO-Idaho/UIXp one. Finally, we can notice that the total χ^2 values at both center-of-mass energies are very similar, whatever two- plus three-nucleon Hamiltonian model is used.

4. Conclusions

We have presented a first study of $N-d$ elastic scattering problem below the deuteron breakup threshold using either local or non-local realistic Hamiltonian models. In particular, we have presented results for the zero-energy scattering lengths, and $n-d$ $E_{cm} = 1.33$ MeV and 2 MeV ($p-d$ $E_{cm} = 1.66$ MeV and 2 MeV) differential cross section, vector and tensor analyzing powers. Among the considered potential models, the N3LO-Idaho/N2LO seems to describe the experimental data slightly better than the previously widely used Hamiltonian models, like the AV18/UIX. A thorough study of $n-d$ and $p-d$ elastic scattering observables at several E_{cm} values, below the deuteron breakup threshold, with different local and non-local Hamiltonian models is required and currently underway [3].

Table 3. χ^2/datum of the $p-d$ elastic scattering observables at $E_{cm} = 1.66$ MeV and $E_{cm} = 2$ MeV calculated with the different Hamiltonian models considered in the present work. The different number N of experimental data of Ref. [23] for each observable is also indicated.

$E_{cm} = 1.66$ MeV	$d\sigma/d\Omega$	A_y	iT_{11}	T_{20}	T_{21}	T_{22}	total
N	37	44	50	50	50	50	
AV18	29.1	267.1	121.5	2.1	3.4	6.7	69.4
N3LO-Idaho	21.5	227.9	92.6	1.0	2.2	2.7	56.1
AV18/UIX	3.1	198.2	68.7	3.9	6.7	2.2	43.9
N3LO-Idaho/UIXp	4.3	185.9	67.0	2.0	3.1	3.2	43.1
N3LO-Idaho/N2LO	6.6	159.4	84.3	2.1	3.9	2.8	42.4
$E_{cm} = 2$ MeV	$d\sigma/d\Omega$	A_y	iT_{11}	T_{20}	T_{21}	T_{22}	total
N	37	38	51	51	51	51	
AV18	29.5	236.3	148.7	3.7	5.0	12.5	67.1
N3LO-Idaho	21.7	186.0	108.3	1.1	2.8	4.4	49.7
AV18/UIX	3.7	198.5	99.1	4.9	11.9	2.5	49.2
N3LO-Idaho/UIXp	4.3	152.4	81.8	3.1	5.5	1.6	38.1
N3LO-Idaho/N2LO	7.5	114.0	85.8	3.6	8.3	1.6	34.7

References

[1] Kievsky A *et al* 2008 *J. Phys. G: Nucl. Part. Phys.* **35** 063101
[2] Viviani M *et al* 2006 *Few-Body Syst.* **39** 159
[3] Marcucci L E *et al* 2009 in preparation
[4] Fabre de la Ripelle M 1983 *Ann. Phys.* **147** 281
[5] Raynal J and Revai J 1970 *Il Nuovo Cim.* **68** 612
[6] Wiringa R B, Stoks V G J and Schiavilla R 1995 *Phys. Rev.* C **51** 38
[7] Pudliner B S *et al* 1995 *Phys. Rev. Lett.* **74** 4396
[8] Machleidt R 2001 *Phys. Rev.* C **63** 024001
[9] Coon S A and Han H K 2001 *Few-Body Syst.* **30** 131
[10] Entem D R and Machleidt R 2003 *Phys. Rev.* C **68** 041001
[11] Navrátil P 2007 *Few-Body Syst.* **41** 117
[12] Kohn W 1948 *Phys. Rev.* **74** 1763
[13] Kievsky A, Viviani M and Rosati S 1994 *Nucl. Phys.* A **577** 511
[14] Marcucci L E *et al* 2008 *Few-Body Syst.* **44** 207
[15] Dilg W *et al* 1971 *Phys. Lett.* **36B** 208
[16] Schoen K *et al* 2003 *Phys. Rev.* C **67** 044005
[17] Kievsky A *et al* 2009 contribution in these proceedings
[18] Adair R K *et al* 1953 *Phys. Rev.* **89** 1165
[19] Weber J *et al* 1981 *Helv. Phys. Acta* **54** 657
[20] Neidel E M *et al* 2003 *Phys. Lett.* B **552** 29
[21] Schwarz P *et al* 1983 *Nucl. Phys.* A **398** 1
[22] McAninch J E *et al* 1994 *Phys. Rev.* C **50** 589
[23] Shimizu S *et al* 1995 *Phys. Rev.* C **52** 1193
[24] Kievsky A *et al* 1996 *Nucl. Phys.* A **607** 402
[25] Kievsky A *et al* 2001 *Phys. Rev.* C **63** 024005

12th Conference on 'Theoretical Nuclear Physics in Italy' IOP Publishing
Journal of Physics: Conference Series **168** (2009) 012006 doi:10.1088/1742-6596/168/1/012006

Nuclear matter and nuclear dynamics

M Colonna

Laboratori Nazionali del Sud, INFN, via Santa Sofia 62, I-95123, Catania, Italy

E-mail: colonna@lns.infn.it

Abstract. Highlights on the recent research activity, carried out by the Italian Community involved in the "Nuclear matter and nuclear dynamics" field, will be presented. In particular, we discuss new results on the Equation of State (EOS) of asymmetric nuclear matter, from microscopic many-body approaches on one side and from the study of sensitive observables in dissipative nuclear reactions on the other. Considering a wide range of beam energies, it is possible to explore the behavior of the EOS symmetry term in several conditions of density and temperature. New results concerning the structure and the life of neutron stars, that are affected by the high density behavior of asymmetric nuclear matter, will also be reported.

1. Introduction

The EOS of nuclear matter plays a fundamental role in the understanding of many aspects of nuclear physics and astrophysics. Transient states of nuclear matter far from normal conditions can be created in terrestrial laboratories and many experimental and theoretical efforts have been devoted to the study of nuclear reactions, from low to intermediate energies, as a possible tool to learn about the behavior of nuclear matter and its EOS. In particular, the availability of exotic beams has opened the way to explore, in laboratory conditions, new aspects of nuclear structure and dynamics up to extreme ratios of neutron (N) to proton (Z) numbers. Over the past years, measurements of isoscalar collective vibrations, collective flows and meson production have contributed to constrain the EOS for symmetric matter for densities up to five time the saturation value [1]. However, the EOS of asymmetric matter has comparatively few experimental constraints: The isovector part of the nuclear effective interaction (Iso-EOS) and the corresponding symmetry energy are largely unknown far from normal density.

The knowledge of the EOS of asymmetric matter is very important at low densities, for studies concerning neutron skins, pigmy resonances, nuclear structure at the drip lines, neutron star formation and crust, as well as at high densities, for investigations of neutron star mass-radius relation, cooling, hybrid structure, transition to a deconfined phase, formation of black holes. Hence a large variety of phenomena, involving an enormous range of scales in size, characteristic time and energy, but all based on nuclear processes at fundamental level, are linked by the concept of EOS.

From one side, this has stimulated new thorough studies of the ground state energy of (asymmetric) nuclear systems, based on microscopic many-body approaches, from which EOS and effective interactions can be extracted, that are essential for modeling heavy ion collisions (HIC) and the structure of neutron stars [2, 3, 4, 5]. Predictions of different many-body techniques will be compared in section 2.

On the other side, several observables which are sensitive to the Iso-EOS and testable experimentally, have been suggested [6, 7, 8, 9]. Taking advantage of new opportunities in theory (development of rather reliable microscopic transport codes for HIC) and in experiments (availability of very asymmetric radioactive beams, improved possibility of measuring event-by-event correlations), in the following sections 3-5 we will review new studies aiming to constrain the existing effective interaction models. We will discuss dissipative collisions in a wide range of beam energies, from just above the Coulomb barrier up to the $AGeV$ range. Low to Fermi energies will bring information on the symmetry term around (below) normal density, while intermediate energies will probe high density regions. As far as the high density behavior of asymmetric nuclear matter is concerned, neutron stars, where densities ten times higher than the densities inside atomic nuclei may be reached in the core, appear as natural astrophysical laboratories that allow for a wide range of physical studies linked to the properties of ultra-dense matter and of its EOS. We will review recent developments in the field also in connection with the possibility that other degrees of freedom, besides the nucleonic ones, (hyperons, quarks), may appear, see section 6. Finally, the link to the possible appearance of new phases of matter in charge asymmetric HIC, already at intermediate energies, is discussed in section 7.

2. Nuclear matter EOS and microscopic approaches

The renewed interest in nuclear matter properties and EOS, that play an important role in the understanding of nuclear structure and astrophysical phenomena, has led to the development of more detailed investigations of the ground state of nuclear many-body systems, where the energy functional is evaluated starting from the bare nucleon-nucleon (NN) interaction. Therefore, from one side one is facing the problem of finding a fundamental scheme for the description of nuclear forces, valid from light nuclei up to dense matter, that is still an open fundamental problem, and, on the other, that of solving a complex many-body problem, that involves strongly spin-isospin dependent forces.

It is well known that all non-relativistic many-body theories need at least three-body forces (TBF) in the Hamiltonian. This is essential to reproduce the correct saturation point of the nuclear EOS, extracted from phenomenological data [10]. However, TBF cannot be constrained in a stringent way, due to the rather limited experimental data on three-nucleon systems. Phenomenological TBF have been devised to reproduce the properties of light nuclei and to correct the nuclear matter salturation point, leading to a new behavior of the whole EOS of both symmetric and asymmetric nuclear matter [11]. However, a unique TBF, which is able to describe accurately light nuclei, as well as the nuclear matter saturation point, is not available yet. Hence, it is interesting to study to which extent the nuclear matter EOS, in a wide density range, is actually dependent on the choice of two- and three-body forces, once the saturation point has been reproduced, i.e. to test to what extent the reproduction of the saturation point constrains the behavior of the whole EOS.

In the following we will review recent results obtained mostly in the context of the many-body Bethe-Brueckner-Goldstone (BBG) theory [4, 5], without the pretension to be exhaustive. Among the most accurate two-body forces, one has to consider the latest one of the Bonn potential series, that is more explicitely constructed from meson exchange processes [12]. Also the Argonne v_{18} NN interaction, that is constructed by a set of two-body operators which arise naturally in meson exchange processes, but with partially phenomenological form factors, can be considered as a very accurate one [13]. Concerning the phenomenological TBF (Urbana model), it contains a term that accounts for the so-called two-pion exchange contribution, with the creation of an intermediate excited Δ state, and a phenomenological repulsive part (Ropler excitation) [14]. As shown in Ref.[4], using exactly the same values of parameters of the TBF for the two choices of NN two-body interaction indicated above, the difference between the corresponding EOS's, which are apparent in the case without TBF, are strongly reduced when

TBF are introduced, not only around saturation, but for the whole density range. Hence, the overall effect of the same TBF on the EOS can be different according to the two-body force adopted, being less repulsive in the case when the two-body force is less attractive and pointing in the direction of making the EOS's closer to each other, see figure 1. This result is far from being trivial and it would be desirable to understand on physical grounds what is the interplay of two- and three-body forces in determining the nuclear EOS.

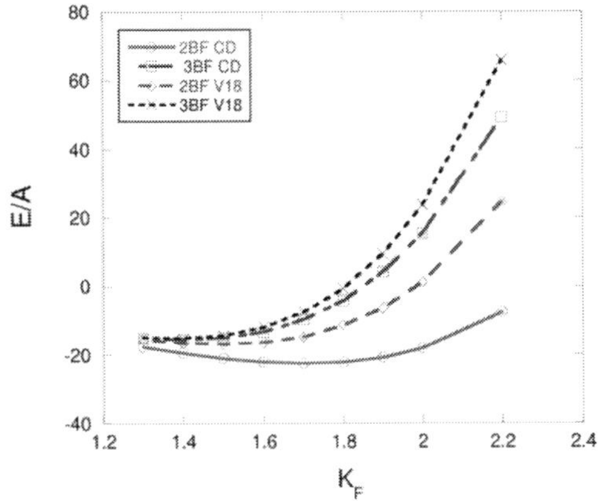

Figure 1. Equation of state of symmetric nuclear matter for the CD Bonn and the v_{18} two-body interactions and with the inclusion of three-body forces (TBF). The energy per particle E/A is in MeV and the Fermi momentum k_F is in fm^{-1}. Taken from Ref.[4].

A different strategy is followed in Ref.[5]: TBF can be evaluated on the basis of a meson-exchange approach. Starting from two-body one-boson exchange (OBE) potentials, like the Bonn or Nijmegen potentials (see Ref.[5] and references therein), the meson exchange parameters of the TBF can be chosen completely consistent with the given NN potential, i.e. the same parameters are used in two and three-body forces. This kind of TBF involves the intermediate excitation of Δ, Roper, and nucleon-antinucleon states by the exchange of π, ρ, σ, and ω mesons in the TBF diagrams. However, one must be aware of the fact that some parameters are not given by the two-body force, in particular the ones related to Δ and Roper excitation, and are evaluated according to recent prescriptions [5]. Contrary to what is done in Ref.[4], within this kind of approach, the nuclear matter saturation point is not fitted with the help of a phenomenological TBF, but it naturally arises from the full calculation, including two- and three-body forces and has to be checked against the experimental value. In figure 2 (middle panel), we report the EOS obtained with only two-body forces and with the full calculations, for different NN potentials. Also in this case one can see that the effect of the TBF is more repulsive when the two-body force is more attractive. This is the kind of compensation discussed before. The result corresponding to the Urbana (phenomenological) TBF, coupled to the Argonne two-body interaction, is also shown on the figure (dot-dashed curve). Comparing dashed and dot-dashed lines, it appears that a very different behavior is obtained for the high density EOS and for the symmetry energy (middle panel) when using different TBF's, inspite of the fact that the saturation point is reasonably reproduced in all cases. One may notice, in particular, the rather stiff high density behavior obtained with the Bonn B (BOB) potential, for which the reproduction of the saturation point properties is excellent. The corresponding symmetry energy exhibits an interesting behavior, that is soft at low density and rather stiff at high density (see top panel).

From the results discussed here, one may conclude that the density behavior of the symmetric matter EOS and symmetry energy is still a rather open problem and it is thus very appealing to try to get hints from the phenomenology of nuclear reactions and nuclear astrophysics. We will review this kind of studies in the following, with a special emphasis on the information one can get on the density dependence of the symmetry energy.

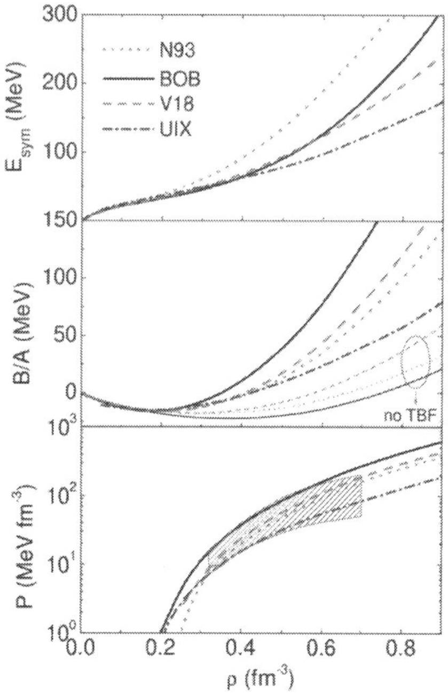

Figure 2. Symmetry energy (upper panel), binding energy per nucleon of symmetric nuclear matter (central panel), and pressure of symmetric matter (lower panel), employing different interactions. The shaded region indicates the constraints of Ref.[1]. Taken from Ref.[15].

3. Nuclear reactions and symmetry energy

Heavy ion reactions with neutron rich nuclei, from low up to relativistic energies, can be used to study the properties of the symmetry term of the nuclear interaction in a wide range of densities.

Nuclear reactions are modeled by solving transport equations based on mean field theories, with correlations included via hard nucleon-nucleon elastic and inelastic collisions and via stochastic forces, selfconsistently evaluated from the mean phase-space trajectory, see [16, 7]. Stochasticity is essential in order to get distributions as well as to allow for the growth of dynamical instabilities.

In the energy range up to a few hundred $AMeV$, the appropriate tool is the so-called Boltzmann-Langevin equation (BLE) [16]:

$$\frac{df}{dt} = \frac{\partial f}{\partial t} + \{f, H\} = I_{coll}[f] + \delta I[f], \tag{1}$$

where $f(\mathbf{r}, \mathbf{p}, t)$ is the one-body distribution function, or Wigner transform of the one-body density, $H(\mathbf{r}, \mathbf{p}, t)$ the mean field Hamiltonian, I_{coll} the two-body collision term incorporating the Fermi statistics of the particles, and $\delta I[f]$ the fluctuating part of the collision integral. For higher beam energies, a covariant formulation of transport equations will be considered, see section 5.2. Hence effective interactions and the nuclear matter EOS can be considered as ingredients of the transport codes and from the comparison with experimental data one can finally get some hints on nuclear matter properties.

We recall that the symmetry energy E_{sym} appears in the energy density $\epsilon(\rho, \rho_3) \equiv \epsilon(\rho) + \rho E_{sym}(\rho_3/\rho)^2 + O(\rho_3/\rho)^4 + ..$, expressed in terms of total ($\rho = \rho_p + \rho_n$) and isospin ($\rho_3 = \rho_p - \rho_n$) densities. E_{sym} gets a kinetic contribution directly from basic Pauli correlations and a potential part, $C(\rho)$, from the highly controversial isospin dependence of the effective interactions:

$$\frac{E_{sym}}{A} = \frac{E_{sym}}{A}(kin) + \frac{E_{sym}}{A}(pot) \equiv \frac{\epsilon_F}{3} + \frac{C(\rho)}{2\rho_0}\rho. \tag{2}$$

The sensitivity of the simulation results is tested against different choices of the density and momentum dependence of the isovector part of the Equation of State. In the non-relativistic frame, one can consider for the potential part of the symmetry energy, $C(\rho)$, two opposite density parametrizations (Iso-EOS) of the mean field [17, 7]: i) $\frac{C(\rho)}{\rho_0} = 482 - 1638\rho\ (MeV fm^3)$ for Asysoft EOS: $E_{sym}/A(pot)$ has a weak density dependence close to the saturation, with an almost flat behavior below the saturation density ρ_0 and even decreasing at suprasaturation; ii) a constant coefficient, $C = 32MeV$, or a $C(\rho)$ coefficient linearly increasing with density, for the Asystiff EOS choice: the interaction part of the symmetry term displays a faster decrease at lower densities and much stiffer above saturation, with respect to the Asysoft case, see figure 3. The isoscalar section of the EOS is the same in both cases, fixed requiring that the saturation properties of symmetric nuclear matter with a compressibility around $220MeV$ are reproduced.

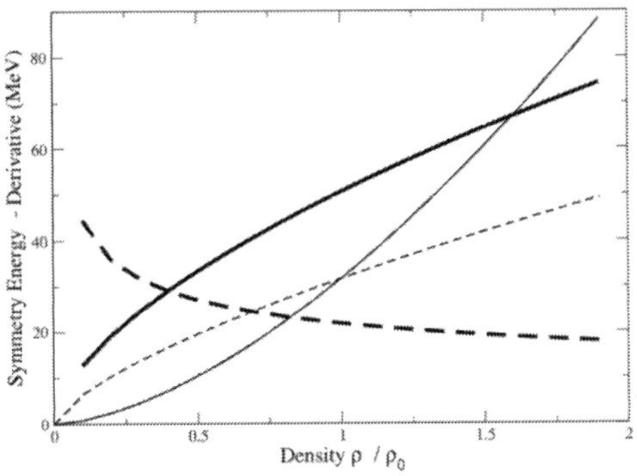

Figure 3. Two representative effective parameterizations of the symmetry energy (thin lines): Asystiff (full line) and Asysoft (dashed line). The tick lines show the corresponding derivatives.

The role of the isospin degree of freedom on the dynamics of a nuclear reaction can be discussed in a compact way by means of the chemical potentials for protons and neutrons as a function of density ρ and asymmetry $I = (N - Z)/A$ [18]. The p/n currents can be expressed as

$$\mathbf{j}_{p/n} = D^{\rho}_{p/n}\nabla\rho - D^{I}_{p/n}\nabla I, \tag{3}$$

with $D^{\rho}_{p/n}$ the drift, and $D^{I}_{p/n}$ the diffusion coefficients for transport, which are given explicitely in Ref. [18]. Of interest for the study of isospin effects are the differences of currents between protons and neutrons which have a simple relation to the density dependence of the symmetry energy

$$\begin{aligned} D^{\rho}_{n} - D^{\rho}_{p} &\propto 4I\frac{\partial E_{sym}}{\partial\rho}, \\ D^{I}_{n} - D^{I}_{p} &\propto 4\rho E_{sym}. \end{aligned} \tag{4}$$

Thus the isospin transport due to density gradients, i.e. isospin migration, depends on the slope of the symmetry energy, or the symmetry pressure, while the transport due to isospin concentration gradients, i.e. isospin diffusion, depends on the absolute value of the symmetry energy. Hence transport phenomena in nuclear reactions are directly linked to the EOS properties.

4. Symmetry energy at low density

At sub-saturation density, the symmetry energy of nuclear matter seems to be under control from the theoretical point of view since different microscopic many-body calculations agree among each other and TBF have a negligible effects [10, 19]. According to these calculations, the symmetry energy can be approximately described by $E_{sym}/A = 31.3(\rho/\rho_0)^{0.6}$. The existing data and analyses extracted from nuclear reactions [20, 21] seem to point to a behavior of the type $E_{sym}/A \propto (\rho/\rho_0)^{\gamma}$, with γ in the range $0.6 - 1$. Hence it is of great interest to pursue this investigation and devise new observables to test the behavior of the symmetry energy at low density.

4.1. The prompt dipole γ-ray emission in dissipative collisions

The low-density behavior of the symmetry energy can be explored looking at dipole excitations in dissipative charge asymmetric reactions around 10 $AMeV$. The possibility of an entrance channel bremsstrahlung dipole radiation due to an initial different N/Z distribution was suggested at the beginning of the nineties [22]. After several experimental evidences, in fusion as well as in deep-inelastic reactions, [23, 24] and references therein, the process is now well understood and stimulating new perspectives are coming from the use of radioactive beams.

During the charge equilibration process taking place in the first stages of dissipative reactions between colliding ions with different N/Z ratios, a large amplitude dipole collective motion develops in the composite dinuclear system, the so-called Dynamical Dipole mode. This collective dipole gives rise to a prompt γ-ray emission which depends on the absolute value of the initial amplitude, $D(t=0)$, on the fusion/deep-inelastic dynamics and on the symmetry term, below saturation, that is acting as a restoring force. Indeed this oscillation develops in the low density interface between the two colliding ions (neck region).

A detailed description is obtained in mean field transport approaches [25]. One can follow the time evolution of the dipole moment in the r-space, $D(t) = \frac{NZ}{A}(R_Z - R_N)$ and in p−space, $DK(t) = (\frac{P_p}{Z} - \frac{P_n}{N})$, being R_p, P_p (R_n, P_n) the centers of mass in coordinate and momentum space for protons (neutrons). A nice "spiral-correlation" clearly denotes the collective nature of the mode, see figure 4. The "prompt" photon emission probability, with energy $E_\gamma = \hbar\omega$, can be estimated applying a bremsstrahlung approach to the dipole evolution given from the BLE approach [25]:

$$\frac{dP}{dE_\gamma} = \frac{2e^2}{3\pi\hbar c^3 E_\gamma}|D''(\omega)|^2, \tag{5}$$

where $D''(\omega)$ is the Fourier transform of the dipole acceleration $D''(t)$. We remark that in this way it is possible to evaluate, in *absolute* values, the corresponding pre-equilibrium photon emission.

We must add a couple of comments of interest for the experimental selection of the Dynamical Dipole: i) The centroid is always shifted to lower energies (large deformation of the dinucleus); ii) A clear angular anisotropy should be present since the prompt mode has a definite axis of oscillation (on the reaction plane) at variance with the statistical giant dipole resonance (GDR). These features have been observed in recent experiments [24].

The use of unstable neutron rich projectiles would largely increase the effect, due to the possibility of larger entrance channel asymmetries. This can be observed in figures 4,5, where the results of the reactions $^{132}Sn+^{58}Ni$ and $^{124}Sn+^{58}Ni$ are compared [26].

One can notice in figure 4 the large amplitude of the first oscillation for the "132" system, but also the delayed dynamics for the Asystiff EOS related to a weaker isovector restoring force. In figure 5 (left panel) we report the power spectrum, $| D''(\omega) |^2$ in semicentral "132" reactions, for different Iso-EOS choices. The gamma multiplicity is simply related to it, see Eq.(5). The corresponding results for the stable "124" system are drawn in the right panel. As expected from

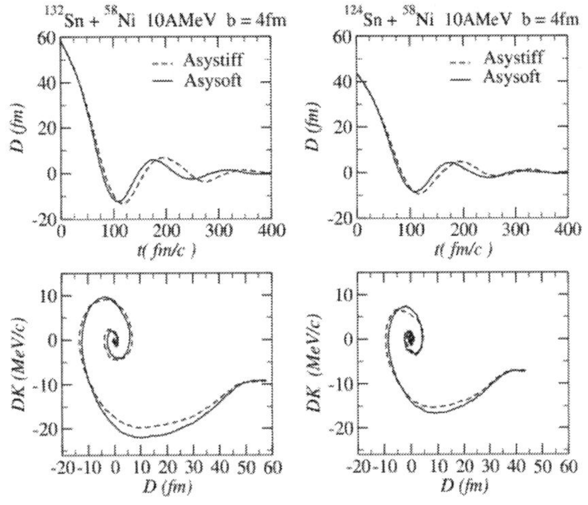

Figure 4. Dipole Dynamics at 10 $AMeV$, $b = 4fm$ centrality. Left panels: exotic "132" system. Upper panel: time evolution of dipole moment D(t) in real space; Lower panel: dipole phase-space correlation (see text). Right panels: same as before for the stable "124" system. Solid lines correspond to Asysoft EOS, the dashed to Asystiff EOS.

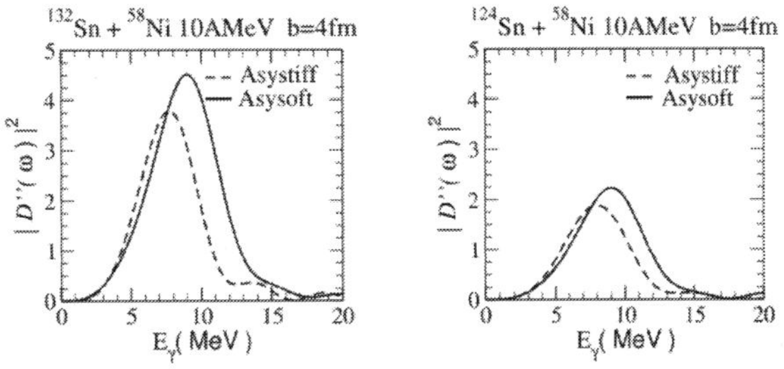

Figure 5. Left panel: exotic "132" system. Power spectra of the dipole acceleration at $b = 4$fm (in c^2 units). Right panel: corresponding results for the stable "124" system. Solid lines correspond to Asysoft EOS, the dashed to Asystiff EOS.

the larger initial charge asymmetry, we clearly see an increase of the Prompt Dipole Emission for the exotic n-rich beam. Such entrance channel effect allows also for a better observation of the Iso-EOS dependence. We remind that in the Asystiff case we have a weaker restoring force for the dynamical dipole in the dilute "neck" region, where the symmetry energy is smaller, see figure 3. This is reflected in the lower value of the centroid, ω_0, as well as in the reduced total yield, as shown in figure 5. The sensitivity of ω_0 to the stiffness of the symmetry energy is amplified by the increase of $D(t_0)$ using exotic, more asymmetric beams.

The prompt dipole radiation angular distribution is the result of the interplay between the collective oscillation life-time and the dinuclear rotation. In this sense one expects also a sensitivity to the Iso-EOS of the anisotropy, in particular for high spin event selections [26].

4.2. Isospin equilibration at the Fermi energies

In this energy range the doorway state mechanism of the Dynamical Dipole will disappear and so one can study a direct isospin transport in binary events. This process also involves the low density neck region and is sensitive to the low density behavior of E_{sym}, see Refs.[20, 27] and references therein.

It is interesting to look at the asymmetries of the various parts of the reaction system in the exit channel: emitted particles, projectile-like (PLF) and target-like fragments (TLF), and in the case of ternary events, intermediate mass fragments (IMF). In particular, one can study the so-called Imbalance Ratio, which is defined as

$$R^x_{P,T} = \frac{2(x^M - x^{eq})}{(x^H - x^L)} \;, \tag{6}$$

with $x^{eq} = \frac{1}{2}(x^H + x^L)$. Here, x is an isospin sensitive quantity that has to be investigated with respect to equilibration. We consider primarily the asymmetry $\beta = I = (N - Z)/A$, but also other quantities, such as isoscaling coefficients, ratios of production of light fragments, etc, can be of interest [8]. The indices H and L refer to the symmetric reaction between the heavy (n-rich) and the light (n-poor) systems, while M refers to the mixed reaction. P, T denote the rapidity region, in which this quantity is measured, in particular the PLF and TLF rapidity regions. Clearly, this ratio is ± 1 in the projectile and target regions, respectively, for complete transparency, and oppositely for complete rebound, while it is zero for complete equilibration.

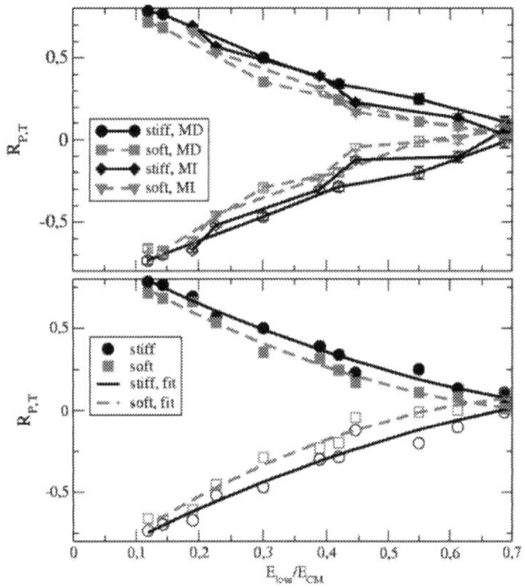

Figure 6. Imbalance ratios as a function of relative energy loss. Upper panel: separately for stiff (solid) and soft (dashed) Iso-EOS, and for two parameterizations of the isoscalar part of the interaction: MD (circles and squares) and MI (diamonds and triangles), in the projectile region (full symbols) and the target region (open symbols). Lower panel: quadratic fit to all points for the stiff (solid), resp. soft (dashed) Iso-EOS.

In a simple model one can show that the imbalance ratio mainly depends on two quantities: the strength of the symmetry energy and the interaction time between the two reaction partners. Let us take, for instance, the asymmetry β of the PLF (or TLF) as the quantity x. At a first order approximation, in the mixed reaction this quantity relaxes towards its complete equilibration value, $\beta_{eq} = (\beta_H + \beta_L)/2$, as

$$\beta^M_{P,T} = \beta^{eq} + (\beta^{H,L} - \beta^{eq})\; e^{-t/\tau}, \tag{7}$$

where t is the time elapsed while the reaction partners stay in contact (interaction time) and the damping τ is mainly connected to the strength of the symmetry energy [27]. Inserting this expression into Eq.(6), one obtains $R^\beta_{P,T} = \pm e^{-t/\tau}$ for the PLF and TLF regions, respectively. Hence the imbalance ratio can be considered as a good observable to trace back the strength of the symmetry energy from the reaction dynamics provided a suitable selection of the interaction time is performed.

The centrality dependence of the Imbalance Ratio, for (Sn,Sn) collisions at 35 and 50 $AMeV$, has been investigated in experiments as well as in theory [18, 20]. We report here a new analysis which appears experimentally more selective [27]. Longer interaction times should be correlated to a larger dissipation. It is then natural to look at the correlation between the imbalance ratio and the total kinetic energy loss. In this way one can also better disentangle dynamical effects of the isoscalar and isovector part of the EOS, see [27].

It is seen in figure 6 (top) that the curves for the Asysoft EOS (dashed) are generally lower in the projectile region (and oppositely for the target region), i.e. show more equilibration, than those for the Asystiff EOS, due to the higher value of the symmetry energy at low density. To emphasize this trend, all the values for the stiff (circles) and the soft (squares) Iso-EOS, corresponding to different impact parameters, beam energies and also to two possible parameterizations of the isoscalar part of the nuclear interaction (MD and MI) are collected together in the bottom part of the figure. One can see that all the points essentially follow a given line, depending only on the symmetry energy parameterization adopted. It is seen, that there is a systematic effect of the symmetry energy of the order of about 20 percent, which should be measurable. The correlation suggested in figure 6 should represent a general feature of isospin diffusion, and it would be of great interest to verify it experimentally.

4.3. Dipole emission in peripheral collisions at relativistic energies

Here we briefly review results obtained with more mascroscopic models on dipole excitation in peripheral collisions, such as ^{40}Ca + ^{24}O at 500 $AMeV$. In Ref.[28] a relativistic treatment is elaborated, to derive the equations of motion of (projectile and target) neutron and proton centroids, under the action of the restoring isovector nuclear force, taking into account in the excitation of the dipole oscillations, apart from Coulomb effects, also the contribution of the nuclear potential, treated with a Wood-Saxon potential well. It is observed that mean-field effects give a strong contribution to the excitation of the dipole oscillation, that is connected to the neutron skin of the target nucleus, leading to a different attraction between the target protons or neutrons and the projectile nucleons. The actual cross section for one- or two-phonon excitations is then evaluated according to the expression $\sigma_\alpha = 2\pi \int_0^\infty P_\alpha(b)T(b)bdb$, where $P_\alpha(b)$ is the probability, at the impact paramter b, for a given reaction channel α. This is related to the number of phonons, $N_{ph} = E_\infty/(\hbar\omega)$ excited by the driving force, being E_∞ the energy associated with the dipole oscillations at asymptotic distance. $T(b)$ are attenuation factors that take into account the depopulation of the reaction channels due to couplings not explicitely included in the truncated working space.

Comparing the obtained cross sections to experimental data, from these studies one can get information on the neutron skin structure of neutron-rich systems that, in turn, depends on the low density behavior of the symmetry energy.

4.4. Isospin distillation in central collisions

In central collisions at 30-50 $AMeV$, where the full disassembly of the system into many fragments is observed, one can study specifically properties of liquid-gas phase transitions occurring in asymmetric matter [29, 16, 7, 9, 30]. For instance, in neutron-rich matter, phase co-existence leads to a different asymmetry in the liquid and gaseous phase: fragments (liquid) appear more symmetric with respect to the initial matter, while light particles (gas) are more neutron-rich. The amplitude of this "distillation" effect, that is connected to the presence of density gradients in the system, depends on specific properties of the isovector part of the nuclear interaction, namely on the value and the derivative of the symmetry energy at low density (see also the discussion in section 3). Indeed, since the symmetry energy increases with density, it is energetically more convenient for asymmetric systems to transfer the neutron excess to the gas phase, while fragments (liquid phase) become more symmetric. These features can be easily

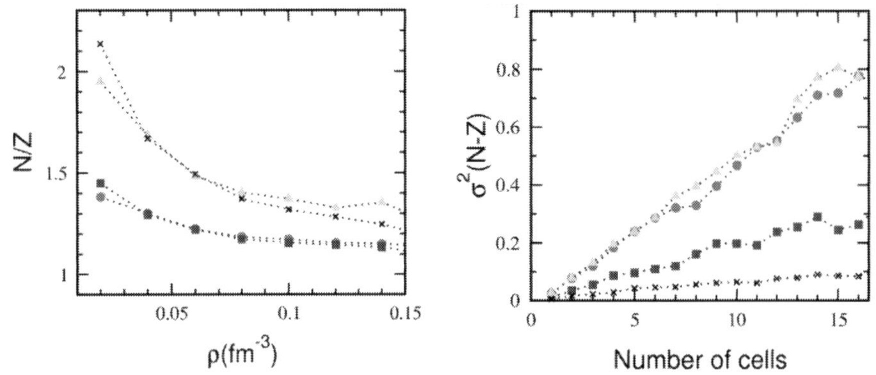

Figure 7. Left panel: correlation between the N/Z and the density of the domains, obtained with the two asy-EOS in the decomposition of unstable nuclear matter, with asymmetry $\beta = 0.1$ (squares and circles) and $\beta = 0.2$ (triangles and stars). Right panel: variance of the (N-Z) content of the density domains as a function of their volume [31].

evidenced by studying fragment formation in unstable asymmetric nuclear matter inside a box. In figure 7 (left) the N/Z of the domains formed inside the box is reported as a function of the density, for systems with two initial charge asymmetry values and for two Iso-EOS employed. The variance of the (N-Z) difference inside the density domains is displayed on the right part of the figure, as a function of the volume of the domain considered. One can see that, both the average N/Z and the isovector variance decrease as the density increases, leading to a more symmetric and less fluctuating liquid phase. However, fluctuations are rather large in the Asysoft case (circles and triangles) [31].

The investigation of isospin distillation is interesting in a more general context: In heavy ion collisions the dilute phase appears during the expansion of the interacting matter. Thus we study effects of the coupling of expansion, fragmentation and distillation in a two-component (neutron-proton) system [32]. Let us focus now on actual collisions, considering symmetric reactions between systems having three different initial asymmetry: $^{112}Sn + ^{112}Sn$,$^{124}Sn + ^{124}Sn$, $^{132}Sn + ^{132}Sn$, with $(N/Z)_{in} = 1.24,1.48,1.64$, respectively. The considered beam energy is 50 $AMeV$. The average N/Z of emitted nucleons (gas phase) and IMF's is presented in figure 8 (left panel) as a function of the initial $(N/Z)_{in}$ of the three colliding systems. One observes a clear Isospin-Distillation effect, i.e. the gas phase (circles) is more neutron-rich than the IMF's (squares). This is particularly evident in the Asysoft case due to the larger value of the symmetry energy and of its derivative at low density [7].

Fragmentation originates from the break-up of a composite source that expands with a given velocity field. Since neutrons and protons experience different forces, one may expect a different radial flow for the two species. In this case, the N/Z composition of the source would not be uniform, but would depend on the radial distance from the center or mass or, equivalently, on the local velocity. It has been recently proposed that this trend should then be reflected in a clear correlation between isospin content and kinetic energy of the formed IMF's [32].

This observable is plotted in figure 8 (right panel) for the three reactions. The behaviour observed is rather sensitive to the Iso-EOS. For the proton-rich system, the N/Z decreases with the fragment kinetic energy, expecially in the Asystiff case (left), where the symmetry energy is relatively small at low density. In this case, the Coulomb repulsion pushes the protons towards the surface of the system. Hence, more symmetric fragments acquire larger velocity. The decreasing trend is less pronounced in the Asysoft case (right) because Coulomb effects on protons are counterbalanced by the larger attraction of the symmetry potential at low density.

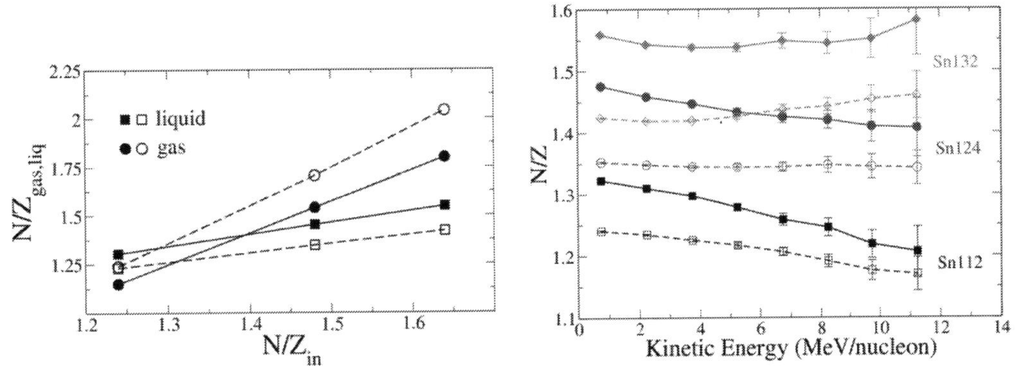

Figure 8. Left panel: the N/Z of the liquid (squares) and of the gas (circles) phase is displayed as a function of the system initial N/Z. Full lines and symbols refer to the Asystiff parameterization. Dashed lines and open symbols are for Asysoft. Right panel: the fragment N/Z as a function of the kinetic energy. Full lines: Asystiff; Dashed lines: Asysoft.

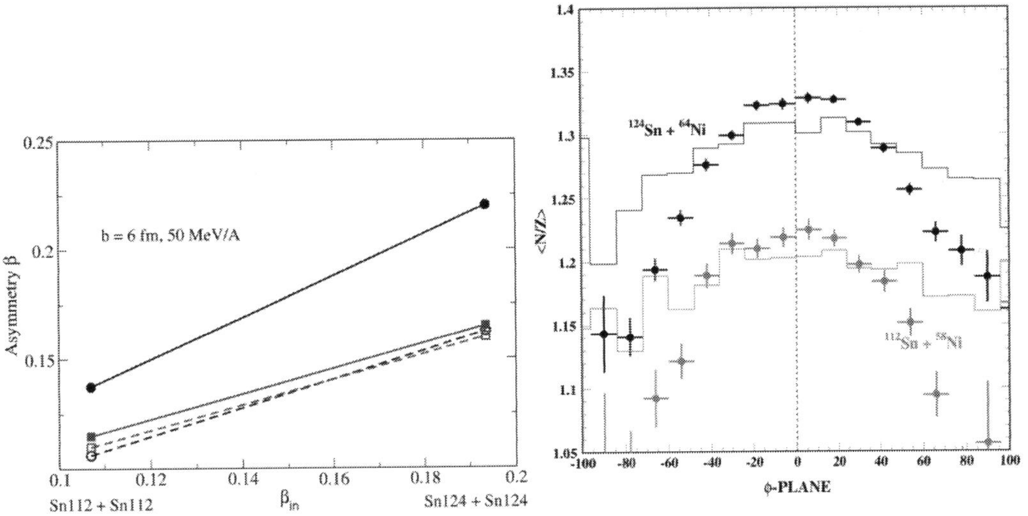

Figure 9. Left panel: asymmetry of IMF's (circles) and PLF-TLF (squares), as a function of the system initial asymmetry, for two Iso-EOS choices: Asystiff (full lines) and Asysoft (dashed lines). Right panel: experimental results on correlation between N/Z of IMF and *alignement* in ternary events of the $^{124}Sn+^{64}Ni$ and $^{112}Sn+^{58}Ni$ reactions at 35 $AMeV$. Points correspond to fast formed IMFs; histograms to all IMFs at mid-rapidity (including statistical emissions). From Ref.[33].

In systems with higher initial asymmetry, the decreasing trend is inversed, due to the larger neutron repulsion in neutron-rich systems. This analysis reveals the existence of significant, EOS-dependent correlations between the N/Z and the kinetic energy of IMF's produced in central collisions. This appears as a promising experimental observable to be investigated, though fragment secondary effects are expected to reduce the sensitivity to the Iso-EOS [32].

4.5. Isospin dynamics in neck fragmentation at Fermi energies
It is now quite well established that the largest part of the reaction cross section for dissipative collisions at Fermi energies goes through the *Neck Fragmentation* channel, with IMFs directly produced in the interacting zone in semiperipheral collisions on very short time scales [34]. It is possible to predict interesting isospin transport effects for this fragmentation mechanism since clusters are formed still in a dilute asymmetric matter but always in contact with the regions of the projectile-like and target-like remnants almost at normal densities. As discussed in section 3, in presence of density gradients the isospin transport is mainly ruled by drift coefficients and so we expect a larger neutron flow to the neck clusters for a stiffer symmetry energy around saturation [7]. This is shown in figure 9 (left), where the asymmetry of the neck region of charge asymmetric reactions is plotted for two Iso-EOS choices.

A very nice new analysis has been performed on the Sn+Ni data at 35 $AMeV$ by the Chimera Collab.[33], see figure 9 right panel. A strong correlation between neutron enrichement and fragment alignement (when the short emission time selection is enforced) is seen, that points to a stiff behavior of the symmetry energy, for which a large neutron enrichment of neck fragments is seen (left). This represents a clear evidence in favor of a relatively large slope (symmetry pressure) around saturation. We note a recent confirmation from structure data, i.e. from monopole resonances in Sn-isotopes [35].

5. Iso-EOS at supra-saturation density
5.1. Effective mass splitting and collective flows
The problem of Momentum Dependence in the Isovector channel ($Iso-MD$) of the nuclear interaction is still very controversial and it would be extremely important to get more definite experimental information, see the recent Refs. [36, 37]. Exotic Beams at intermediate energies (100-500 $AMeV$) are of interest in order to have high momentum particles and to test regions of high baryon (isoscalar) and isospin (isovector) density during the reaction dynamics. Transport codes are usually implemented with different (n,p) momentum dependences, see [37, 36]. This allows one to follow the dynamical effect of opposite n/p effective mass (m^*) splitting while keeping the same density dependence of the symmetry energy [27].

For central collisions in the interacting zone baryon densities about $1.7-1.8\rho_0$ can be reached in a transient time of the order of 15-20 fm/c. The system is quickly expanding and the Freeze-Out time is around 50fm/c. Here it is very interesting to study again the collective response of the system. Collective flows are very good candidates since they are expected to be rather sensitive to the momentum dependence of the mean field, see [38, 7]. The transverse flow, $V_1(y,p_t) = \langle \frac{p_x}{p_t} \rangle$, provides information on the anisotropy of nucleon emission on the reaction plane. Very important for the reaction dynamics is the elliptic flow, $V_2(y,p_t) = \langle \frac{p_x^2-p_y^2}{p_t^2} \rangle$. The sign of V_2 indicates the azimuthal anisotropy of emission: on the reaction plane ($V_2 > 0$) or out-of-plane ($squeeze-out$, $V_2 < 0$) [38]. The $Iso-MD$ of the fields can be tested just evaluating the difference of neutron/proton transverse and elliptic flows $V_{1,2}^{(n-p)}(y,p_t) \equiv V_{1,2}^n(y,p_t) - V_{1,2}^p(y,p_t)$ at various rapidities and transverse momenta in semicentral ($b/b_{max} = 0.5$) collisions, such as ^{197}Au+^{197}Au at 400 $AMeV$. For the nucleon elliptic flows the mass splitting effect is evident at all rapidities, and nicely increasing at larger rapidities and transverse momenta, with more neutron flow when $m_n^* < m_p^*$. From figure 10 it is clear how at, mid-rapidity, the mass splitting effects are more evident for higher tranverse momentum selections, i.e. for high density sources. In particular the elliptic flow difference becomes negative when $m_n^* < m_p^*$, revealing a faster neutron emission and so more neutron squeeze out (more spectator shadowing). In correspondance the proton flow is more negative (more proton squeeeze out) when $m_n^* > m_p^*$. Due to the difficulties in measuring neutrons, one could investigate the difference between light isobar flows, like ^{3}H vs. ^{3}He. We still expect to see effective mass splitting effects.

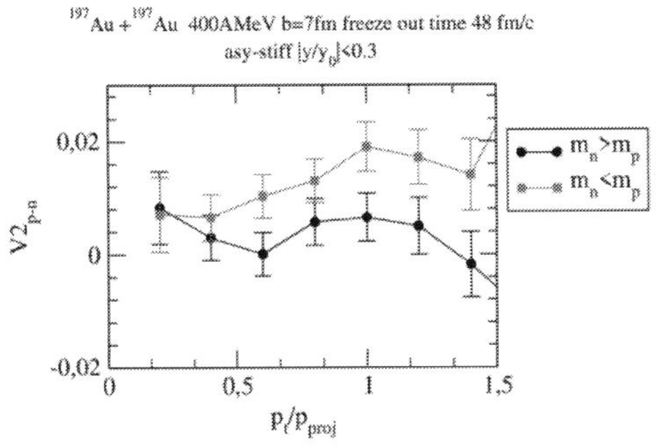

Figure 10. Transverse momentum dependence of the difference between proton and neutron V_2 flows, at mid-rapidity, in a semi-central reaction Au+Au at 400 $AMeV$. Taken from Refs.[39, 40].

5.2. *Meson production in relativistic heavy ion collisions*

The phenomenology of isospin effects on heavy ion reactions at relativistic energies (few $AGeV$ range) is extremely rich and can allow a "direct" study of the covariant structure of the isovector interaction in a high density hadron medium. In this energy range, one has to work within a relativistic transport frame, beyond the cascade picture, consistently derived from effective Lagrangians, where isospin effects are accounted for in the mean field and collision terms. Heavy ion collisions are described solving the covariant transport equation of the Boltzmann type (RBUU) [41], and applying a Monte-Carlo procedure for the hard hadron collisions. The collision term includes elastic and inelastic processes involving the production/absorption of the $\Delta(1232MeV)$ and $N^*(1440MeV)$ resonances as well as their decays into pion channels [42].

The effective Lagrangian approach to the hadron interacting system can be extended to the isospin degree of freedom: within the same frame equilibrium properties (EOS, [43]) and transport dynamics can be consistently derived. Within a covariant picture of the nuclear mean field, for the description of the symmetry energy at saturation (a) only the Lorentz vector ρ mesonic field, and (b) both, the vector ρ (repulsive) and scalar δ (attractive) effective fields are included. In the latter case a rather intuitive form of the symmetry energy can be obtained [7]

$$E_{sym} = \frac{1}{6}\frac{k_F^2}{E_F} + \frac{1}{2}\left[f_\rho - f_\delta\left(\frac{m^*}{E_F}\right)^2\right]\rho_B. \tag{8}$$

The competition between scalar and vector fields leads to a stiffer symmetry term at high density [7].

To discuss meson production, the starting point is a simple phenomenological version of the Non-Linear (with respect to the iso-scalar, Lorentz scalar σ field) effective nucleon-boson field theory, the Quantum-Hadro-Dynamics [43]. According to this picture the presence of the hadronic medium leads to effective masses and momenta $M^* = M + \Sigma_s$, $k^{*\mu} = k^\mu - \Sigma^\mu$, with Σ_s, Σ^μ scalar and vector self-energies. For asymmetric matter the self-energies are different for protons and neutrons, depending on the isovector meson contributions. The corresponding models are named $NL\rho$ and $NL\rho\delta$, respectively, and just NL for the case without isovector interactions. For the more general $NL\rho\delta$ case the self-energies of protons and neutrons read:

$$\begin{aligned}\Sigma_s(p,n) &= -f_\sigma\sigma(\rho_s) \pm f_\delta\rho_{s3},\\ \Sigma^\mu(p,n) &= f_\omega j^\mu \mp f_\rho j_3^\mu,\end{aligned} \tag{9}$$

(upper signs for neutrons), where $\rho_s = \rho_{sp} + \rho_{sn}$, $j^\alpha = j_p^\alpha + j_n^\alpha$, $\rho_{s3} = \rho_{sp} - \rho_{sn}$, $j_3^\alpha = j_p^\alpha - j_n^\alpha$ are the total and isospin scalar densities and currents and $f_{\sigma,\omega,\rho,\delta}$ are the coupling constants of the various mesonic fields. $\sigma(\rho_s)$ is the solution of the non linear equation for the σ field [7]. From the form of the scalar self-energies we note that in n-rich environment the neutron effective masses are definitely below the proton ones.

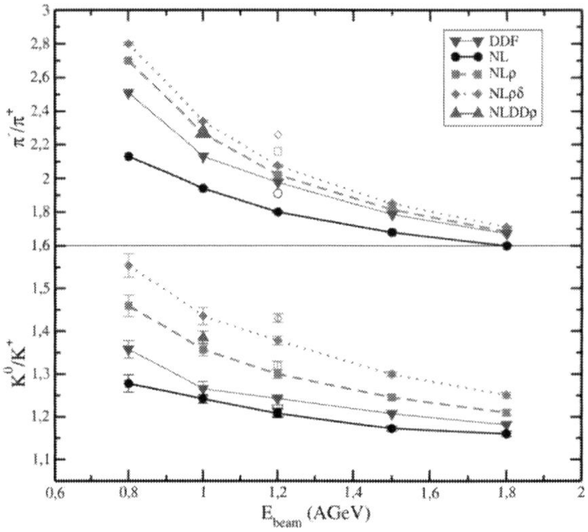

Figure 11. π^-/π^+ (top) and K^0/K^+ (bottom) ratios as a function of the incident energy for central Au + Au collisions, for several parameterizations of the isovector nuclear interaction. The open symbols at 1.2 $AGeV$ show the corresponding results for a ^{132}Sn + ^{124}Sn collision, more neutron rich.

Kaon production has been proven to be a reliable observable for the high density EOS in the isoscalar sector [44]. Here we show that the $K^{0,+}$ production (in particular the K^0/K^+ yield ratio) can be also used to probe the isovector part of the EOS [42]. Pion and kaon production are analyzed in central ^{197}Au+^{197}Au collisions in the $0.8 - 1.8$ $AGeV$ beam energy range [42].

When isovector fields are included the symmetry potential energy in neutron-rich matter is repulsive for neutrons and attractive for protons. In a heavy ion collision this leads to a fast, pre-equilibrium, emission of neutrons. Such a *mean field* mechanism, often referred to as isospin fractionation [7], is responsible for a reduction of the neutron to proton ratio during the high density phase, with direct consequences on particle production in inelastic NN collisions. On the other hand, the self-energy contributions, Eqs.9, will influence the particle production at the level of thresholds as well as of the phase space available in the final channel. As a matter of fact, the threshold effect is dominant and consequently the results are nicely sensitive to the covariant structure of the isovector fields, see figure 11. The beam energy dependence of the π^-/π^+ (top) and K^0/K^+ (bottom) ratios is shown on the figure. At each beam energy we see an increase of the π^-/π^+ and K^0/K^+ yield ratios with the models $NL \rightarrow NL\rho \rightarrow NL\rho\delta$. The effect is larger for the K^0/K^+ compared to the π^-/π^+ ratio. This is due to the subthreshold production and to the fact that the isospin effect enters twice in the two-step production of kaons, see [42]. Interestingly the Iso-EOS effect for pions is increasing at lower energies, when approaching the production threshold.

6. Hints from neutron stars

Neutron stars appear as natural astrophysical laboratories to study the behavior of the EOS at high density. In particular, the structure and composition of neutron stars is affected by the density dependence of the symmetry energy. Neutron stars are long-lived systems in β equilibrium, hence in rather different conditions with respect to the transient systems that are formed in nuclear reactions. The concept of EOS is very appropriate for these objects, but astrophysical observations of their properties are more difficult.

It is generally believed that a neutron star (NS) is formed as a result of the gravitational

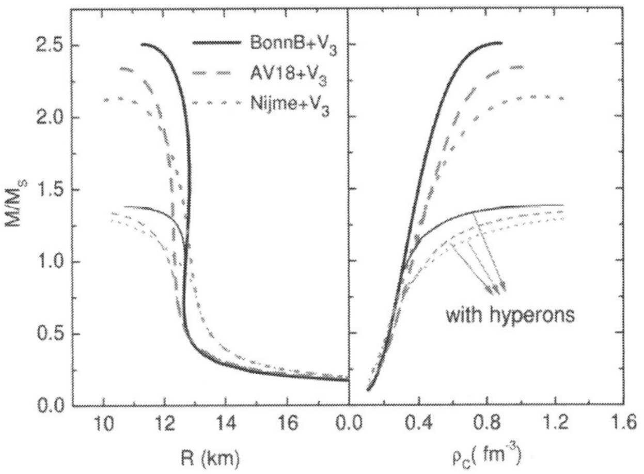

Figure 12. Mass-radius (left panel) and mass-central density (right panel) relations of neutron stars evaluated with different nucleonic EOS. Taken from Ref.[15].

collapse of a massive star in a type-II supernova. Hence a protoneutron star (PNS) appears, a very hot and lepton-rich object, where neutrinos are temporarily trapped. The following evolution of the PNS is dominated by neutrino diffusion, which eventually results in cooling. The star stabilizes at practically zero temperature, and no trapped neutrinos are left.

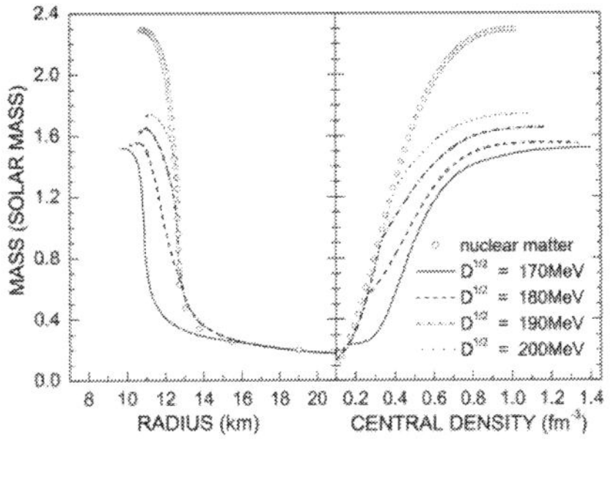

Figure 13. Mass-radius (left panel) and mass-central density (right panel) relations of hybrid stars for four values of the confinement parameter D. The nucleonic EOS is the one corresponding to the BOB interaction, see figure 2. Taken from Ref.[45].

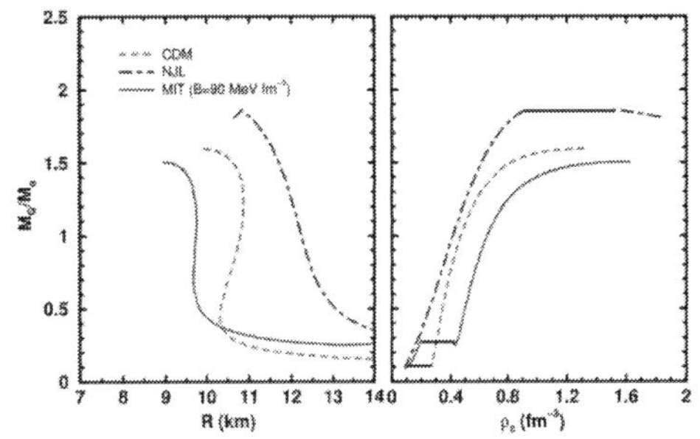

Figure 14. Mass-radius (left panel) and mass-central density (right panel) relations of hybrid stars, corresponding to three different models of the quark phase. The hadronic EOS includes hyperons. Adapted from Refs.[46, 48].

The dynamical transformation of a PNS into a NS could be strongly influenced by a phase transition in the central region of the star. In fact, the central particle density of a massive PNS may reach values larger than $1/fm^3$. In this density range the nucleon cores start to touch each other, and it is likely that quark degrees of freedom will play a role.

For stars in which the strongly interacting particles are only baryons, the composition at given baryon density ρ_B is determined by the requirements of β equilibrium and charge neutrality, involving nucleons, hyperons, leptons (electrons, muons) and neutrinos [45, 46]. These conditions fix, at the density ρ_B, and for given muon and electron numbers, the relative densities of all species. From the knowledge of these densities, using the Fermi-gas model for leptons, (non-interacting) hyperons and for neutrinos and the usual thermodynamical relation $P = \rho^2 \frac{\partial(f/\rho)}{\partial\rho}$ (being f the free-energy density) for neutrons and protons, one finally obtains the EOS of the matter that composes the star, i.e. the pressure corresponding to ρ_B, $P(\rho_B)$. In the above relation, the partial derivative of the free energy is taken at constant temperature and constant baryon number. The resulting EOS clearly depends on the nucleonic EOS adopted, usually taken from microscopic calculations, see section 2. Nucleon-hyperon interactions can also be included [46, 15].

To investigate the possible phase transition to quark matter in neutron stars, it is essential to know also the EOS of the quark matter. One usually deals with ordinary strange quark matter (SQM) in β equilibrium. The special problem in studying the EOS of ordinary quark matter is to treat quark confinement in a proper way. The conventional standard approach is to add an extra constant term to the quark energy, the bag constant B (MIT bag model), that leads to a negative contribution to the pressure, to confine quarks in a finite volume [46]. Alternatively, in Ref.[45] the quark confinement is treated using the chirally dependent quark mass scaling, where the quark mass scales with density. From the weak-equilibrium conditions of SQM, for given baryon and charge densities, it is possible to get the the relative densities of quarks (u,d,s) and electrons and finally derive the EOS of the quark matter [45].

The possible co-existence between quark and hadronic matter is studied imposing the Gibbs conditions [47]. Requiring the equality of the chemical potentials in the two phases, only two chemical potentials remain independent, such as (μ_n, μ_p) for instance. Their corresponding values as well as the quark fraction, are determined by imposing to have the same pressure in the two phases at the given total ρ_B and electric charge.

Calculations show that the transition from the hadron to the mixed phase occurs at a density a bit less than 0.15 fm^{-3}, close to the saturation density. Of course, this is not the case in terrestrial laboratories, since the nuclear matter so far realized in exotic nuclei or heavy ion collisions is much less neutron-rich than neutron stars. However, the transition to the pure quark phase occurs at much higher density, of the order of 0.6 fm^{-3}. The transition density depends on the MIT bag parameter or, in the case of the density dependent quark masses, on the parameter usually named D [45]. One can note that, increasing the bag parameter B, the pressure associated with the quark phase decreases and the transition starts at higher density. It is worth mentioning that the asymmetry of the quark phase is higher than the asymmetry of the co-existing hadronic phase.

With the EOS which has the mixed and/or quark phases at hand, the structure of hybrid stars can be studied by solving the Tolmann-Opennheimer-Volkoff (TOV) equation:

$$\frac{dP}{dr} = -\frac{GmE}{r^2} \frac{(1 + P/E)(1 + 4\pi r^3 P/m)}{1 - 2Gm/r}, \tag{10}$$

where G is the gravitational constant, r is the distance from the center of the star, $E = E(r)$ and $P = P(r)$ are the energy density and pressure at the radius r, respectively. The EOS $P = P(E)$ is essentially the input of the equation. The subsidiary condition is:

$$dm/dr = 4\pi r^2 E, \tag{11}$$

where $m(r)$ denotes the gravitational mass of the star and, for the considered unit choice, the speed of light in vacuum is $c = 1$. Starting with a central mass density E_C, Eq.(10) is integrated out until the pressure on the surface equals the one corresponding to the density of iron. This gives the stellar radius R and the gravitational mass M_G. In particular, one can investigate the maximum radius of stable stars.

As one can see from figure 12, a remarkable reduction of the large range of the maximum mass obtained with the nucleonic EOS's (1.8-2.5 $M_\odot$) to the nearly unique value of 1.3-1.4 $M_\odot$ is observed when only hyperons are included in the calculation of the nuclear EOS, i.e. excluding the possible transition to the quark matter, but implementing the nucleon-hyperon interaction in the EOS. The hyperons provide a self-regulating softening effect on the EOS: The stiffer the original nucleonic EOS, the more it is softened by the earlier appearance at higher concentration of hyperons. Due to the softening of the EOS, the maximum stellar mass decreases.

In any case, the maximum masses of hadronic (hyperonic) neutron stars are much too low in order to cover all present observational values. Within the present descriptions, the introduction of nonhadronic "quark" matter is thus necessary and heavier neutron stars can only be hybrid stars. Some calculations including the possibility of a transition (or co-existence) to the quark phase are shown in figures 13,14. Results obtained with different possibilities to model the quark phase and different nucleonic EOS are presented. With the inclusion of quarks, one obtains a maximum mass in the range 1.4-1.8 $M_\odot$, that depends on the model considered, but is closer to the experimental observations. The largest value is obtained adopting the Nambu-Jona Lasinio (NJL) model for the quark phase [48]. Indeed, in this case, the co-existence with the quark phase starts at rather high density and no transition at all is observed if hyperons are also included. Moreover, the onset of the pure quark phase at the center of the PNS as the mass increases marks an instability of the star, i.e. the PNS collapses to a black hole at the transition point since the quark EOS is unable to sustain the increasing central pressure due to gravity.

From the body of these calculations, one can conclude that limiting masses for PNS smaller than 2 solar masses are generally confirmed in all models.

All that shows clearly the great relevance of observations on NS to our knowledge of the high density nuclear EOS. The astrophysical observations could be able to rule out definite EOS or put constraints on them. However, the results discussed above indicate a larger sensitivity to the model adopted for the quark phase than to the nucleonic EOS.

7. On the transition to a mixed hadron-quark phase in HIC at relativistic energies

The possibility of the transition to a mixed hadron-quark phase, at high baryon and isospin density, has been suggested also for the dense and hot matter that is formed in HIC of charge asymmetric systems at relativistic energies, such as for instance Au + Au or $^{238}U+^{238}U$ (average proton fraction $Z/A = 0.39$) at 1 $AGeV$.

In figure 15 (left) we discuss the evolution of temperature, baryon density, energy density and Z/A in a space cell located in the c.m. of the system. We note that a rather exotic nuclear matter is formed in a transient time of the order of 10 fm/c, with baryon density around $3-4\rho_0$, temperature $50-60$ MeV, energy density 500 MeV fm^{-3} and proton fraction between 0.35 and 0.40, likely inside the estimated mixed phase region.

In fact the transition densities are largely isospin dependent and, as seen in the previous section in the case of neutron stars, reach rather low values in very asymmetric matter.

In the hadronic phase the charge chemical potential is given by $\mu_3 = 2E_{sym}(\rho_B)\frac{\rho_3}{\rho_B}$. Thus, we expect critical densities rather sensitive to the isovector channel in the hadronic EOS.

In figure 15 (right) we show the crossing density ρ_{cr} separating nuclear matter from the quark-nucleon mixed phase, as a function of the proton fraction Z/A, for different parameterizations of E_{sym}, see section 5.2. The MIT bag model is adopted for the quark phase. We can see the effect of the δ-coupling towards an *earlier* crossing due to the larger symmetry repulsion at high baryon densities. In the same figure the paths in the $(\rho, Z/A)$ plane followed in the c.m. region during the collision of the n-rich $^{132}Sn+^{132}Sn$ system, at different beam energies, are also reported. One can see that already at 300 $AMeV$ the systems may be on the border of the

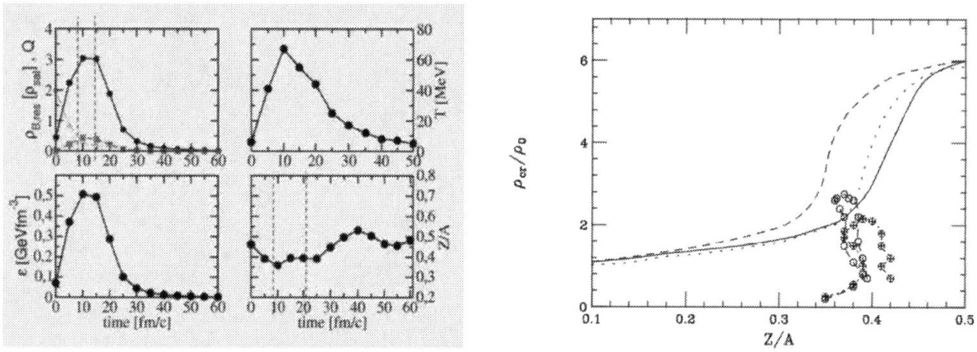

Figure 15. Left panel: time evolution of density, temperature, energy density and Z/A content reached in a central region of the reaction Au + Au at 1 $AGeV$, as obtained in RBUU calculations. Right panel: variation of the transition density with proton fraction for various hadronic EOS parameterizations. Dotted line: $GM3$ RMF-model [49]; dashed line: $NL\rho$; solid line: $NL\rho\delta$. For the quark EOS: MIT bag model with $B^{1/4}$=150 MeV. The points represent the path followed in the interaction zone during a semi-central ^{132}Sn+^{132}Sn collision at 1 $AGeV$ (circles) and at 300 $AMeV$ (crosses). From Ref.[50].

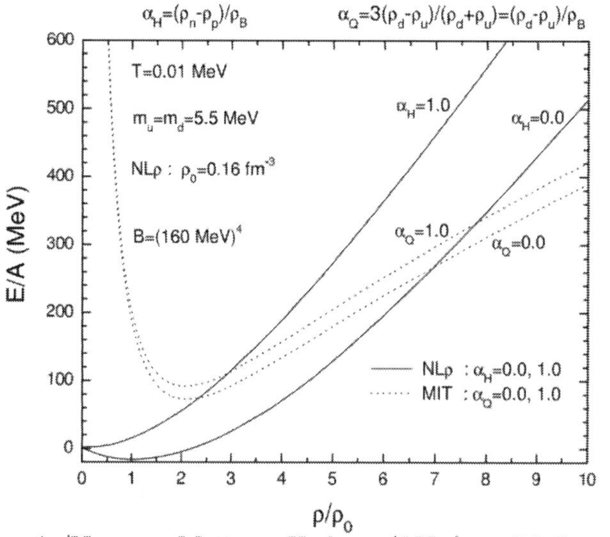

Figure 16. EOS of Symmetric/Neutron Matter: Hadron ($NL\rho$), solid lines, vs. Quark (MIT-Bag), dashed lines. $\alpha_{H,Q}$ represent the isospin asymmetry parameters respectively of the hadron,quark matter: $\alpha_{H,Q} = 0$, Symmetric Matter; $\alpha_{H,Q} = 1$, Neutron Matter. Taken from Ref.[40].

mixed phase, and well inside it at 1 $AGeV$ [50].

As a signature of co-existence with quark matter in HIC, one can expect a *neutron trapping* effect, supported by statistical fluctuations as well as by a symmetry energy difference in the two phases. In fact while in the hadron phase we have a large neutron potential repulsion (in particular in the $NL\rho\delta$ case), in the quark phase we only have the much smaller kinetic contribution. Observables related to such neutron "trapping" could be an inversion in the trend of the formation of neutron rich fragments and/or of the π^-/π^+, K^0/K^+ yield ratios for reaction products coming from high density regions, i.e. with large transverse momenta.

From the above discussion and from the results reported in section 6, it is clear that the possibility of a transition to the quark phase crucially depends on the way to describe the quark EOS. From this point of view, it appears extremely important to investigate the possible inclusion of the Isospin degree of freedom in the effective approaches to the QCD dynamics [40]. In figure 16 we report the EOS of symmetric and neutron matter for hadronic (full lines) and quark (dashed lines) phases. One can see that the transition to the quark phase is particularly convenient because, in the quark models considered so far, the symmetry energy of the quark matter takes contribution only from the kinetic energy and it is rather small. In fact, the two dashed lines in figure 16 are very close, while in the hadronic case the potential part of the symmetry energy is large.

Of course, these considerations are relevant also for the description of the properties of neutron stars.

8. Perspectives

We have reviewed some aspects of the rich phenomenology associated with nuclear reactions and astrophysical observations, from which interesting hints are emerging to constrain the nuclear EOS and, in particular, the largely debated density behavior of the symmetry energy. The greatest theoretical uncertainties concerns the high density domain, that has the largest impact on the understanding of the properties of neutron stars. This regime can be explored in terrestrial laboratories by using relativistic heavy ion collisions of charge asymmetric nuclei. Differential collective flows and meson production are promising observables. On the other hand, the behavior of the symmetry energy at low density can be accessed in reactions from low to intermediate energies, where collective excitations and fragmentation mechanisms are dominant. A considerable amount of work has already been done in this domain. In the near future, thanks to the availability of both stable and rare isotope beams, more selective analyses, also based on new exclusive observables, are expected to provide further stringent constraints.

References

[1] Danielewicz P, Lacey R and Lynch WG 2002 *Science* **298** 1592
[2] Fuchs C and Wolter HH 2006 *Eur. Phys. Jour.* A **30** (2006)
[3] Fantoni S *et al* 2008 *AIP Conf.Proc.* **1056** 233 and *Preprint* arXiv:0807.5043[nucl-th]
[4] Baldo M and Shaban AE 2008 *Phys. Lett.* B **661** 373
[5] Li ZH, Lombardo U, Schulze H-J *et al* 2008 *Phys. Rev.* C **77** 034316
[6] *Isospin Physics in Heavy-ion Collisions at Intermediate Energies*, Eds. Li BA and Schröder WU, Nova Science Publishers (2001, New York)
[7] Baran V, Colonna M, Greco V and Di Toro M 2005 *Phys. Rep.* **410** 335
[8] Colonna M and Tsang MB 2006 *Eur. Phys. J* A **30** 165, and references therein
[9] Li BA, Chen LW and Ko CM 2008 *Phys. Rep.* **465** 113
[10] Baldo M and Maieron C 2007 *J. Phys. G:Nucl. Part. Phys.* **34** R243; Baldo M *et al* 2004 *Nucl. Phys.* A **736** 241
[11] Zhou XR, Burgio GF, Lombardo U, Schulze H-J and Zuo W 2004 *Phys. Rev.* C **69** 018801, and references therein
[12] Machleidt R 2001 *Phys. Rev.* C **63** 024001
[13] Wiringa RB, Stoks VGJ and Schiavilla R 1995 *Phys. Rev.* C **51** 38
[14] Carlson J, Pandharipande VR and Wiringa RB, 1983 *Nucl. Phys* A **401** 59
[15] Li ZH, Lombardo U and Schulze H-J, submitted
[16] Chomaz P, Colonna M and Randrup J 2004 *Phys. Rep.* **389** 263
[17] Baran V *et al* 2002 *Nucl. Phys.* A **703** 603
[18] Baran V *et al* 2005 *Phys.Rev.* C **72** 064620

Journal of Physics: Conference Series **168** (2009) 012006 doi:10.1088/1742-6596/168/1/012006

[19] Baldo M and Maieron C 2008 *Phys. Rev.* C **77** 015801
[20] Tsang MB *et al* 2004 *Phys. Rev. Lett.* **92** 062701
[21] Xu HS *et al* 2000 *Phys. Rev. Lett.* **85**, 716; Famiano *et al Phys. Rev. Lett.* **97** 052701; Iglio J *et al* 2006 *Phys. Rev.* C **74** 024605
[22] Chomaz P, Di Toro M and Smerzi A 1993 *Nucl. Phys.* A **563** 509
[23] Pierroutsakou D *et al* 2005 *Phys. Rev.* C **71** 054605
[24] Martin B, Pierroutsakou D *et al* (Medea Collab.) 2008 *Phys. Lett.* B **664** 47
[25] Baran V, Brink DM, Colonna M and Di Toro M 2001 *Phys. Rev. Lett.* **87** 182501
[26] Baran V, Rizzo C, Colonna M, Di Toro M and Pierroutsakou D 2009 *Phys. Rev.* C **79** 021603
[27] Rizzo J *et al* 2008 *Nucl.Phys.* A **806** 79
[28] Dasso CH, Gallardo MI, Lanza EG *et al* 2008 *Nucl. Phys.* A **801** 129
[29] Li BA and Ko CM 1997 *Nucl.Phys.* A **618** 498
[30] Bonasera A *et al* 2008 *Phys.Rev.Lett.* **101** 122702
[31] Colonna M and Matera F 2008 *Phys. Rev.* C **77** 064606
[32] Colonna M *et al* 2008 *Phys. Rev.* C **78** 064618
[33] De Filippo E *et al* (Chimera Collab.), *Time scales and isospin effects on reaction dynamics*, NN06 Conf., Rio de Janeiro, August 2006, and *Isospin signals in reaction dynamics*, Int.Conf.on Nuclear Fragmentation, Antalya 2007
[34] Di Toro M, Olmi A and Roy R 2006 *Eur. Phys. Jour.* **A30** 65
[35] Li T, Garg U *et al* 2007 *Phys. Rev. Lett.* **99** 162503
[36] Li BA, Das Champak B, Das Gupta S and Gale C 2004 *Nucl. Phys.* A **735** 563
[37] Rizzo J, Colonna M and Di Toro M 2005 *Phys. Rev.* C**72** 064609
[38] Danielewicz P 2000 *Nucl. Phys.* A **673** 375
[39] Giordano V, Master Thesis, Univ. of Catania 2008
[40] Di Toro M *et al* 2009 *Progress in Particle and Nuclear Physics*, doi:10.1016/j.ppnp.2008.12.038
[41] Fuchs C, Wolter HH 1995 *Nucl. Phys.* A **589** 732
[42] Ferini G, Gaitanos T, Colonna M, Di Toro M and Wolter HH 2006 *Phys. Rev. Lett.* **97** 202301
[43] Serot BD and Walecka JD 1986 *Advances in Nuclear Physics* **16** 1, eds. J. W. Negele, E. Vogt, (Plenum, N.Y.)
[44] Fuchs C 2006 *Prog.Part.Nucl.Phys.* **56** 1
[45] Peng GX, Li A and Lombardo U 2008 *Phys. Rev.* C **77** 065807
[46] Nicotra OE, Baldo M, Burgio GF *et al* 2006 *Phys. Rev.* D **74** 123001
[47] Landau LD, Lifshitz L, *Statistical Physics*, Pergamon Press, Oxford 1969
[48] Burgio GF and Plumari S 2008 *Phys. Rev.* D **77** 085022
[49] Glendenning NK and Moszkowski SA 1991 *Phys. Rev. Lett.* **67** 2414
[50] Di Toro M, Drago A, Gaitanos T, Greco V and Lavagno A 2006 *Nucl. Phys.* A **775** 102

Scale and chiral symmetry in a effective lagrangian for nuclear physics

Luca Bonanno and Alessandro Drago
INFN Sezione di Ferrara, 44100 Ferrara, Italy
E-mail: bonanno@fe.infn.it,drago@fe.infn.it

Abstract. We study the restoration of chiral symmetry and scale invariance at high baryon density and/or energy-density through a effective chiral lagrangian for nuclear physics. In this lagrangian the breaking of scale invariance is regulated by the expectation value of a scalar field, called dilaton. A peculiar feature of this model is that the masses of the ω and ρ mesons initially slightly decrease with the density but then they increase again for densities larger than $\sim 3\,\rho_0$.

1. Introduction

In the next years several Heavy Ion Collisions (HICs) experiments at energies of the order of a few ten A GeV (as e.g. the ones proposed at facility FAIR at GSI [1], at RHIC (Brookhaven) and at the Nuclatron in Dubna) will probably start their activity. In these experiments the Equation of State (EOS) of matter will be tested at large density and/or temperature. It is therefore very important to provide, through theoretical investigations, a map of the "new" regions which will likely be explored by the new experiments.

Lattice QCD calculations have been crucial in providing a rather precise hint about what can happen at large temperature [2, 3]. One important result of lattice calculations is that, at least at zero density, chiral symmetry restoration and deconfinement take place at the same temperature. An important question is if the two symmetries are restored together also when working at finite density. It is therefore interesting to explore models in which chiral symmetry is embedded in the hadronic lagrangian, so that one can study chiral symmetry restoration independently of quark deconfinement.

In this work, based on our recent paper [4], we use an effective model based on a chiral invariant lagrangian for nuclear physics. In that model chiral fields are present together with a dilaton field which reproduces the breaking of scale invariance in QCD, but at a mean field level. This feature is by itself extremely interesting because it allows us to explore, within this effective model, the interplay between chiral symmetry and scale invariance.

2. The Chiral Dilaton Model

Spontaneously chiral symmetry breaking has long been studied in several microscopic models because chiral symmetry is a fundamental feature of low-energy effective lagrangians. At finite temperature the restoration of chiral symmetry in the linear and non-linear sigma model has been discussed but the attempt to describe nuclear dynamics fails due to the impossibility to reproduce basic properties of nuclei [5]. More sophisticated approaches have been proposed

Journal of Physics: Conference Series **168** (2009) 012007 doi:10.1088/1742-6596/168/1/012007

in the literature, both within a SU(2) chiral symmetric models [6, 7] and also extending the symmetry to the strange sector [8, 9, 10, 11, 12]. Here we use the model introduced by the Minnesota group [13, 14, 15]. In that model chiral fields are present together with a dilaton field which reproduces at a mean field level the breaking of scale invariance which takes place in QCD. In [13, 14, 15] it has been developed a formalism (which we adopt) allowing resummations of the thermal contributions. This is important when studying a strongly non-perturbative problem as the restoration of chiral symmetry. The lagrangian of the model reads:

$$\begin{aligned}\mathcal{L} &= \tfrac{1}{2}\partial_\mu\sigma\partial^\mu\sigma + \tfrac{1}{2}\partial_\mu\boldsymbol{\pi}\cdot\partial^\mu\boldsymbol{\pi} + \tfrac{1}{2}\partial_\mu\phi\partial^\mu\phi - \tfrac{1}{4}\omega_{\mu\nu}\omega^{\mu\nu} \\ &- \tfrac{1}{4}\boldsymbol{B_{\mu\nu}}\cdot\boldsymbol{B_{\mu\nu}} + \tfrac{1}{2}G_{\omega\phi}\phi^2\omega_\mu\omega^\mu + \tfrac{1}{2}G_{b\phi}\phi^2\boldsymbol{b}_\mu\cdot\boldsymbol{b}^\mu \\ &+ [(G_4)^2\omega_\mu\omega^\mu]^2 - \mathcal{V} \\ &+ \bar{N}\left[\gamma^\mu(i\partial_\mu - g_\omega\omega_\mu - \tfrac{1}{2}g_\rho\boldsymbol{b}_\mu\cdot\boldsymbol{\tau}) - g\sqrt{\sigma^2+\boldsymbol{\pi}^2}\right]N\end{aligned} \tag{1}$$

where the potential reads as the following:

$$\begin{aligned}\mathcal{V} &= B\phi^4\left(\ln\frac{\phi}{\phi_0} - \frac{1}{4}\right) - \tfrac{1}{2}B\delta\phi^4\ln\frac{\sigma^2+\boldsymbol{\pi}^2}{\sigma_0^2} \\ &+ \tfrac{1}{2}B\delta\zeta^2\phi^2\left[\sigma^2+\boldsymbol{\pi}^2 - \frac{\phi^2}{2\zeta^2}\right] - \tfrac{3}{4}\epsilon_1' \\ &- \tfrac{1}{4}\epsilon_1'\left(\frac{\phi}{\phi_0}\right)^2\left[\frac{4\sigma}{\sigma_0} - 2\left(\frac{\sigma^2+\boldsymbol{\pi}^2}{\sigma_0^2}\right) - \left(\frac{\phi}{\phi_0}\right)^2\right].\end{aligned} \tag{2}$$

Here σ and $\boldsymbol{\pi}$ are the chiral fields, ϕ the dilaton field, ω_μ the vector meson field and $\mathbf{b}_\mu$ the vector-isovector meson field, introduced in order to study asymmetric nuclear matter. The field strength tensors are defined by $F_{\mu\nu} = \partial_\mu\omega_\nu - \partial_\nu\omega_\mu$, $\mathbf{B}_{\mu\nu} = \partial_\mu\mathbf{b}_\nu - \partial_\nu\mathbf{b}_\mu$. In the vacuum $\phi = \phi_0$, $\sigma = \sigma_0$ and $\boldsymbol{\pi} = 0$. The ω and ρ vacuum masses are generated by their couplings with the dilaton field so that $m_\omega = G_{\omega\phi}^{1/2}\phi_0$ and $m_\rho = G_{\rho\phi}^{1/2}\phi_0$. Moreover $\zeta = \phi_0/\sigma_0$, B and δ are constants and ϵ_1' is a term that breaks explicitly the chiral invariance of the lagrangian.

The form of the potential $\mathcal{V}$ in eq. (2) comes from the necessity to reproduce the same divergence of the scale current as in QCD [4]. That divergence does not vanish due to the presence of the dimensional parameters ϕ_0 and σ_0 [4]. Moreover the potential $\mathcal{V}$ plays the role of the well known "Mexican hat" potential in generating the spontaneous breaking of the chiral symmetry. Notice also that when the mean value of the dilaton field drops to zero also the radius of the chiral circle drops to zero (see figure 1). In other words, when the scale invariance is restored also the chiral symmetry is automatically restored. This result is crucial since the restoration of scale invariance means the vanishing of all the condensates, including the chiral one. Finally, the values of the parameters used in calculations (listed in table 1 of Ref. [4]) were determined in Ref. [14] by fitting the properties of nuclear matter and finite nuclei.

3. Phase diagrams

The advantage of using a sophisticated model which includes both the chiral dynamics and the gluon condensate dynamics, is that one can make previsions on the restorations of both the chiral and the scale symmetry. In the following we study the restoration of both these symmetries.

3.1. Chiral symmetry restoration

Concerning the restoration of chiral symmetry, the Chiral Dilaton Model (CDM) gives different previsions depending on the presence or the absence of a symmetry breaking term into the lagrangian.

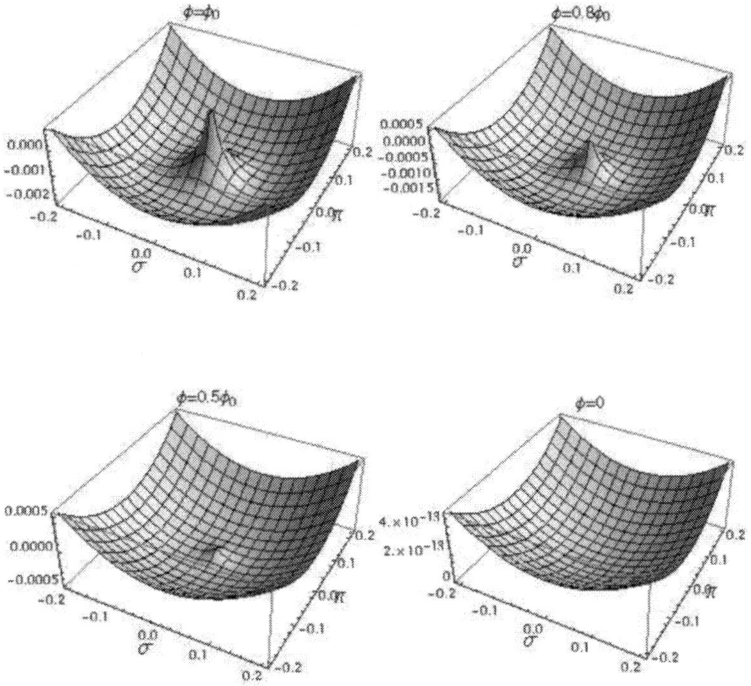

Figure 1. Dilaton potential (in GeV^4) as a function of the mean values of σ and π (in GeV) plotted for different values of the dilaton mean field. Here we consider $\epsilon_1' = 0$.

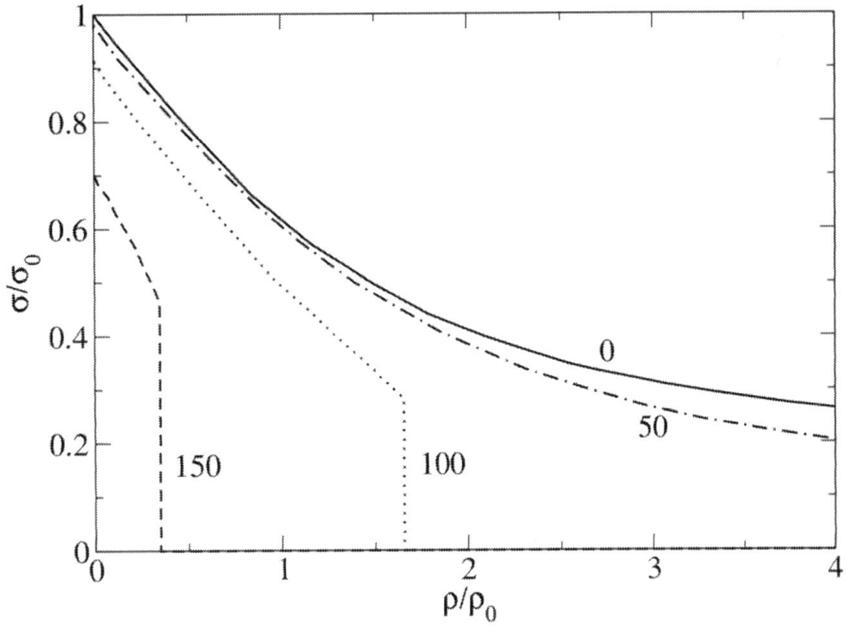

Figure 2. Mean value of the sigma field as a function of the baryon density for different values of the temperature. Here we are in the case where $\epsilon_1' = 0$.

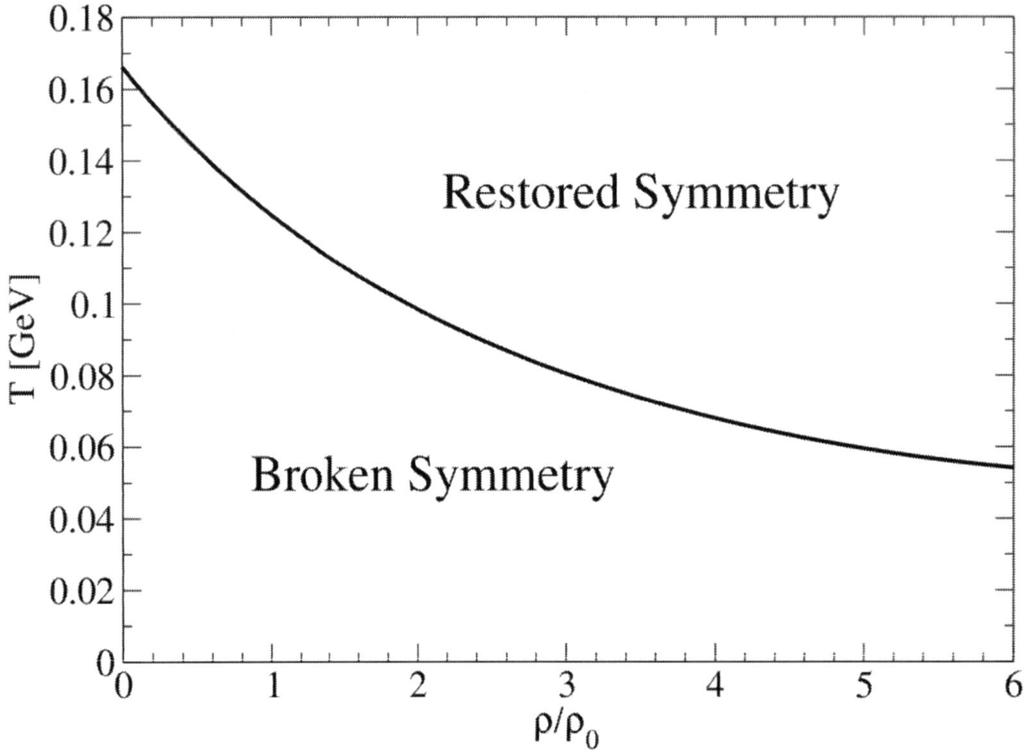

Figure 3. Chiral phases diagram for symmetric nuclear matter on the temperature versus baryon density plane. Here the explicit symmetry breaking term $\epsilon_1' = 0$.

In particular the restoration of chiral symmetry at finite density and temperature occurs only when the chiral invariance of the lagrangian is exact, or equivalently, only when the explicit symmetry breaking term $\epsilon_1' = 0$. In this case, as shown in figure 2, at a given temperature we observe that the mean field value of σ drops to zero after a critical density is reached. The transition turns out to be first order, due to the discontinuous behavior of the sigma field. The chiral phase diagram for isospin symmetric nuclear matter is shown in figure 3 in the temperature versus baryon density plane. Notice that the critical temperature at zero baryon density is $T_C = 165$ MeV, which is rather close to the QCD critical temperature estimated by lattice QCD [2, 3]. Obviously, in the case of QCD the restoration of chiral symmetry comes together with the appearance of the quark-gluon plasma phase.

As it can be observed, the critical temperature decreases with increasing density, but it never reaches zero, not even at very large densities. It can be interesting to notice that some (very preliminary) lattice calculations at finite density seem to support this result [16].

The chiral phase restoration in the chiral limit is also represented in figure 5, on the energy-density vs. baryon-density plane, by two bold lines, very close one to the other. These lines are separated by a small energy gap, which takes place since the chiral transition is first order.

Let us concentrate now in the more realistic case where $\epsilon_1' \neq 0$ and the pion has a finite mass. As shown in figure 4, in this case the σ mean field is continuous everywhere and it never drops to zero but decreases continuously with the density. This behavior, called "cross over", is predicted by the QCD lattice calculations with two flavors. Although a phase transition does not occur, there is a region where the σ mean field drops rapidly with the temperature, reaching very small values. This region, ranging from T=150 MeV to T$\sim$ 190 MeV, is indicated by a dashed

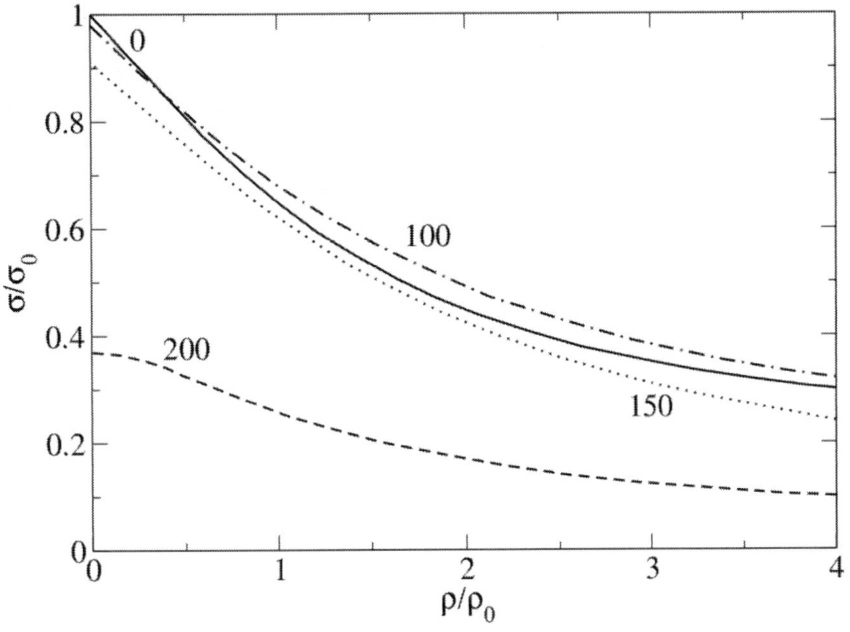

Figure 4. Same as in figure 2, but for the case where the chiral symmetry is explicitly broken ($\epsilon_1' \neq 0$).

curve in figure 5, where a sort of "phases map" for symmetric nuclear matter is shown on the energy-density versus baryon-density (ρ_B, ϵ) plane. In the same figure we have indicated with asterisks a rough estimates of the maximum energies and densities reached at the moment of maximum compression during a HIC experiment (the numbers above the marks indicate the beam energy in units of A GeV). These estimates, for a given beam energy, are extracted by two simple formulae provided in Ref.[17]:

$$\epsilon = 2\gamma_{cm}^2 m_N \rho_0 \; ; \qquad \rho = 2\gamma_{cm}\rho_0 \, , \tag{3}$$

where m_N is the nucleon mass, $\rho_0 = 0.15 fm^{-3}$ is the saturation density and $\gamma_{cm}^2 = 1 + E_0/2m_N$, with E_0 the beam energy. These formulae are based on the idea that the maximum baryon density can be estimated by considering the overlapping of two Lorentz-contracted nuclei. Notice that when the energy is increased from $E_0 = 2$ A GeV to $E_0 = 10$ A GeV the system crosses the dashed line and therefore the effect of the (at least partial) restoration of the chiral symmetry should be observable. In Ref. [18] we discussed the softening of the EOS due to chiral symmetry restoration and an attempt was made to connect our results with a analysis of the experimental data [19, 20].

3.2. Scale invariance restoration

Let us first recall that the order of a phase transition cannot be computed at a mean field level. Therefore our result, indicating a first order transition for scale invariance restoration, probably just indicates a rapid drop of the dilaton condensate at large temperatures [4]. As shown in figure 5, a large energy gap ($\sim$ 1 GeV) appears between the broken symmetry phase and the restored symmetry phase (the energy gap is indicated in figure 5 by a dark-shaded region). As previously discussed, the scale invariance restoration induces the restoration of the chiral symmetry. At zero baryon density, we obtain a critical energy density $\epsilon_c(0) = 523$ MeV/fm^3, which corresponds to T$\sim$ 200 MeV. Notice also that the critical energy density $\epsilon_c(\rho_B)$ increases

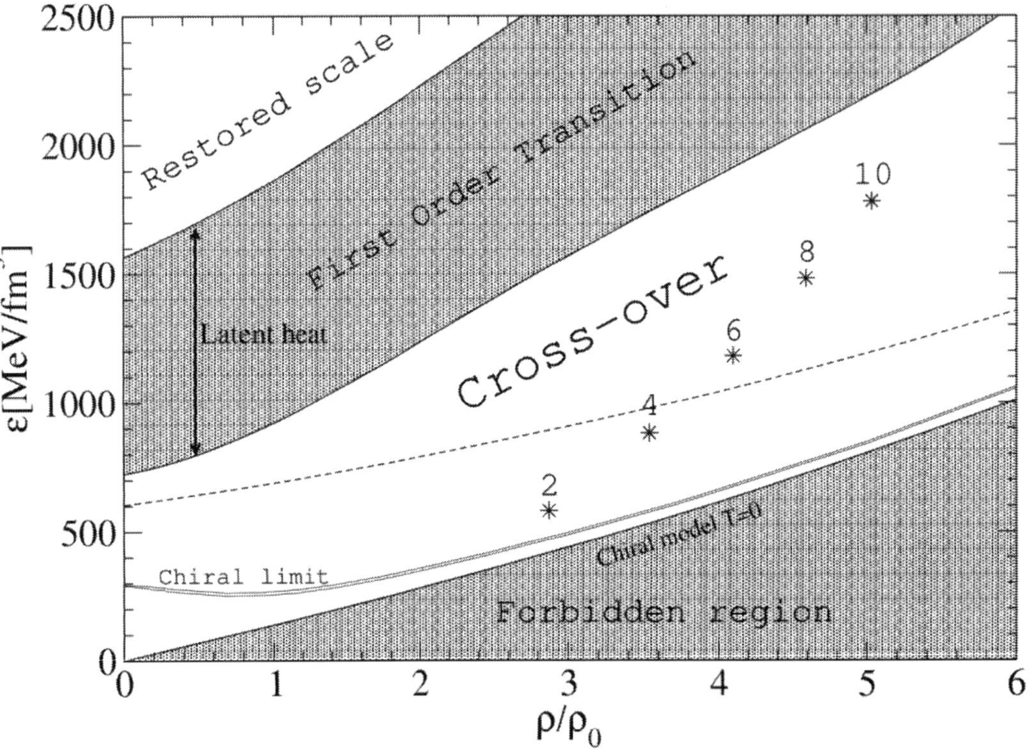

Figure 5. Phase diagram for symmetric nuclear matter on the energy-density versus baryon-density plane. Bold lines separate the (low-lying) region where chiral symmetry is broken from the (upper-lying) region where the chiral symmetry is restored in the chiral limit case $\epsilon_1' = 0$. All other graphical signs refer to the broken symmetry case $\epsilon_1' \neq 0$. The lower shaded area indicates the forbidden region under the T=0 EOS. The upper shaded area separates the (low-lying) region where scale symmetry is broken from the region where scale symmetry is restored. The dashed line indicates the maximum (negative) variation of the sigma mean value with the temperature (see text). The asterisks indicate the maximum energies and densities reached in the initial phase of a HICs experiment. The numbers above the marks indicate the beam energy in units of A GeV.

with the baryon density (it is possible to check that also the critical temperature for scale invariance restoration increases with the baryon density). This interesting effect arises from the repulsive contribution of the ω vector meson, which contributes to the energy density as $E_\omega = (1/2)(g_\omega/m_\omega^*)^2\rho_B^2$ (here for simplicity we have assumed $G_4 = 0$). Since the ω meson mass scales with the dilaton field, E_ω increases when the dilaton field decreases. Therefore at large densities it is less convenient to restore scale symmetry.

4. Vector meson masses

The experimental determination of the in-medium modification of the light vector mesons (ω, ρ and Φ) is still uncertain. Recent experiments, based on the dilepton spectroscopy, have advanced to a level that in-medium spectral information on vector mesons can be deduced. Within the CDM the vector meson masses are generated by coupling the vector mesons to the dilaton field (see the lagrangian density), so that the in the medium:

$$m_\omega^* = m_\omega\,(\phi/\phi_0) \tag{4}$$

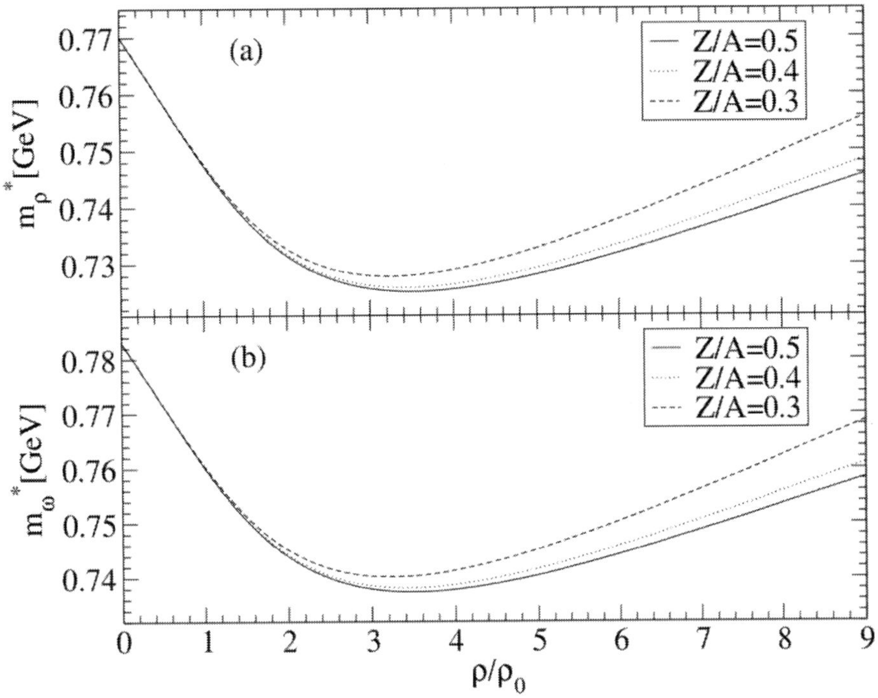

Figure 6. Vector meson masses as a function of the baryon density at T=0 and for different proton fractions.

$$m_\rho^* = m_\rho\,(\phi/\phi_0)\ .$$

In figure 6 the peculiar behavior of m_ω^* and m_ρ^* with the density is shown at T=0 and for different proton ratios: at low densities the masses initially drop, reaching a minimum in the neighbor of $\rho = 3.5\rho_0$, while for higher values of ρ, the masses increase again. This effect, due to the difficulty in restoring the scale invariance at large densities (see previous section), seems to be in agreement with several experimental analysis indicating a moderate reduction of the masses of the vector mesons at densities of the order of ρ_0 and no significant reduction when the masses are tested at very large densities [21, 22, 23, 24, 25, 26]. These experimental results presumably discards the scenario predicted by Brown and Rho [23, 27], where the vector meson masses drop continuously with the density [28].

Notice also that the effect of a finite isospin density is to increase the vector meson masses as it can be observed in figure 6, although at $\rho = 9\rho_0$ the increment is only of about 1%.

5. Conclusions

The main aim of our work was to investigate the behavior of matter at large densities and temperatures by using a effective lagrangian with broken scale invariance and chiral symmetry. In this contribution, extracted from a recent work [4], we presented two important results:

– we have provided a phase diagram mapping the regions in which chiral symmetry and/or scale invariance are restored. We have shown that scale invariance restoration is more difficult at large densities, because the repulsive effect of the exchange of vector mesons is enhanced if their mass is reduced;

– the masses of the vector mesons first reduce at finite density, but at larger densities they even increase.

Journal of Physics: Conference Series **168** (2009) 012007 doi:10.1088/1742-6596/168/1/012007

References

[1] Senger P 2004 *J. Phys.* G **30** S1087
[2] Cheng M *et al* 2007 *Phys. Rev.* D **75** 034506
[3] Cheng M *et al* 2006 *Phys. Rev.* D **74** 054507
[4] Bonanno L and Drago A 2009 *Phys. Rev.* C **79** 045801
[5] Furnstahl R J, Serot B D and Tang H B 1996 *Nucl. Phys.* A **598** 539
[6] Mishustin I, Bondorf J and Rho M 1993 *Nucl. Phys.* A **555** 215
[7] Furnstahl R J, Tang H B and Serot B D 1995 *Phys. Rev.* C **52** 1368
[8] Papazoglou P, Schramm S, Schaffner-Bielich J, Stoecker H and Greiner W 1998 *Phys. Rev.* C **57** 2576
[9] Papazoglou P *et al* 1999 *Phys. Rev.* C **59** 411
[10] Wang P, Lyubovitskij V E, Gutsche T and Faessler A 2004 *Phys. Rev.* C **70** 015202
[11] Dexheimer V, Schramm S and Stoecker H 2008 *J. Phys.* G **35** 014060
[12] Dexheimer V and Schramm S 2008 *Preprint* `arXiv:0802.1999`
[13] Heide E K, Rudaz S and Ellis P J 1994 *Nucl. Phys.* A **571** 713
[14] Carter G W, Ellis P J and Rudaz S 1996 *Nucl. Phys.* A **603** 367
[15] Carter G W and Ellis P J 1998 *Nucl. Phys.* A **628** 325
[16] Blum T C, Hetrick J E and Toussaint D 1996 *Phys. Rev. Lett.* **76** 1019
[17] Arsene I C *et al* 2007 *Phys. Rev.* C **75** 034902
[18] Bonanno L, Drago A and Lavagno A 2007 *Phys. Rev. Lett.* **99** 242301
[19] Ivanov Y B, Russkikh V N and Toneev V D 2006 *Phys. Rev.* C **73** 044904
[20] Russkikh V N and Ivanov Y B 2006 *Phys. Rev.* C **74** 034904
[21] Trnka D *et al* (CBELSA/TAPS) 2005 *Phys. Rev. Lett.* **94** 192303
[22] Muto R *et al* (KEK-PS-E325) 2007 *Phys. Rev. Lett.* **98** 042501
[23] Metag V 2007 *J. Phys.* G **34** S397
[24] Nasseripour R *et al* (CLAS) 2007 *Phys. Rev. Lett.* **99** 262302
[25] Adamova D *et al* (CERES/NA45) 2003 *Phys. Rev. Lett.* **91** 042301
[26] Arnaldi R *et al* (NA60) 2006 *Phys. Rev. Lett.* **96** 162302
[27] van Hees H and Rapp R 2006 *Preprint* `arXiv:hep-ph/0604269`
[28] Brown G E and Rho M 1991 *Phys. Rev. Lett.* **66** 2720

Magnetic fields generated by r-modes in accreting millisecond pulsars

Carmine Cuofano and Alessandro Drago

Dipartimento di Fisica, Universitá di Ferrara
and INFN sezione di Ferrara, 44100 Ferrara, Italy

E-mail: `cuofano@fe.infn.it`

Abstract. In millisecond pulsars the existence of the Coriolis force allows the development of the so-called Rossby oscillations (r-modes) which are know to be unstable to emission of gravitational waves. These instabilities are mainly damped by the viscosity of the star or by the existence of a strong magnetic field. A fraction of the observed millisecond pulsars are known to be inside Low Mass X-ray Binaries (LMXBs), systems in which a neutron star (or a black hole) is accreting from a donor whose mass is smaller than 1 $M_{\odot}$. Here we show that the r-mode instabilities can generate strong toroidal magnetic fields by inducing differential rotation. In this way we also provide an alternative scenario for the origin of the magnetars.

1. Introduction

The r-mode oscillations in all rotating stars are unstable for emission of gravitational waves [1]. These modes play therefore a very important role in the astrophysics of compact stars and in the search for gravitational waves. On the other hand the existence of millisecond pulsars implies the presence of damping mechanisms of the r-modes. Damping mechanisms are associated with bulk and shear viscosity and with the possible existence of the so called Ekman layer. The latter is located at the interface between the solid crust and the fluid of the inner core and in this region friction is significantly enhanced respect to friction in a purely fluid component. All these mechanisms are strongly temperature dependent.

An important class of rapidly rotating neutron stars are the accreting millisecond pulsars associated with Low Mass X-ray Binaries (LMXBs). For these objects the internal temperature is estimated to be in the range 10^8-$10^{8.5}$ K [2, 3] and their frequencies can be as large as ~ 650 Hz. In this range of temperatures and in the case of a purely nucleonic star, bulk and shear viscosities alone cannot stabilize stars whose frequency exceeds ~ 100 Hz. A possible explanation of the stability of stars rotating at higher frequencies is based on the Ekman layer, but recent calculations show that this explanation holds only for rather extreme values of the parameters [4]. In this contribution we propose a new damping mechanism based on the generation inside the star of strong magnetic fields produced by r-mode instabilities. This same mechanism has been proposed in the case of rapidly rotating, isolated and newly born neutron stars in [5, 6]. In that paper the mechanism which generates the magnetic field is investigated only during the relatively short period in which the star remains always in the instability region. In our work we consider accreting stars and we investigate the interplay between r-modes and magnetic field on an extremely long period and we show that in this scenario a very strong magnetic field can be produced.

2. R-mode equations in the presence of magnetic field

R-mode instabilities are associated to kinematical secular effects which generate differential rotation in the star and large scale mass drifts, particularly in the azimuthal direction. Differential rotation in turn can produce very strong toroidal magnetic fields in the nucleus and these fields damp the instabilities extracting angular momentum from the modes. In order to derive the equations regulating the evolution of r-modes in the presence of a pre-existent poloidal magnetic field we have modified the equations derived in [7], taking into account also the magnetic damping.

We use the estimate given in [1] for the gravitational radiation reaction rate due to the $l = m = 2$ current multipole

$$F_g = \frac{1}{47} M_{1.4} R_{10}^4 P_{-3}^{-6} \quad \mathrm{s} \tag{1}$$

as well as the bulk and shear viscosity damping rates

$$\begin{aligned} F_b &= \frac{1}{2.7 \times 10^{11}} M_{1.4}^{-1} R_{10} P_{-3}^{-2} T_9^6 \quad \mathrm{s} \\ F_s &= \frac{1}{6.7 \times 10^7} M_{1.4}^{5/4} R_{10}^{-23/4} T_9^{-2} \quad \mathrm{s} \end{aligned} \tag{2}$$

where we have used the notation $M_{1.4} = M/1.4M_\odot$, $R_{10} = R/10$ Km, $P_{-3} = P/1$ ms and $T_9 = T/10^9$ K.

The total angular momentum J of a star can be decomposed into a equilibrium angular momentum J_* and a perturbation proportional to the canonical angular momentum of a r-mode J_c:

$$J = J_*(M, \Omega) + (1 - K_j) J_c, \qquad J_c = -K_c \alpha^2 J_* \tag{3}$$

where $K_{(j,c)}$ are dimensionless constants and $J_* \cong I_* \Omega$.

Following Ref.[8] the canonical angular momentum obeys the following equation:

$$dJ_c/dt = 2J_c\{F_g(M,\Omega) - [F_v(M,\Omega,T_v) + F_{m_i}(M,\Omega,B_p)]\} \tag{4}$$

where $F_v = F_s + F_b$ is the viscous damping rate and we have introduced the magnetic damping rate F_{m_i} that we discuss in the next section.

The total angular momentum satisfies instead the equation:

$$dJ/dt = 2J_c F_g + \dot{J}_a(t) - I_* \Omega F_{m_e} \tag{5}$$

where $\dot{J}_a$ is the rate of accretion of angular momentum and we have assumed it to be $\dot{J}_a = \dot{M}(GMR)^{1/2}$, and F_{m_e} is the magnetic braking rate associated to the poloidal magnetic field.

Combining the equations (4) and (5) than we give the dynamical evolution relation

$$\begin{aligned} \frac{d\alpha}{dt} &= \alpha(F_g - F_v - F_{m_i}) + \alpha[K_j F_g + (1 - K_j)(F_v + F_{m_i})] K_c \alpha^2 - \frac{\alpha \dot{M}}{2\tilde{I}\Omega} \left(\frac{G}{MR^3}\right)^{1/2} + \frac{\alpha F_{m_e}}{2} \\ \frac{d\Omega}{dt} &= -2K_c \Omega \alpha^2 [K_j F_g + (1 - K_j)(F_v + F_{m_i})] - \frac{\dot{M}\Omega}{M} + \frac{\dot{M}}{\tilde{I}} \left(\frac{G}{MR^3}\right)^{1/2} - \Omega F_{m_e} \end{aligned} \tag{6}$$

where $I = \tilde{I} M R^2$ with $\tilde{I} = 0.261$ for an n=1 polytrope. The previous equations can be simplified when the star is close to the instability region identified by the condition $F_g - F_v - F_m \approx 0$ and we obtain:

$$\begin{aligned} \frac{d\Omega}{dt} &= -2\Omega\alpha^2 K_c (F_v + F_m) - \frac{\dot{M}}{M}\Omega + \frac{\dot{M}}{\tilde{I}}\left(\frac{G}{MR^3}\right)^{1/2} \\ \frac{d\alpha}{dt} &= \alpha F_g - \alpha (F_v + F_m)(1 - \alpha^2 K_c) - \frac{\alpha \dot{M}}{2\tilde{I}\Omega}\left(\frac{G}{MR^3}\right)^{1/2} \end{aligned} \tag{7}$$

We note that $K_c = 9.4 \times 10^{-2}$ and that in this approximation the value of K_j is unimportant. Moreover the magnetic breaking is negligible for magnetic fields typical of the LMXBs (10^8-10^9G) and for accretion rates considered (10^{-8}-10^{-9} $M_\odot$ yr^{-1}).
Viscosity depends critically on temperature. We include three factors in modelling the temperature evolution: modified URCA cooling, shear viscosity reheating, and accretion heating. The cooling rate due to the modified URCA reactions, $\dot{\epsilon}_u$, is given in [9]

$$\dot{\epsilon}_u = 7.5 \times 10^{39} M_{1.4}^{2/3} T_9^8 \quad \text{erg s}^{-1} \tag{8}$$

The neutron star will be heated by the action of shear viscosity on the r-mode oscillations. The heating rate due to shear viscosity, $\dot{\epsilon}_s$, is given by [1]

$$\dot{\epsilon}_s = 2\alpha^2\Omega^2 MR^2 \tilde{J} F_s = 8.3 \times 10^{37} \alpha^2 \Omega^2 \tilde{J} M_{1.4}^{9/4} R_{10}^{-15/4} T_9^{-2} \quad \text{erg s}^{-1} \tag{9}$$

where $\tilde{J} = 1.635 \times 10^{-2}$.
Accretion heating have two components. We use the estimates given in [10]. The first contribution arises when accreting matter undergoes nuclear burning at the surface of the star

$$\dot{\epsilon}_n = \frac{\dot{M}}{m_B} \times 1.5\text{MeV} = 4 \times 10^{51} \dot{M}_{1.4} \quad \text{erg s}^{-1} \tag{10}$$

where m_B is the mass of a barion.
The second contribution arise because the flow is assumed to be advection dominated. Matter falling in towards a star liberate $\sim GM/R$ of potential energy per unit mass. In a non-advection dominated flow most of this would be dissipated as heat in accretion disk, while in an advection dominated flow this energy is carried in with the flow of matter. The heating rate is

$$\dot{\epsilon}_h \sim \frac{R}{\lambda}\frac{GM\dot{M}}{R} = 8 \times 10^{51} M_{1.4}^{13/6} \dot{M}_{1.4} \quad \text{erg s}^{-1} \tag{11}$$

Finally we use the estimate of the heat capacity C_v given in [10]

$$C_v = 1.6 \times 10^{39} M_{1.4}^{1/3} T_9 \quad \text{erg K}^{-1} \tag{12}$$

3. Magnetic damping

The crucial ingredient introduced in the previous Section is the magnetic damping rate, which we have inserted in the evolution of the r-modes. This specific modification of the r-mode equations is at the basis of the phenomenology we are going to discuss and was never considered in previous calculations. The expression of the magnetic damping rate has been derived in [5, 6], where it has been shown that while the star remains in the instability region, the r-modes generate a differential rotation which can greatly amplify a pre-existent magnetic field. More specifically, if a poloidal magnetic field was originally present, a strong toroidal field is generated inside the star. The energy of the modes is therefore transfered to the magnetic field and the instability is damped.

The expression of the magnetic damping rate reads:

$$F_m \equiv \frac{1}{\tau_m} \simeq \frac{4(1-p)}{9\pi p \cdot (8.2 \times 10^{-3})} \frac{B_p^2 R \Lambda' \int_0^t \alpha^2(t')\Omega(t')dt'}{M\Omega}, \tag{13}$$

where p and Λ' are dimensionless parameters of order unit. The time integral over the r-mode amplitude α takes contribution from the period during which the star is inside the instability region. The crucial point is if the toroidal magnetic field generated during an instability phase can be stabilized or unwind on a Alfvén timescale.

4. Results

We consider a scenario in which the mass accretion spin up a initially slowly rotating neutron star and we investigate the case in which the star cannot find a state of thermal equilibrium, then it enters a regime of thermo-gravitational runaway within a few hundred years of crossing the r-mode stability curve. In figure 1 and figure 2 we show the evolution of temperature, spin frequency and r-mode amplitude obtained without magnetic field. At the instant $\boldsymbol{A}$ the accreting neutron star enters the r-modes instability window. Until the star is close to equilibrium the instability grows slowly and the star moves to higher temperatures mainly because of non thermal equilibrium. After the instant $\boldsymbol{B}$ the r-mode instability grows exponentially due to the decrease of the shear viscosity with increasing temperature. As a consequence the r-modes amplitude rapidly reaches the saturation value and the viscosity heats significantly the star. At this stage the star loses angular momentum by emission of gravitational waves and within few hundred years goes out of the instability region (instant $\boldsymbol{C}$).

Taking into account the magnetic field, the evolutionary scenario for the star is quite different. After the instant $\boldsymbol{B}$, the secular effects are extremely large and the toroidal magnetic field is either produced or amplified by the wrapping of the poloidal magnetic field produced by the (mostly) toroidal secular velocity field. Our results obtained solving Eqs.(7) and reported in figure 3 show how the new magnetic field so generated reaches a value $B_{tor} \sim 10^{14}$ G in about one hundred years and stabilizes the star when it is still in the instability region.

The subsequent evolution of the system is difficult to control because it strongly depends on the evolution of the magnetic field. Analytical and numerical simulations show that strong purely toroidal fields are unstable. We will limit ourselves to outline a possible scenario that we discuss in a forthcoming paper. Following the results of some numerical simulations, the magnetic field can evolve on an Alfvén timescale into a stable configuration of a mixed poloidal-toroidal twisted-torus shape [11]. In this scenario the newly formed magnetic field can provide a effective damping of the r-modes and rapidly rotating stars can be stabilized also in the absence of the Ekman layer.

It is a pleasure to thank Giuseppe Pagliara and Andreas Reisenegger for many useful and stimulating discussions.

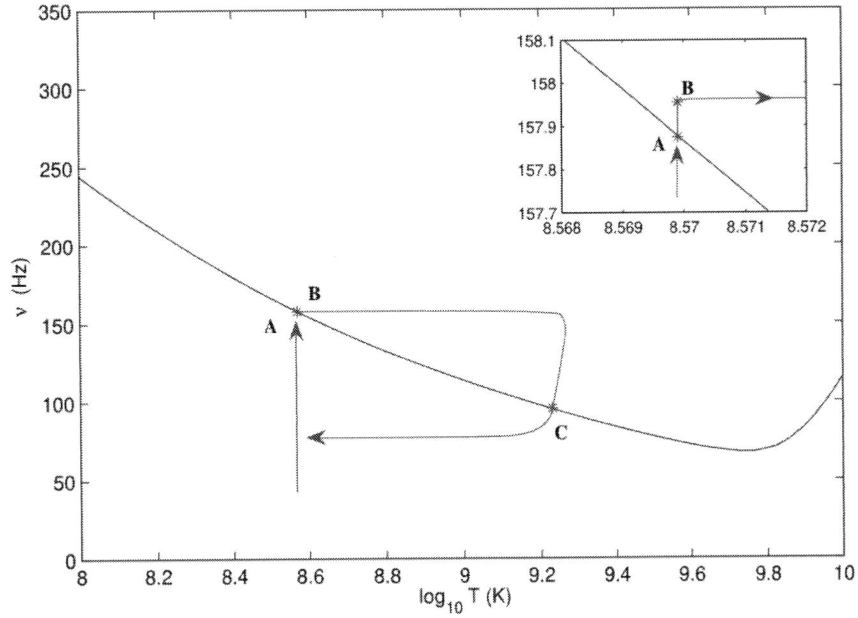

Figure 1. Path followed by the accreting neutron star, without toroidal magnetic field, in the Temperature vs Frequency plane. Here $\dot{M} = 10^{-8}\ M_\odot\ yr^{-1}$ and poloidal magnetic field $B_{pol} = 10^8$ G.

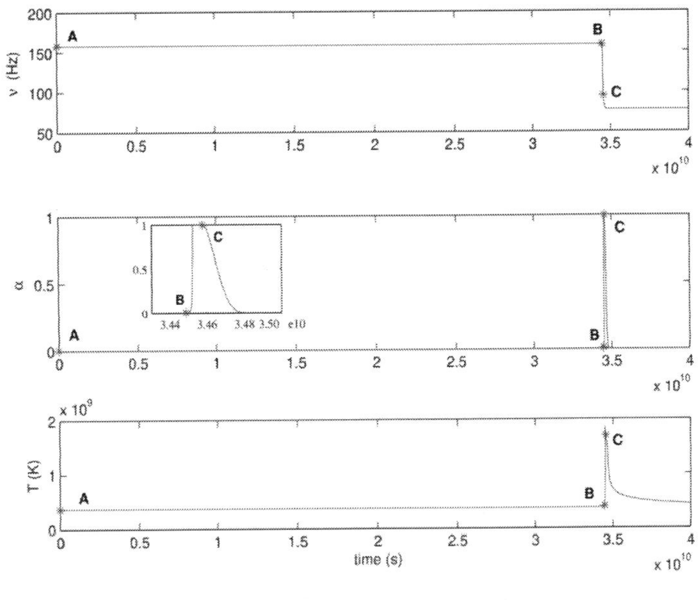

Figure 2. Temporal evolution, without toroidal magnetic field, of frequency, r-mode amplitude and temperature of the accreting neutron star.

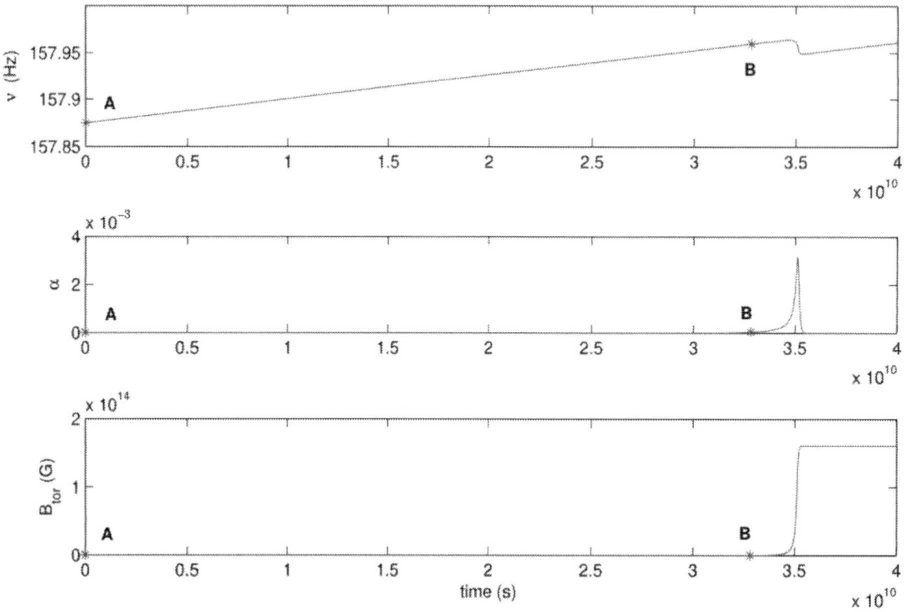

Figure 3. Temporal evolution, with toroidal magnetic field, of frequency, r-mode amplitude and temperature of the accreting neutron star.

References

[1] Andersson N and Kokkotas K D 2001 *Int. J. Mod. Phys.* D **10** 381
[2] Brown E F and Bildsten L 1998 *Astrophys. J.* **496** 915
[3] Bhattacharyya S 2001 *Preprint* `arXiv:astro-ph/0112178`
[4] Bondarescu R, Teukolsky S A and Wasserman I 2007 *Phys. Rev.* D **76**
[5] Rezzolla L, Lamb F K, Markovic D and Shapiro S L 2001 *Phys. Rev.* D **64** 104013
[6] Rezzolla L, Lamb F K, Markovic D and Shapiro S L 2001 *Phys. Rev.* D **64** 104014
[7] Wagoner R V 2002 *Astrophys. J.* **578** L63
[8] Friedman J L and Schutz B F 1978 *Astrophys. J.* **222** 281
[9] Shapiro S and Teukolsky S A 1983 *Black Holes White Dwarfs and Neutron Stars, Wiley*
[10] Watts A L and Andersson N 2002 *Mon. Not. Roy. Astron. Soc.* **333**
[11] Braithwaite J and Spruit H C 2005 *Preprint* `arXiv:astro-ph/0510287`

Journal of Physics: Conference Series **168** (2009) 012009 doi:10.1088/1742-6596/168/1/012009

Nuclear structure

Silvia M Lenzi

Dipartimento di Fisica, Università degli Studi di Padova,
and INFN, Sezione di Padova,
via F. Marzolo 8, 35131 Padova, Italy

E-mail: `lenzi@pd.infn.it`

Abstract. A review of the research activity of the Italian community on theoretical nuclear structure is presented and discussed. These studies cover a broad field of nuclear structure phenomena and a large variety of theoretical models and methods to treat the many-body systems. Most of this activity is developed within international collaborations that in many cases include experimental groups. An increasing interest on nuclei far from the valley of stability is evidenced.

1. Introduction

Nuclei present a large variety of phenomena that can be encountered in other systems, such as pairing and superfluidity, shell structure, collective motion, symmetries and phase transitions. Nuclear structure thus provides an excellent laboratory to develop models and methods that can help to understand Nature from a wide perspective. The research in nuclear structure theory is a rich field that spans from *ab initio* theories, that describe the many-body system starting from the nucleon–nucleon (NN) force, to the description of the nuclear properties with algebraic methods, including microscopic formalisms such as shell model, mean-field and beyond mean-field theories. All these research lines are pursued by the different Italian groups and constitute an extremely live activity. In this report I will give an overview of the progress in nuclear structure theory achieved in the last two years by the Italian community. This is the sixth report of a series started ten years ago [1, 2, 3, 4, 5].

There is a great variety of nuclear phenomena that can be treated at very different theoretical levels. And although many aspects and properties of the atomic nucleus are well understood, modern studies still face fundamental questions that await for definite answers both for stable and unstable matter such as the nature of the effective nucleon–nucleon interaction, the properties of exotic nuclei at the limits of binding, the location of the drip lines, how symmetries in the nuclear structure emerge and the existence of new collective motions and nuclear shapes.

This research is further challenged by the new data arising from the large-scale facilities and from the new accelerators of exotic beams. Nuclei far from stability allow amplifying and isolating particular aspects of the nuclear interaction and dynamics that probe the nuclear models developed near stability. Future exotic beams will allow to approach and map the drip-line regions and to explore nuclear stability at the very limits. The construction of high intensity stable and more powerful radioactive beams not only in Europe, and in particular in Italy with the SPES project, but also in Japan and USA, will produce in the next future a large amount of high quality data. This will be possible also due to the progress in sensitivity and efficiency of experimental tools built in ambitious international collaborations. Among them, it is important to mention the gamma-ray spectrometer Agata, based on a new detection concept that will have an important impact on the nuclear-structure investigations. An initial prototype of this array, composed of only 15 out of the 180 detectors of the full project, the Agata Demonstrator, is being now installed at LNL and will start taking data by the fall this year.

The collaboration with the experimental groups is very active in the Italian nuclear structure community and has produced, in the period covered by this report, many excellent results that are documented by publications in the most relevant scientific journals. This interaction is essential to allow a deep understanding of the different nuclear structure aspects and to trigger new investigations from both theoretical and experimental sides. Another fact that contributes to enlarge our way of approaching the research is the increasing connection with hadron physics and nuclear astrophysics. Last, but not least, the growing computational capabilities are modifying our approach to nuclear structure research in the last years. They allow to take into account more degrees of freedom of the physical systems under analysis, to improve their description and to increase the degree of predictibility of the results.

The main groups active in nuclear structure research in Italy are in Bologna, Catania, Lecce, Milano, Napoli and Padova. On reviewing the activity progress of the last two years of the research on theoretical nuclear structure, the variety, quantity and quality of the results are quite impressive. Although most of the activity is carried out within important international collaborations, the leading role of the Italian groups is remarkable.

In this contribution I will present some of the results achieved during this period starting from the problems related to the nucleon-nucleon force and the related effective interaction, describing the different microscopic approaches to study ground-state properties and collective excitations, to arrive finally to different methods to describe nuclear structure based on the underlying symmetries of the nuclear system.

2. From the bare to the effective NN interaction

Understanding the properties of nuclei starting from the forces between nucleons is one of the fundamental goals of nuclear physics. *Ab initio* approaches have been developed in the last years which start from a nucleon-nucleon force as input and compute the equation of state for nuclear matter. For the case of finite nuclei, *ab initio* approaches such as no-core shell model [6], coupled-cluster [7] and Green's function Monte Carlo methods [8] have improved considerably during the last years. Extensive no-core shell model calculations can now be performed using NN and three-body interactions derived from chiral effective field theory, including explicit two-pion exchange dynamics. (The importance of three-body forces has been discussed by several authors and significant efforts have been devoted to these studies.) Despite the important progress in this field and in the computing power, these fully microscopic calculations are limited presently to nuclei with mass $A \leq 20$.

For larger masses several methods are used. Among the different microscopic nuclear models, the shell model aims at a fully correlated description of ground state and excitations up to few MeV above the yrast line. These calculations make use of an effective nuclear interaction deduced from the NN force. As the bare nucleon-nucleon force shows strong repulsion at short distance and prevents a straightforward treatment in the many-body theory [9, 10, 11], effective interactions are used which behave smoothly and can be used when treating finite nuclei. In deriving the effective interaction one can follow microscopic or more phenomenological approaches. In the shell model framework, the problem is solved in a limited space, the so-called model space, subspace of the full Hilbert space. This is done by means of the Brueckner theory, which obtains the so-called G-matrix by summing "ladder" diagrams. The contribution from the excluded space by means of expansions of the effective interactions can be also obtained with the $\hat{Q}$-box method that includes diagrams to a certain order in perturbation theory [12]. In general, single particle energies are taken from the experimental data. Short-range correlations can be taken into account by resorting to the Correlated Basis Function approach, which is based on a variational method. Alternative approaches are the self-consistent mean field theories. The goal is to construct an energy-density functional and to produce directly the appropriate single-particle potential corresponding to the actual density distribution for a given nucleus. Aside non-relativistic methods such as the Skyrme-Hartree-Fock, in the last years relativistic mean field models have been successfully applied to the description of nuclei. The idea is to link the relativistic mean field to QCD observables. In this chapter I will briefly describe the related activity in progress within the Italian community.

2.1. *Relativistic energy density functional and the QCD constrains*

The Energy Density Functional (EDF) theory, originally applied in condensed matter physics and chemistry, has been successfully extended to the description of the nuclear many-body problem. Rather than calculating the many-body wave function, the density functional theory is based on finding the response of the ground-state energy to changes in the density. This allows to describe many important properties of systems across the whole nuclear chart and one of the major goals is to build an energy density functional which is universal, in the sense that the same functional is used for all nuclei, with the same set of parameters. This framework should then provide a reliable microscopic description of infinite nuclear and neutron matter, ground-state properties of bound nuclei, rotational and vibrational motion, etc. The accurate energy density functionals presently used are of mainly phenomenological origin.

The Bologna group has developed a relativistic nuclear energy density functional with the goal of linking the parameters of the EDF to QCD observables. A specific contribution by Paolo Finelli is included in this volume. The formalism is based on two important features that establish connections with chiral dynamics and the symmetry breaking pattern of low-energy QCD: i) the nuclear ground state is characterised by strong scalar and vector fields related to in-medium changes of QCD vacuum condensates, and ii) the long- and intermediate-range interactions generated by one- and two-pion exchange are derived from in-medium chiral perturbation theory, with explicit inclusion of $\Delta(1232)$ excitations. By means of relativistic Hartree-Bogoliubov calculations, this group has constructed nuclear energy density functional using the effective FKVW interaction (derived by the Bologna group) in the p-h channel plus the Gogni pairing in the p-p channel. This provides a consistent microscopic framework not only for ground-state properties, but also for complex nuclear excitations.

In particular, the formalism has been applied to the description of binding energies, radii of proton and neutron distributions and other observables over a wide range of spherical and deformed nuclei from ^{16}O to ^{210}Po. Moreover, isotopic chains of Sn and Pb nuclei have been studied as test cases for the isospin dependence of the underlying interactions and to the study of isobaric analog states and GamowTeller resonances. [13, 14, 15]. Although not as good as the phenomenological descriptions, the accuracy of these calculations is very good and provides an excellent microscopic framework for further investigations. Moreover, it is remarkable to note that the different terms of the FKVW interaction are in good agreement with the chiral perturbation theory predictions. The perspectives of this research line for the next future are the description of unnatural parity states and the spectroscopy of hypernuclei. Important developments are also pursued to link perturbative in-medium chiral dynamics and $V_{\mathrm{low-k}}$ in the low-density limit.

2.2. *The $V_{\mathrm{low-k}}$ and shell-model effective interaction*

The problem of deducing an effective two-body interaction form the free nucleon–nucleon potential when dealing with the interacting shell model started more than forty years ago [16, 17]. The strong repulsion at short distances, related to the high-momentum components of the interaction, precludes the straightforward use of a realistic NN interaction in the solution of the many-body system. This problem can be tackled by resorting to the Brueckner G-matrix method, which has been widely employed to construct effective interactions in different mass regions. A disadvantage of the G matrix is that it is model-space and energy dependent. An alternative formalism that has been introduced in the last few years is the $V_{\rm low-k}$ approach [18]. This is based on the fact that in nuclear physics there are different energy scales: Inspired by the Effective Field Theory, the $V_{\rm low-k}$ is derived by renormalization group techniques. The decoupling between the low- and high-momentum spaces is achieved through the construction of a low-momentum NN potential obtained by integrating out the high-momentum components of the original NN potential down to a cutoff momentum Λ. The resulting potential $V_{\rm low-k}$ preserves the physical properties of the NN potential and, having a smooth behaviour, is suitable to be used directly in nuclear structure calculations. In recent years this approach, which constitutes an advantageous alternative to the G-matrix method, has been the subject of great theoretical interest.

The Napoli group has used the $V_{\rm low-k}$ approach in the framework of the Goldstone expansion for doubly closed-shell nuclei [19, 20] as well as in the derivation of the shell-model effective interactions by means of the linked-cluster expansion. The development and current status of shell-model calculations with effective two-body interactions derived from the NN potential constitute the subject of a recent review article [21]. This group has performed several realistic shell-model calculations for a number of nuclei near the doubly magic ^{16}O, ^{100}Sn, ^{132}Sn, and ^{208}Pb. The two-body effective interactions are derived from different realistic NN potentials and renormalized through the $V_{\rm low-k}$ approach (see [21] and references therein). It has been shown that the results do not significantly depend on the particular choice of the NN potential. In particular, in the case of the nucleus ^{134}Te, the three potentials CD-Bonn, Nijmegen and Argonne V18, which fit well the NN phase shifts and the deuteron properties, produce energy spectra and transition probabilities in very good agreement with experiment. The rms deviations from the experimental spectrum up to 5 MeV excitation energy are $\sigma_{\rm CD-Bonn}$=115 keV, $\sigma_{\rm Nij}$=143 keV and $\sigma_{\rm V18}$=128 keV, a striking result which shows the nearly "universal" character of $V_{\rm low-k}$ [21].

Recently, the group has put a considerable effort in the description of exotic, neutron-rich nuclei near ^{132}Sn [22, 23, 24, 25, 26, 27]. New data is becoming available in this mass region, which will be further investigated at future radioactive beam facilities. In contrast to some theoretical attempts to reproduce the experimental data by introducing shell-structure modifications, the group has been able to give a very good description of the available data without invoking any modifications. All these calculations have been performed using single particle energies taken from experiment and an effective interaction derived from the $V_{\rm low-k}$ obtained renormalizing the CD-Bonn potential with a cutoff $\Lambda = 2.2$ fm^{-1}. In particular, attention has been focused on ^{134}Te, ^{134}Sn and ^{134}Sb, which give direct information on the proton-proton, neutron-neutron, and proton-neutron effective interaction. The nucleus ^{136}Te with two protons and two neutrons above ^{132}Sn also provides a stringent test of the interaction. The spectrum, as well as the transition probabilities recently measured in Coulomb excitation of radioactive ion beams, are very well reproduced [28].

In this volume there is a contribution from Angela Gargano, where some details on the derivation of the realistic effective interaction are given and some applications are presented. The successful application of $V_{\rm low-k}$, derived by renormalizing realistic NN interactions, to recent data near shell closures gives confidence on the predictive power of this approach and opens new perspectives on the construction of effective two-body interactions.

2.3. Correlated Basis Function

When going to higher energies, the short-range correlations need to be included explicitly and the $V_{\rm low-k}$ approach results inadequate. The problem of treating the hard core of the bare NN potential and its correlations can be overcome by acting directly on the trial wave functions in solving the many-body Schroedinger equation with a variational method. This is the basic idea in the Correlated Basis Functions (CBF). The trial wave functions of a system of A interacting particles used to minimise the energy functional are of the form:

$$\psi(1, ..., A) = F(1, ..., A)\ \Phi(1, ..., A) \tag{1}$$

where $\Phi(1...A)$ is a Slater determinant and $F(1, ..., A)$ is the correlation function that in its simplest form can be written as:

$$F(1, ..., A) = \prod_{j>i=1}^{A} f(r_{i,j}). \tag{2}$$

The technique used to solve the variational problem uses the cluster expansion method. The particles, correlated by the function f, form clusters and a set of integral equations is constructed that allows to sum, in a closed form, the contributions of all the clusters with some specific topological properties. This set of integral equations is called hypernetted chain equations [29] and can be used to describe both classical and bosonic systems. Originally derived for infinite systems, this method was extended to finite systems of fermions (after including the Pauli principle) by Fantoni and Rosati [30].

The group of Lecce and Pisa has extended for the first time this method to finite nuclei [31, 32]. More recently, a formulation of the equations in a general form in order to handle separately protons and neutrons was introduced. In Ref. [33], the ^{12}C, ^{16}O, ^{40}Ca, ^{48}Ca and ^{208}Pb binding energies and density distributions have been calculated using realistic microscopic interactions (of the Argonne and Urbana type) with tensor and spin-orbit terms. The Hamiltonian included also three-body interactions. From the analysis of the different contributions to the binding energy, it emerges that the three-body term is repulsive. This is at odd with what happens in few-body and light nuclei where the contribution of this part of the Hamiltonian is attractive. This puzzling result is presently being investigated. It could be due, for example, to the fact that the inclusion of a three-nucleon correlation function is necessary or that the three-body interactions used are suited for light nuclei, but they are not adequate for heavier nuclear systems. A review article where an exhaustive description of the method and its applications can be found has been published recently by the Lecce-Pisa group [34]. Several perspectives for the application of the CBF are in program, as for example the description of deformed nuclei, excited states and hypernuclei.

2.4. The evolution of the nuclear shell structure far from stability

An optimum test ground for the nuclear models, and in particular for the effective interactions developed for stable nuclei, are the nuclear systems far from the stability valley. In the last few years, many experimental and theoretical efforts have been put on the study the structure and the reaction mechanisms of these nuclei. In the last years, state of the art experiments have revealed several unexpected effects in nuclei, such as the existence of halos and skins and the modifications of shell closures. One of the current topics of interest is the role of the tensor interaction between protons and neutrons which has been shown to be responsible of the change in the shell structure far from stability. In fact, at very large neutron excess, substantial changes occur: some known magic numbers disappear being replaced by new ones, whereas new regions of deformation develop around nucleon numbers that are magic near stability. In the middle of the Z=20-28, N=28-50 region the increasing neutron excess gives rise to collective effects. These phenomena have interesting consequences for the persistence of the N=40 subshell closure in the middle of the proton $f_{7/2}$ shell, which is of relevance to nucleosynthesys pathways. Theoretical calculations indicate that in nuclei where the proton $1f_{7/2}$ shell is not completely filled, the neutrons excited to the *sdg* shell couple to the *pf* protons and deformation appears [35, 36]. The mechanism of nuclear deformation reflects important properties such as shell effects, pairing correlation, and the long range proton-neutron correlation. Therefore the characterisation of a deformation region can provide new insights into the nuclear structure.

In collaboration with the experimental group of LNL, neutron-rich Fe isotopes have been populated recently in deep-inelastic and multi-nucleon transfer reactions using the γ-ray spectrometer CLARA and the magnetic spectrometer PRISMA. Yrast excited states were populated in $^{62-68}$Fe isotopes [37, 38]. The spectra have been very well reproduced by shell model calculations in the full fp shell using the effective interactions GXPF1A [39] and KB3G [40] up to the ^{64}Fe isotope. Around N=40, which behaves as a subshell closure in ^{68}Ni, an increase of collectivity is observed and calculations do not reproduce the experimental findings. This suggests the need of including the *gds* upper main shell in the model space and the consequent development of a cross-shell interaction. These studies put in evidence the change of the shell structure, in particular, the gap between the neutron $f_{5/2}$ and $g_{9/2}$ decreases considerably making the N=40, Z$\simeq$24-26 a new region of deformation. Further efforts in the shell model calculations are under development.

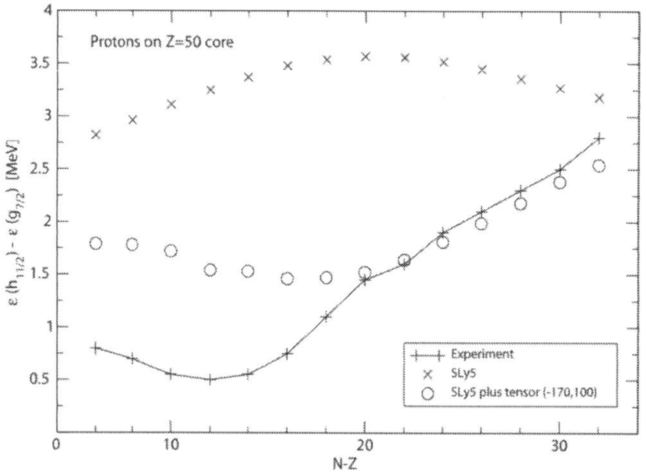

Figure 1. Energy differences between the $1h_{11/2}$ and $1g_{7/2}$ single-proton states along the Z = 50 isotopes. Calculations are performed without (crosses) and with (circles) the tensor term in the spinorbit potential on top of SLy5. See the text for details.

Mean-field approaches can also shed light on the contribution of the tensor interaction to the evolution of the single-particle orbitals as a function of the isospin. Here, the tensor term of the Hamiltonian is not always present in the effective interaction. In particular, the Skyrme parameter sets which are widely used in nuclear structure calculations do not include the tensor contribution. This contribution was included only in the so-called Skyrme-Landau parametrisations of Ref. [41]. The group of Milano has recently studied the contribution of the zero-range tensor force to the spin-orbit splittings in neutron-rich nuclei. The calculations have been performed in the Hartree-Fock plus BCS framework, using the SLy5 effective interaction and adding a tensor force whose parameters have been fitted. Changes in the single particle energies have been investigated through the Z=50 isotopic and N=82 isotonic chains, taking into account the effects of the triplet-even and triplet-odd tensor contributions separately [42]. When comparing with the experimental data, calculations including the tensor term improve considerably the agreement with respect to those with the bare SLy5 interaction (see figure 1). More recently, the group has extended the study of the tensor correlations to lighter systems in the fp shell (Ca isotopes and N=28 isotones). In this case, the trend of the proton single-particle states along the Ca isotopes and the reduction of the neutron spin-orbit splittings on going from ^{48}Ca to ^{46}Ar is qualitatively explained as an effect of the tensor force [43]. These two examples show the important role of the proton-neutron tensor interaction both in medium-mass and in heavy nuclei. On the other hand, the choice of the sign of the proton-proton or neutron-neutron tensor interaction terms is not clearly understood and deserves further investigation.

2.5. A microscopic study of the pairing field

One of the fundamental blocks of the NN interaction are the pairing correlations. Their effect on the nuclear spectrum and dynamics is clearly manifested. The pairing gap in superfluid nuclei is often calculated making use of a simple, phenomenological parametrisation of the pairing interaction in terms of density-dependent, contact interactions. In the last years, the Milano group has extensively studied pairing correlations following a more microscopic approach, starting from the v_{14} NN Argonne interaction which reproduce the experimental NN phase shifts, and then adding the main medium polarisation effects. It has been found that the exchange of surface vibrations between pairs of single-particle states in time reversal gives rise to an induced interaction which enhances in an important way the pairing gap obtained from the bare interaction [44, 45]. The total pairing field, resulting from the contribution of the bare and induced pairing interactions, is strongly peaked at the nuclear surface. It is to be noted that the absolute values of the pairing gap depends not only on the adotped pairing interaction, but also on the properties of the mean field, and in particular on the density of single-particle levels. A consistent study therefore must take into account the medium effects not only in the particle-particle but also on the particle-hole channel [46].

In a recent work, the Milano group has shown that, although the induced-pairing interaction is non-local and energy dependent, it is possible to introduce a density-dependent, contact approximation which reproduces the essential features of the pairing gap obtained in the full quantal solution [47]. Interestingly, the new parametrisation found for the contact interaction turns out to be quite different to those commonly used in the phenomenological approaches. For a more detailed discussion of this issue the reader is addressed to Alessandro Pastore's contribution to this volume.

3. Description of collective modes

3.1. Excited states and correlations in finite systems

While the description of the nuclear systems starting from the NN interaction, as we saw in the previous Section, provides a good description of ground or low energy states, the description of collective or high-lying states can be better approached by means of beyond mean-filed models. The Hartree-Fock plus Random Phase Approximation (HF+RPA) is a microscopic method to study the collective vibrations in nuclei. Although this remains a very good tool to understand some physical properties, it is not well suited for some quantitative descriptions as it is based on some approximations, like the quasiboson approximation, using an uncorrelated HF ground state, in place of the correlated one. This substitution introduces a visible inconsistency since the correlated g.s. is used in the formal derivation of the equations of motion. Several improvements and renormalisations have been introduce to this method. In particular, the Catania group has proposed an extension of this method in Refs. [48, 49]. In a recent work [50], a more general approach has been introduced that avoids the quasiboson approximation and includes g.s. correlations in a self-consistent way. The linearisation of the equation of motion is performed in a way that all matrix elements are expressed in terms of the One Body Density Matrix and the non-linear eigenvalue problem is solved in an iterative procedure. The method is applied to describe the excitation spectrum of dipole strength in Na metal clusters within the jellium approximation with bare Coulomb interaction both for the electrons with the jellium and for the electrons among themselves. Although this approximation does not allow a full quantitative description of the data, it provides a simplified picture to compare different theoretical approaches. It is found that the extended RPA is in better agreement with the experimental findings than the standard RPA results (see figure 2). In this volume, the contribution of Danilo Gambacurta treats further extensions such as the Second RPA that includes 2p-2h excitations. The idea is to explore different extensions of the RPA to be then applied to more complex systems as the nuclear excitations. The Catania group has also calculated, by means of self consistent Hartree-Fock plus RPA with two Skyrme effective interactions (SGII and SLY4), the excitation of the pygmy resonance as a function of the neutron excess. The full description can be found in the contribution of Edoardo Lanza to these proceedings.

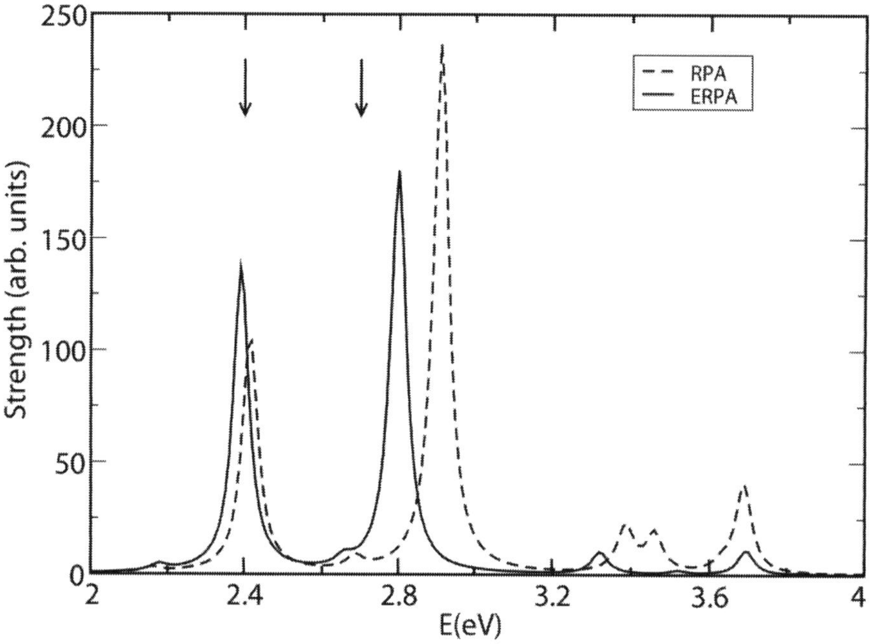

Figure 2. Spin=0 dipole strength for Na_{40} metal cluster. The solid line refers to calculations performed within the extended RPA (ERPA) approach,while the dashed line refers to standard RPA calculations. The arrows roughly indicate the positions of the experimental peaks.

Beyond mean field calculations in the framework of the Quasi-particle Phonon Model (QPM) constitute an accurate microscopic approach that allows the inclusion of a large number of configurations. A multiphonon basis is constructed from phonons generated in a QRPA calculation. A Hamiltonian of general separable form is diagonalised in a space spanned by states composed of 1, 2 and 3 phonons. In a recent work, the Napoli group has studied with this method the structure of low-lying states in the N = 80 isotones ^{134}Xe, ^{136}Ba, and ^{138}Ce [51]. Both levels and electromagnetic transition strengths are in good agreement with data. Due to its phonon structure, QPM can be naturally related to algebraic approaches as the Interacting Boson Model (IBM) and, in this context, the QPM multiphonon basis states can be viewed as the microscopic counterparts of the IBM bosonic states. QPM, however, can give additional information: An important capability of the QPM, is that it can disclose the shell structure of the multiphonon states. In fact, in the case of ^{138}Ce the splitting of the M1 strength is shown to be due to the proton subshell closure which magnifies the role of pairing in the excitation mechanism. It is also shown that the low-lying states observed in the N = 80 isotones can be well classified according to the *np* symmetry in agreement with the F-spin IBM-2 scheme.

3.2. Energy spectrum and de-excitation of superdeformed nuclei

Beyond mean field calculations are an ideal tool to study the thermal excitations of rapidly rotating nuclei, such as the phenomena of superdeformation (SD). Fusion-evaporation reactions, followed by neutron evaporation, can be used to produce excited, rapidly rotating deformed nuclei, that decay by $\gamma-$emission towards the yrast line. The experimental study of the decay, based on $\gamma-\gamma$ correlation spectra, gives important information concerning the collective behaviour of the nucleus as a function of angular momentum and thermal energy. In the de-excitation process towards the ground state, the nucleus passes through different regimes, from a chaotic behaviour at high excitation energy, to the ordered cold structure near the yrast line. The chaotic regime is characterised by a statistical spectrum with transitions governed by Porter-Thomas fluctuations, while regular transitions are observed in the spectrum of a decaying, cold ordered system. Starting from high excited, non-yrast states, the γ- decay transitions of E2 character, do not follow a definite band due to a high degree of band mixing. The γ flow is therefore fragmented. This process, known as rotational damping, is translated into a broad B(E2) strength and can be studied theoretically within the framework of the interacting cranked shell model [52]. Here, the mean field states of a deformed potential, on which unperturbed rotational bands are built, are mixed by a residual interaction. With increasing excitation energy and level density, the mixing leads to complex states and the spreading in the frequencies displayed by the individual unperturbed bands, can then produce a fragmentation of the E2 transitions from the mixed states, leading eventually to rotational damping. A quantitative calculation of the decay process, however, also involves the cooling of the nucleus by statistical E1 transitions.

Recent experimental studies have investigated the population and the decay of the superdeformed nuclei ^{151}Tb and ^{196}Pb, measuring several independent observables that provide new constraints to the dynamics of the decay process and represent a challenge for a quantitative interpretation. For this purpose, the Milano group has developed a detailed Monte Carlo simulation of the $\gamma-$decay of a superdeformed nucleus, which also describes the tunnelling between the super- and normal-deformed potential energy wells, based on microscopic penetration probabilities. These simulations include, as basic ingredients, microscopically calculated levels and E2 transition energies and statistical E1 transitions. Good agreement with experiment is found provided one introduces an enhancement in the E1 strength around 1-2 MeV. This could be related to the role played by octupole vibrations in the $A\sim150$ and $A\sim190$ mass regions [53].

In another recent experimental study of the decay of ^{194}Hg, an intermediate ergodic regime has been observed where the Porter-Thomas fluctuations dominate the spectrum, with the exception of collective E2 transitions along preserved rotational bands. These nearly parallel ergodic bands present very similar moments of inertia. The configuration mixing among excited superdeformed states does not lead to a spread in the rotational strength and therefore no rotational damping is observed. This is a unique case where this phenomena has been observed. The Milano group has performed a cranked shell model calculation with pairing and a surface delta interaction [54]. These calculations show that the wave functions contain many mean-field configurations, so that a chaotic regime is approached, yet conserving a rotational coherence and preserving the decay flow. This phenomenon can be explained in terms of the combination of shell effects and motional narrowing and gives new insight into the transition from order to chaos.

The softness towards octupole deformation at the SD well in this mass regions can be explained in terms of the presence of $\Delta l = 3$ intruder states near the Fermi surface. This has been studied by the Napoli group [55] in the framework of the cranked Nilsson plus QRPA calculations. This formalism has been applied to the study of SD structures in ^{152}Dy and the isotope chain ^{190,192,194}Hg. The calculated dynamical moment of inertia are in agreement with the experimental data and with the octupole character interpretation of the low-lying negative parity SD bands, in agreement with the data and with other theoretical studies [55, 56]. The analysis is extended to the effect of the rapid rotation and high deformation on the magnetic dipole mode M1 [57]. It is observed that the orbital M1 strength is enhanced giving a dominant scissors-like character to the low-energy M1 excitations.

4. Symmetries, dynamical symmetries and phase transitions

The study of symmetries and its relation to nuclear structure properties allows to give a simple description of complex phenomena involving a many-body system. Many symmetries have been identified in nuclear physics and there are several groups in the Italian community that study these features. In this chapter I will discuss recent results on the study of symmetries related to the nuclear shape and its phase transitions, the pairing modes and the isospin symmetry.

4.1. Shape-phase transitions in nuclei

The nuclear shape of a nucleus is usually characterised by the two well known deformation parameter β and the angle variable γ that accounts for deviations from axial symmetry. On solving the Bohr Hamiltonian there are three analytical solutions which correspond to the axially deformed rotor, the spherical vibrator and the so called γ-soft solution. These three limits allow a particularly simple description in terms of dynamical symmetries and supersymmetries: $U(5)$ (spherical shape), $SU(3)$ (axially deformed shape) and $SO(6)$ (deformed γ-soft). They are the three vertex of the Casten triangle of figure 3 and provide explicit analytic forms for the observables. However, most of the known nuclei do not fit exactly into one of these classes and present an intermediate behaviour. By changing the number of nucleons, for example along an isotopic chain, nuclei move from one limit to another. This can be thought as a quantum phase transition. At the critical point of the phase transitions, F. Iachello has found particularly simple solutions to the Bohr Hamiltonian. They are the E(5) for the transition from $U(5)$ to $SO(6)$ [58], the X(5) for the critical point in the transition from $U(5)$ to $SU(3)$ [59] and the Y(5) for the axially deformed shapes to triaxial shape transition [60]. In this framework, precise values are predicted not only for the relative energy of the excited states but also for the transition probabilities. These features can also be treated within the Interacting Boson Model. Since the introduction of these dynamical symmetries and supersymmetries, many experimental efforts have been paid to look for nuclei that sit at the critical point and few cases, near stability have already been proposed to fit these conditions. In what follows, we will concentrate on the critical point for the transition between spherical to γ-soft shape.

The Padova group is very much involved in studying phase transitions in both even-even and odd-mass nuclei using algebraic approaches. In collaboration with the experimental group of Padova-LNL, the structure of the nucleus ^{58}Cr has been recently analysed. The yrast structure of this neutron-rich nucleus has been observed using the CLARA-PRISMA experimental setup. The sequence of energy levels follows that consistent with the E(5) dynamical symmetry, indicating ^{58}Cr as a good candidate for the critical point phase-shape transition. This is sustained by theoretical calculations that predict an increase of collectivity in Cr isotopes when approaching N=40. In fact, the excitation energy of the 2^+ state decreases considerably in ^{60,62}Cr and there are several experimental groups that look for the most deformed case that should be ^{64}Cr, not yet observed. Interestingly enough, the level scheme can be reproduced with the IBM and large-scale shell model calculations in the fp shell. This nucleus constitutes therefore

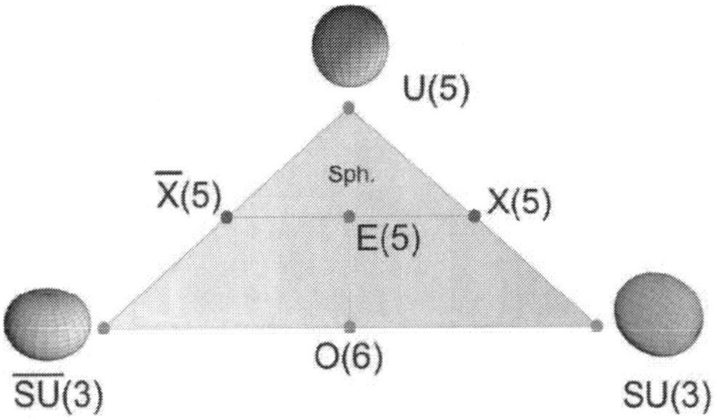

Figure 3. A version of the Casten triangle showing the different shape phases and the related dynamical symmetries.

an excellent benchmark to compare, in the same physical system very different theoretical approaches and build a bridge between shell model, IBM and E(5) [61].

This studies are limited to even-even nuclei, but what happens when we couple the even-even core to an odd particle? When studying the shape transition of the core from spherical to γ-unstable, the coupling to the odd particle generally destroys the situation of γ-instability. A noticeable exception, that preserves the γ-instability along the whole shape transition (including the critical point) results when the odd particle occupies a $j = 3/2$ orbital. F. Iachello has derived a new analytical solution of the Bohr Hamiltonian in the even-odd case. This is the E(5/4) dynamical symmetry [62]. This can be treated with the IBM, as proposed by the Padova group [63] in the framework of the Interacting Boson Fermion Model (IBFM). Details of these developments and applications are presented in Lorenzo Fortunato's contribution to this proceedings.

In a further work, this group has found that another analytical solution to the even-odd nuclei can be deduced from the Bohr Hamiltonian, for a more general case than the E(5/4). The odd particle is assumed to move in single-particle orbitals with $j = 1/2, 3/2, 5/2$. Using an infinite square well potential in the parameter β, the Hamiltonian is analitycally solvable and the new supersymmetry is called E(5/12) [64]. In the framework of the IBFM this problem can be also described. In this case two different Hamiltonians are explored. The first one is constructed to mock up the situation in the E(5/12) model, while the second one leads to the occurrence of the $O^B(6) \otimes U^F(12)$ symmetry when the boson part approaches the O(6) condition. It is shown that both Hamiltonians are in agreement with E(5/12) dynamical symmetry solution [65].

4.2. Phase transitions in the pairing gauge space

A collective model to describe paring correlations has been introduced by Bès and collaborators in the early seventies. The collective variables in the pairing field are the pair deformation α, related to the pairing gap, and the gauge angle ϕ which is the canonical conjugate to the particle-number operator N. In analogy to the Bohr Hamiltonian used to describe the dynamics of the nuclear shape with the β and γ parameters, a collective pairing Hamiltonian has been derived in terms of α and ϕ. It is shown that on solving this Hamiltonian:

$$-\frac{\hbar^2}{2B}\frac{\partial^2\psi}{\partial\alpha^2} - \frac{\hbar^2}{4}\left(\frac{1}{\Im B}\frac{\partial\Im}{\partial\alpha} - \frac{1}{B^2}\frac{\partial B}{\partial\alpha}\right)\frac{\partial\psi}{\partial\alpha} + \left(V(\alpha) + \frac{\hbar^2 M^2}{2\Im} - E\right)\psi = 0 \tag{3}$$

two analytical solutions can be found, corresponding to the normal and superconducting limits. In the above equation, the parameters $\Im$ (inertia) and B (mass) are functions of the pair deformation α, $M = A - A_0$ (number of particles A relative to a reference A_0), and $V(\alpha)$ is the potential. Nuclei with two identical nucleons added or removed from a closed shell core are in the normal regime, without a static deformation of the pairing field. They can undergo fluctuations that give rise to a pairing vibrational spectrum [66]. Nuclei in the middle of the shell, or in general, with many pairs of particles outside of the closed-shell configuration, a static deformation of the pair field arises and the nucleus behaves as a rotor in the gauge space [67]. This corresponds to the superconducting limit. One can then think on a pair-rotational band formed by the sequence of ground states of neighbouring even-even nuclei along an isotopic or isotonic chain. In the vibrational limit, the potential $V(\alpha)$ is parabolic with a minimum at $\alpha = 0$, while in the superconducting limit, the minimum is at a finite value. At the critical point the potential has the typical flat behaviour and can be approximated by a square well. The analytical solution gives a simple form to the energy spectrum of the states in terms of the zeros of Bessel functions of integer order [68]. It is found that the pairing bands of Pb isotopes based on ^{208}Pb are closer to a critical point behaviour than to a pure vibrator character. This description may provide an explanation of the anharmonicities of the pairing vibrational structure observed around ^{208}Pb [69].

4.3. Dynamical chirality

Chiral doublets bands have been proposed to occur in some rapidly rotating nuclei with triaxial deformation. These structures are characterised by pairs of identical $\Delta I = 1$ bands with the same parity in odd-odd nuclei. This can occur when the valence proton is filling the lower part of a high-j (particle-like) subshell and the neutron the upper part of a high-j (hole-like) subshell, or vice versa, on the top of a triaxially deformed core. The angular momenta of the valence particles are aligned along the short and long axes of the triaxial core, while the angular momentum of the rotational core is aligned along the intermediate axis. Therefore, by arranging the three angular momenta, one can form two systems that differ by intrinsic static chirality, a left- and a right-handed system. When the chiral symmetry is broken in the body-fixed frame, this gives rise to degenerate doublet bands. Candidates to the chiral symmetry have been identified in the $A \sim 105$, 130 and 190 mass regions. The underlying symmetry should also manifest on the electromagnetic properties. The B(M1) transition probabilities should exhibit odd-even staggering, while the B(E2) values de-exciting analog states of the chiral twin bands should be almost equal.

Among the theoretical models that have been applied to describe this phenomena are the tilted axis cranking model [70], two quasiparticle plus triaxial rotor model [71] and the core-particle-hole coupling model [72]. The main common assumption of these studies is that of a rigid triaxial core. However, the nuclei to which these calculations are applied are in a region where even-even nuclei (core) are γ-soft and not rigid. This has been recently pointed out by the Bologna group [73] in the framework of the Interacting Boson Fermion Fermion Model (IBFFM). They analyse chiral-bands candidates in the region of mass $A = 130$ showing that they do not fulfil the necessary requirements for the realisation of chirality in a rigid rotor. The authors identify two types of candidates, those in which both the energies and the electromagnetic properties are close to the chiral requirements and another class in which only the excited states of the twin bands are near in energy. From the analysis of the wave functions of both types of chiral candidates it results that both types of nuclei do not sizably differ in geometry. The authors interpret the difference in the structure of the two types of chiral candidates nuclei as due to different fluctuations in β and γ, induced by the exchange boson-fermion interaction of the IBFFM. It is also observed that in all cases the configuration where the angular momenta of the valence proton, neutron and core are in a near orthogonal geometry is present, but not dominant. This gives rise to a weak chirality of dynamical (non rigid) origin.

4.4. Isospin symmetry and the Mirror Energy Differences

The study of isospin symmetry in nuclei as a function of angular momentum has now become established as a very powerful tool to understand nuclear properties in rotating nuclei. These studies have become feasible in the last decade due to recent experimental developments in the identification of proton-rich nuclei produced with very low cross sections. Parallelly, state-of-the-art shell-model codes have been produced for the description of these data. The synergy between theory and experiment for the study of energy differences of mirror and isobaric analogue nuclei in the mass region between $A \sim 30$ and $A \sim 60$ has allowed the investigation of the evolution of the nuclear wave functions with increasing spin. The alignment process along a rotational band and in general nucleon recouplings, changes of the nuclear shape, changes in the intrinsic configuration, together with the evidence of isospin-non-conserving terms of the nuclear interaction are examples of the type of phenomena that can be studied from the analysis of Coulomb energy differences.

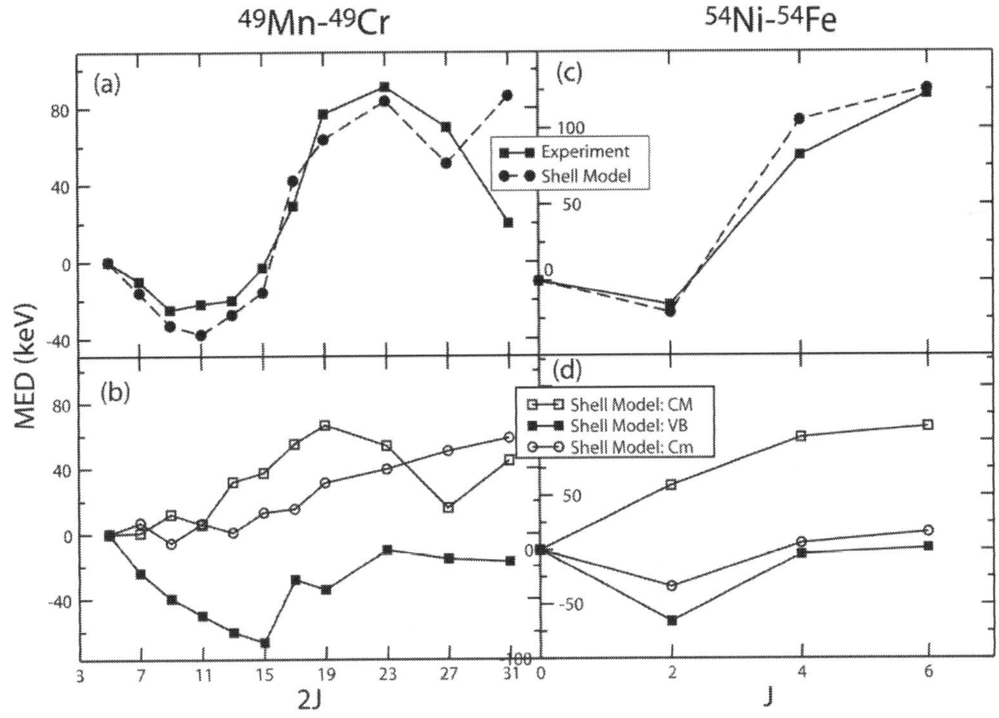

Figure 4. Mirror energy differences between $T = 1/2$ analogue states in $A = 49$ and $T = 1$ analogue states in $A = 54$. The experimental values are compared to the shell-model results in a) and c), respectively. The different contributions to the theoretical results are given in b) and d), where CM corresponds to the multipole Coulomb, Cm to the residual monopole Coulomb and VB is the isospin-breaking interaction.

In the last years, the Padova group, in close collaboration with the experimental colleagues of LNL has explored the so-called Mirror Energy Differences (MED) between excited states in several mirror pairs in the fp and upper sd shells. These energy differences are quite small and amount to tens of keV. Nevertheless, they can be described with good accuracy in a parameter-free shell model calculation at the 10 keV level. The theoretical method has been introduced in Ref. [74, 75]. The main ingredients entering the calculation are the multipole Coulomb interaction, which is calculated in the harmonic oscillator basis and added to the nuclear effective interaction, the electromagnetic spin-orbit and residual monopole Coulomb contributions. Interestingly, all the mirror pairs in the $f_{7/2}$ can be very well described provided an additional matrix element of 100 keV strength in the $T = 1$, $J = 2$ proton-proton interaction is added to the nuclear interaction [75, 76]. This isospin asymmetry in the $J = 2$ matrix element is called the "$J = 2$ Anomaly" and has been suggested to be of *nuclear* origin. Presently, the research is being extended to other mass regions. The nature of this interaction that happens to be in the MED at the same level of importance as the Coulomb one, is not yet understood [77, 78, 79, 80]. The contributions of the different terms to the calculated MED together with the excellent agreement with the experimental data can be seen in figure 4 for some examples of $T = 1/2$ and $T = 1$ analogue states in mirror nuclei. It can be shown that the behaviour of multipole Coulomb (CM) energy gives information on the recoupling and alignment of nucleons along the bands. The monopole Coulomb (Cm), instead is related to changes in the nuclear radius with the spin. Finally, the electromagnetic spin-orbit contribution introduces single-particle energy shifts that put in evidence pure single particle excitations. These latter effects have been clearly observed in the MED of sd-shell nuclei [81]. A detailed description of the experimental and theoretical tools, together with the results for MED from mass $A \sim 30$ to $A \sim 60$ can be found in a recent review article [76]. Evidence of isospin breaking effects of non-Coulomb character (VB) are presently under investigation in different mass regions.

5. Concluding remarks

This review puts in evidence the intense research activity and the leading role of the Italian nuclear structure community. Nuclear aspects and phenomena are studied with very different methods and models and a clear increasing interest on nuclei far from the valley of stability is observed. This is in part encouraged by the new data arising from the radioactive beam facilities and produced with state-of-the-art instrumentation which challenge the theoretical techniques built near stability.

In this contribution I have chosen some of the recent results obtained by the different groups, putting special emphasis on studies done in close collaboration with national and international experimental groups. It is important to note the efforts on developing theoretical techniques that help to explain the richness of phenomena recently observed in systems at the limits of angular momentum, excitation energy, and mainly, of isospin. These extreme situations evidence particular terms of the nucleon-nucleon interaction and help to refine our models and methods in order to improve the predictability of our results when approaching the drip lines. An increasing convergence of research interest with other nuclear communities such as those of nuclear astrophysics and hadron physics is also evinced.

Many challenging facets remain to be analysed and investigated in the next years in which the Italian nuclear-structure theoreticians can continue contributing very actively.

Acknowledgments

This contribution has been prepared in close collaboration with the colleagues in the different Italian research groups in nuclear structure. I would like to thank all of them for the stimulating and fruitful exchange and discussions.

References

[1] Lo Iudice N 1999 *Perspectives in Theoretical Nuclear Physics, proceedings of the VII Convegno su Problemi di Fisica Nucleare Teorica, Cortona, 1998* (Pisa: Edizioni ETS) p 125
[2] Bortignon P F 2001 *Theoretical Nuclear Physics in Italy, proceedings of the 8th Conference on Problems in Theoretical Nuclear Physics, Cortona, 2000* (Singapore: World Scientific) p 3
[3] Gargano A 2003 *Theoretical Nuclear Physics in Italy, proceedings of the 9th Conference on Problems in Theoretical Nuclear Physics, Cortona, 2002* (Singapore: World Scientific) p 23
[4] Colò G 2005 *Theoretical Nuclear Physics in Italy, Proceedings of the 9th Conference on Problems in Theoretical Nuclear Physics, Cortona, 2002* (Singapore: World Scientific) p 1
[5] Andreozzi F 2007 *Theoretical Nuclear Physics in Italy, proceedings of the 9th Conference on Problems in Theoretical Nuclear Physics, Cortona, 2006* (Singapore: World Scientific) p 109
[6] Navràtil P, Vary J P and Barrett B R 2000 *Phys. Rev. Lett.* **84** 5728
[7] Dean D J and Hjorth-Jensen M 2003 *Phys. Rev.* C **69** 054320
[8] Pieper S C, Pandharipande V R, Wiringa R B and Carlson J 2001 *Phys. Rev.* C **64** 014001
[9] Machleidt R, Holinde K and Elster C 1987 *Phys. Rep.* **149** 1
[10] Machleidt R 1989 *Adv. Nucl. Phys.* **19** 189
[11] Stoks V G J, Klomp R A M, Rentmeester M C M and de Swart J 1993 *Phys. Rev.* C **48** 192
[12] Kuo T T S and Osnes E 1990 *Springer Lecture Notes in Physics* vol 364 ed Dunning F B and Hulet R (Berlin: Springer)
[13] Finelli P, Kaiser N, Vretenar D and Weise W 2007 *Nucl. Phys.* A **791** 57
[14] Finelli P, Kaiser N, Vretenar D and Weise W 2006 *Nucl. Phys.* A **770** 1
[15] Finelli P, Kaiser N, Vretenar D and Weise W 2004 *Phys. Rev.* A **735** 449
[16] Dawson J F, Talmi I and Walecka J D 1962 *An.. Phys. (NY)* **18** 339
[17] Kuo T T S and Brown G E 1966 *Nucl. Phys.* **86** 40
[18] Bogner S, Kuo T T S, Coraggio L, Covello A and Itaco N 2002 *Phys. Rev.* C **65** 051301
[19] Coraggio L, Covello A, Gargano A, Itaco N and Kuo T T S 2006 *Phys. Rev.* C **73** 014304
[20] Coraggio L, Covello A, Gargano A, Itaco N and Kuo T T S 2007 *Phys. Rev.* C **75** 057303
[21] Coraggio L, Covello A, Gargano A Itaco N and Kuo T T S 2009 *Prog. Par. Nucl Phys.* **62** 135
[22] Goodin C *et al* 2008 *Phys. Rev.* C **78** 044331
[23] Coraggio L, Covello A, Gargano A, Itaco N, Entem D R, Kuo T T S and Machleidt R 2008 *Nucl. Phys.* A **805** 424
[24] Korgul A *et al* 2007 *Eur. Phys J.* A **32** 25
[25] Simpson G S *et al* 2007 *Phys. Rev.* C **76** 041303
[26] Li K *et al* 2007 *Phys. Rev.* C **75** 044314
[27] Coraggio L, Covello A, Gargano A and Itaco N 2006 *Phys. Rev.* C **73** 031302
[28] Covello A, Coraggio L, Gargano A and Itaco N 2007 *Prog. Part. Nucl. Phys.* **59** 401
[29] van Leeuwen J M J, Groeneveld J and de Boer J 1959 *Physica* C **25** 792
[30] Fantoni S and Rosati S 1979 *Nucl. Phys.* A **328** 478
[31] Cò G, Fabrocini A, Fantoni S and Lagaris I E 1992 *Nucl. Phys.* A **549** 439
[32] Arias de Saavedra F, Cò G, Fabrocini A and Fantoni S 1996 *Nucl. Phys.* A **76** 359
[33] Bisconti C, Arias de Saavedra F, Cò G and Fabrocini A 2006 *Phys. Rev.* C **73** 054304
[34] Arias de Saavedra F, Bisconti C, Cò G and Fabrocini A 2007 *Phys. Rep.* **450** 1
[35] Otsuka T *et al* 2005 *Phys. Rev. Lett.* **95** 232502
[36] Caurier E *et al* 2002 *Eur. Phys. J.* A **15** 145
[37] Lunardi S *et al* 2007 *Phys. Rev.* C **76** 034303
[38] Lenzi S M *et al* 2008 *LNL Ann. Rep. 2007* (INFN-LNL-222) p 15
[39] Honma M *et al* 2005 *Eur. Phys. J.* A **25** 499
[40] Poves A *et al* 2001 *Nucl. Phys.* A **694** 157
[41] Liu K F *et al* 1991 *Nucl. Phys.* A **534** 1
[42] Colò G, Sagawa H, Fracasso S and Bortignon P F 2007 *Phys. Lett.* B **646** 227
[43] Zou W, Colò G, Ma Z, Sagawa H and Bortignon P F 2008 *Phys. Rev.* C **77** 014314
[44] Barranco F, Broglia R A, Gori G, Vigezzi E, Bortignon P F and Terasaki J 1999 *Phys. Rev. Lett.* **83** 2147
[45] Barranco F, Bortignon P F, Broglia R A, Colò G and Vigezzi E 2001 *Eur. Phys. J.* A **11** 385
[46] Barranco F, Bortignon P F, Broglia R A, Colò G, Schuck P, Vigezzi E and Viñas X 2005 *Phys. Rev.* C **72** 054314
[47] Pastore A, Barranco F, Broglia R A and Vigezzi E 2008 *Phys. Rev.* C **78** 024315
[48] Catara F, Piccitto G, Sambataro M and Van Giai N 1996 *Phys. Rev.* B **54** 17536
[49] Catara F, Grasso M, Piccitto G and Sambataro M 1998 *Phys. Rev.* B **58** 16070
[50] Gambacurta D and Catara F 2008 *Phys. Rev.* B **77** 205434

[51] Lo Iudice N, Stoyanov Ch and Tarpanov D 2008 *Phys. Rev.* C **77** 044310
[52] Matsuo M *et al* 1997 *Nucl. Phys.* A **617** 1
[53] Leoni S *et al* 2008 *Phys. Rev. Lett.* **101** 142502
[54] Lopez-Martens A *et al* 2008 *Phys. Rev. Lett.* **100** 102501
[55] Kvasil J, Lo Iudice N, Andreozzi F, Knapp F and Porrino A 2007 *Phys. Rev.* C **75** 034306
[56] Kvasil J, Lo Iudice N, Andreozzi F, Knapp F and Porrino A 2006 *Phys. Rev.* C **73** 034302
[57] Hamamoto I and Nazarewicz W 1992 *Phys. Lett.* B **297** 25
[58] Iachello F 2000 *Phys. Rev. Lett.* **85** 3580
[59] Iachello F 2001 *Phys. Rev. Lett.* **87** 052502
[60] Iachello F 2003 *Phys. Rev. Lett.* **91** 132502
[61] Marginean N *et al* 2006 *Phys. Lett.* B **633** 696
[62] Iachello F 2005 *Phys. Rev. Lett.* **95** 052503
[63] Alonso C E, Arias J M, Fortunato L and Vitturi A 2005 *Phys. Rev.* C **72** 061302
[64] Alonso C E, Arias J M and Vitturi A 2007 *Phys. Rev. Lett.* **98** 052501
[65] Alonso C E, Arias J M and Vitturi A 2007 *Phys. Rev.* C **75** 064316
[66] Bès D R and Broglia R A 1966 *Nucl. Phys.* **80** 289
[67] Broglia R A, Terasaki J and Giovanardi N 2000 *Phys. Rep.* **335** 1
[68] Clark R M, Macchiavelli A O, Fortunato L and Krucken R 2006 *Phys. Rev. Lett.* **96** 032501
[69] Broglia R A, Hansen O and Riedel C 1973 *Adv. Nucl. Phys.* **6** 287
[70] Frauendorf S and Meng J 1997 *Nucl. Phys.* A **617** 131
[71] Dimitrov V I, Frauendorf S and Dö nau F 2000 *Phys. Rev. Lett.* **84** 5732
[72] Starosta K *et al* 2002 *Phys. Rev.* C **65** 044328
[73] Brant S, Tonev D, de Angelis G and Ventura A 2008 *Phys. Rev.* C **78** 034301
[74] Lenzi S M *et al* 2001 *Phys. Rev. Lett.* **87** 122501
[75] Zuker A P, Lenzi S M, Martinez-Pinedo G and Poves A 2002 *Phys. Rev. Lett.* **89** 142502
[76] Bentley M A and Lenzi S M 2007 *Prog. Part. Nucl. Phys.* **59** 497
[77] Gadea A *et al* 2006 *Phys. Rev. Lett.* **97** 152501
[78] Bentley M A *et al* 2006 *Phys. Rev. Lett.* **97** 132501
[79] Rudolph D *et al* 2008 *Phys. Rev.* C **78** 021301
[80] Garrett P E *et al* 2007 *Phys. Rev.* C **75** 014307
[81] Della Vedova F *et al* 2007 *Phys. Rev.* C **73** 034317

Journal of Physics: Conference Series **168** (2009) 012010 doi:10.1088/1742-6596/168/1/012010

Relativistic models for nuclear structure and low-energy QCD phenomenology

P Finelli

Physics Department, University of Bologna and INFN
Via Irnerio 46, 40126 Bologna (Italy)

E-mail: `paolo.finelli@bo.infn.it`
`http://minimafisica.biodec.com`

Abstract. In this contribution we will report about developments on a recently proposed relativistic nuclear energy density functional based on the characteristic features of low-energy QCD and its symmetry breaking pattern.

1. Introduction

The most complete and accurate description of structure phenomena in medium and heavy nuclei is currently provided by self-consistent non-relativistic and relativistic mean-field approaches. By employing phenomenological effective interactions, adjusted to empirical properties of symmetric and asymmetric nuclear matter, and to bulk properties of stable nuclei, self-consistent mean-field methods have achieved a high level of precision in describing ground states and properties of excited states in stable nuclei, exotic nuclei far from β-stability, and in nuclear systems at the nucleon drip-lines.

In the last years, a new approach has been proposed [1, 2, 3, 4] in which the effective interaction is constrained by two essential features of the low-energy (non-perturbative) sector of QCD. It is essentially based on the following conjectures:

(i) The nuclear ground state is characterized by strong scalar and vector mean fields which have short-range origin and can be satisfactorily interpreted as in-medium changes of the scalar quark condensate (the chiral condensate) and of the quark density [5].

(ii) Nuclear binding and saturation arise primarily from chiral (pionic) fluctuations (reminiscent of van der Waals forces) in combination with Pauli blocking effects, superimposed on the condensate background fields and calculated according to the rules of in-medium chiral perturbation theory [6].

2. Medium and long-range interaction: in-medium chiral dynamics

The starting point is the description of nuclear matter based on the chiral effective Lagrangian with pions and nucleons [6]. The relevant *small* scales are the Fermi momentum k_f, the pion mass m_π and the $\Delta - N$ mass difference $\Delta \equiv M_\Delta - M_N \simeq 2.1 m_\pi$, all of which are well separated from the characteristic scale of spontaneous chiral symmetry breaking, $4\pi f_\pi \simeq 1.16$ GeV with the pion decay constant $f_\pi = 92.4$ MeV. The calculations have been performed to three-loop order in the energy density. They incorporate the one-pion exchange Fock term, iterated one-pion exchange and irreducible two-pion exchange, including one or two intermediate Δ's (see the insert in figure 2 for some relavant diagrams).

The only new ingredient in performing new calculations at finite nucleon density is the in-medium nucleon propagator. It expresses the fact that the ground-state of the system has changed from an empty vacuum to a filled Fermi sea of nucleons. As shown in figure 1 the full propagator splits additively into the vacuum nucleon propagator and a *medium insertion*. This allows to organize the diagrammatic calculation according to the number of medium insertions.

$$\mathrm{G}_N = (\not{p} + M_N)[\tfrac{i}{p^2 - M_N^2 + i\epsilon} - 2\pi\delta(p^2 - M_N^2)\theta(k_f - |\vec{p}^2|)\theta(p_0)]$$

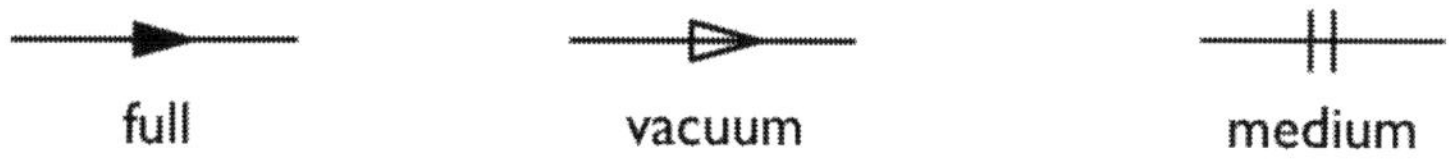

Figure 1. The full nucleon propagator defined as the *vacuum* part plus the *in-medium insertion.*

The pion-nucleon interaction vertices are the pseudovector πNN-vertex and the Tomozawa-Weinberg-contact vertex of the form

$$\frac{g_A}{2f_\pi}\gamma_\mu q_a^\mu \gamma_5 \tau_a \quad \text{and} \quad \frac{1}{4f_\pi^2}(\gamma_\mu q_b^\mu - \gamma_\mu q_a^\mu)\epsilon_{abc}\tau_c \,.$$

As a first application, nuclear matter has been extensively studied [6]: the equation of state for symmetric and asymmetric systems, real and imaginary single particle potentials, the liquid-gas phase transition, the role of Δ-isobar degree of freedom, pairing gaps, the Landau-Migdal parameters and the chiral condensate dynamics. In figure 2 we show only a small selection of the results obtained in this framework, and in particular the very important role of $\Delta(1232)$ excitations in stabilizing the calculations is emphasized by blue thick lines. In the small insert we have written the nuclear matter equation of state as an expansion in powers of the Fermi momentum k_f. The expansion coefficients are functions of k_f/m_π and Δ/m_π, the dimensionless ratios of the relevant small scales. Divergent momentum space loop integrals are regularized by introducing subtraction constants in the spectral representations of these terms. The (few) subtraction constants are the only parameters in this approach. They equivalently correspond to two- and three-nucleon contact interactions (and derivatives thereof), encoding short-distance dynamics not resolved in detail at the characteristic momentum scale $k_f \ll 4\pi f_\pi$. The finite parts of the energy density, written in closed form as functions of k_f/m_π and Δ/m_π, represent long and intermediate range (chiral) dynamics with input fixed entirely in the πN sector. The low-energy constants (contact terms or subtraction constants) are adjusted to reproduce basic properties of symmetric and asymmetric nuclear matter at the saturation point. The same framework has also been used for other purposes: deriving a (non-relativistic) energy density functional [7], extracting the nuclear spin-orbit interaction [8] and the description of hyperons in a nuclear medium [9]. Concerning the spin-orbit interaction, and in particular its microscopic origin, it is worthy mentioning the new remarkable interpretation presented in Ref. [10]: *large* spin-orbit splittings in ordinary nuclei and *almost vanishing* spin-orbit splittings in Λ hypernuclei can naturally arise as different balancing mechanisms between short- and long-range components of the nuclear force without the introduction of any ad-hoc hypothesis.

Journal of Physics: Conference Series **168** (2009) 012010 doi:10.1088/1742-6596/168/1/012010

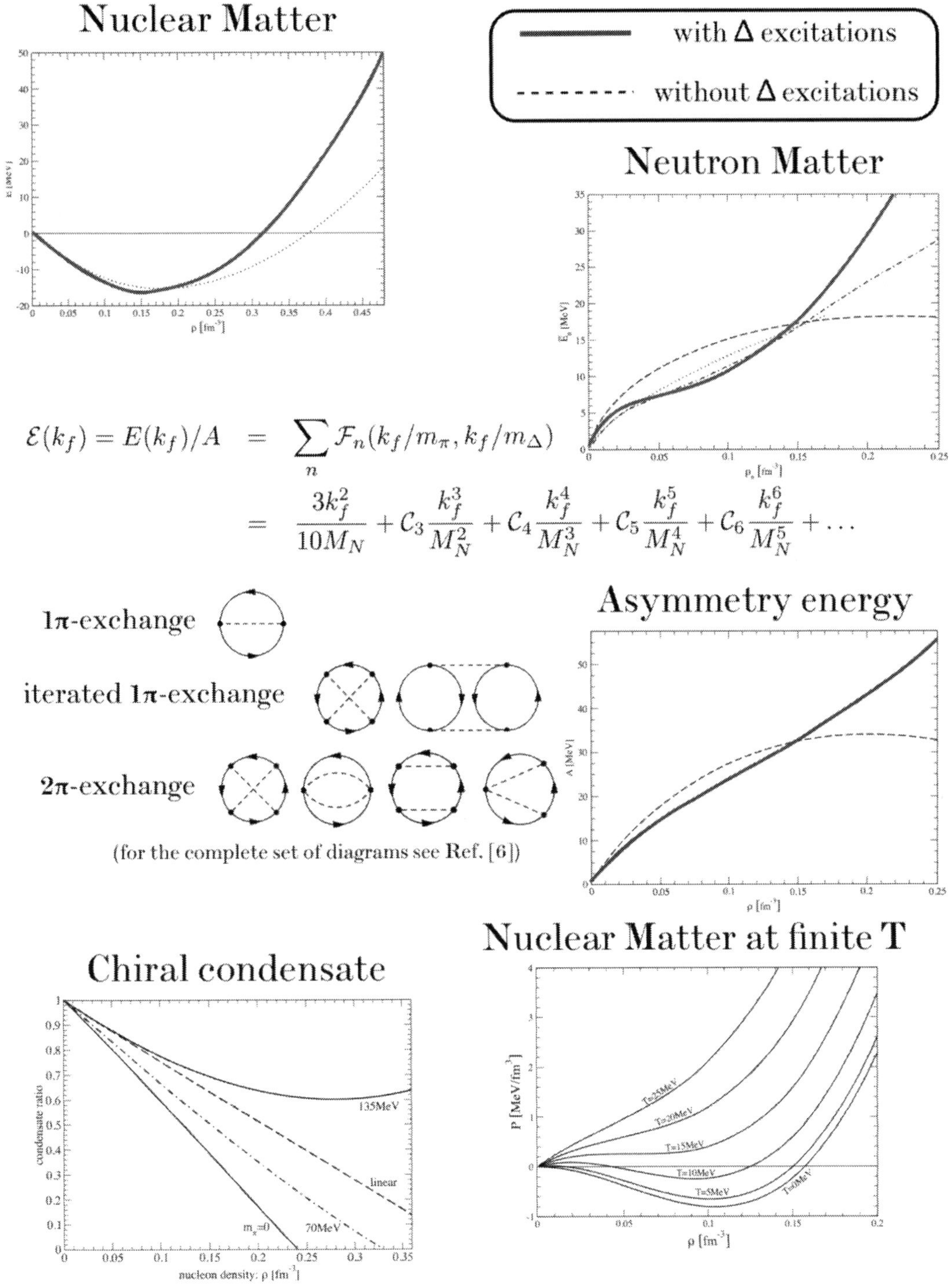

$$\mathcal{E}(k_f) = E(k_f)/A = \sum_n \mathcal{F}_n(k_f/m_\pi, k_f/m_\Delta)$$
$$= \frac{3k_f^2}{10M_N} + \mathcal{C}_3\frac{k_f^3}{M_N^2} + \mathcal{C}_4\frac{k_f^4}{M_N^3} + \mathcal{C}_5\frac{k_f^5}{M_N^4} + \mathcal{C}_6\frac{k_f^6}{M_N^5} + \ldots$$

Figure 2. Some results obtained with in-medium chiral perturbation theory. For details about calculations and figures see Ref. [6].

Journal of Physics: Conference Series **168** (2009) 012010 doi:10.1088/1742-6596/168/1/012010

3. Short-distance dyanmics: in-medium QCD sum rules

The QCD ground state is characterized by condensates of quark-antiquark pairs and gluons, an entirely non-perturbative phenomenon. The quark condensate $\langle\bar{q}q\rangle$, i.e. the ground state expectation value of the scalar quark density, plays a particularly important role as an order parameter of spontaneously broken chiral symmetry. At a renormalization scale of about 1 GeV (with up and down quark masses $m_u + m_d \simeq 12$ MeV) the strength of the chiral vacuum condensate is $\langle\bar{q}q\rangle_0 \simeq -(240\ \text{MeV})^3 \simeq -1.8\ \text{fm}^{-3}$ [11], more than an order of magnitude larger than the nuclear matter density at saturation $\rho_0 \simeq 0.16\ \text{fm}^{-3}$. Hadrons, as well as nuclei, are excitations built on this condensed ground state. The strength of the chiral condensate at normal nuclear densities is reduced by about one third from its vacuum value. The density-dependent changes of the condensate structure in the presence of baryonic matter are a source of strong scalar and vector fields experienced by nucleons (and hyperons). In-medium QCD sum rules relate the leading changes of the scalar quark condensate and quark density at finite baryon density, with the scalar and vector self-energies of a nucleon in the nuclear medium.

The nuclear ground state is characterized by strong scalar U_S and vector U_V mean fields which have their origin in the in-medium changes of the scalar quark condensate (the chiral condensate) and of the quark density (more precisely, nucleons in the medium acquire large scalar and vector self-energies). They can be calculated by QCD sum rules techniques [5] to obtain, at leading order, an approximation valid at nucleon densities below and around saturated nuclear matter,

$$U_S = -\underbrace{\frac{\sigma_N M_N}{m_\pi^2 f_\pi^2}}_{G_S^{(0)}} \rho_S \quad \text{and} \quad U_V = \underbrace{\frac{4(m_u+m_d)M_N}{m_\pi^2 f_\pi^2}}_{G_V^{(0)}} \rho \tag{1}$$

where $\sigma_N = \langle N|m_q\bar{q}q|N\rangle$ is the nucleon sigma term ($\simeq 50$ MeV), m_π the pion mass (138 MeV), M_N the nucleon mass (939 MeV), $m_{u,d}$ the quark masses ($\simeq 5$ MeV), f_π the pion decay constant and ρ and ρ_S are the baryon and the scalar density, respectively. The resulting U_S and U_V are individually of the order of $300 - 400$ MeV in magnitude. $G_S^{(0)}$ and $G_V^{(0)}$ could be interpreted as coupling strengths for an effective contact nucleon-nucleon interaction. It is remarkable that the ratio

$$\frac{U_S}{U_V} = -\frac{\sigma_N}{4(m_u+m_d)}\frac{\rho_S}{\rho} \tag{2}$$

is close to -1. As a result, in the single-nucleon Dirac equation there is an almost complete cancellation in the central potential ($\sim U_V + U_S$) (a feature characteristic of relativistic mean-field phenomenology), giving a negligible contribution to the binding of the system, but, at the same time, a large contribution to the spin-orbit potential

$$V_{LS} \sim \frac{1}{2M_N^2}\frac{1}{r}\left(\frac{\partial}{\partial r}(U_V - U_S)\right)\boldsymbol{l}\cdot\boldsymbol{s}\,. \tag{3}$$

Of course, the constraints implied by Eqs. (1) are not very accurate on a quantitative level. Corrections from condensates of higher dimension and uncertainties in the values of σ_N and m_u+m_d lead to an estimated error for the ratio $\Sigma_S^{(0)}/\Sigma_V^{(0)} \simeq -1$ of about 20%. As a consequence QCD sum rules calculations should be used as theoretical estimates for *a posteriori* comparison with the fitted values for $G_S^{(0)}$ and $G_V^{(0)}$.

4. The model

The relativistic density functional describing the ground-state energy of the system can be written as a sum of four distinct terms:

$$E_0[\hat{\rho}] = E_{\text{free}}[\hat{\rho}] + E_{\text{H}}[\hat{\rho}] + E_{\text{coul}}[\hat{\rho}] + E_\pi[\hat{\rho}]\,, \tag{4}$$

with

$$
\begin{aligned}
E_{\rm free}[\hat{\rho}] &= \int d^3r \, \langle\phi_0|\bar{\psi}[-i\boldsymbol{\gamma}\cdot\boldsymbol{\nabla} + M_N]\psi|\phi_0\rangle \, , & (5)\\
E_{\rm H}[\hat{\rho}] &= \frac{1}{2}\int d^3r \, \{\langle\phi_0|G_S^{(0)}(\bar{\psi}\psi)^2|\phi_0\rangle + \langle\phi_0|G_V^{(0)}(\bar{\psi}\gamma_\mu\psi)^2|\phi_0\rangle\} \, , & (6)\\
E_\pi[\hat{\rho}] &= \frac{1}{2}\int d^3r \, \Big\{\langle\phi_0|G_S^{(\pi)}(\hat{\rho})(\bar{\psi}\psi)^2|\phi_0\rangle + \langle\phi_0|G_V^{(\pi)}(\hat{\rho})(\bar{\psi}\gamma_\mu\psi)^2|\phi_0\rangle \\
&\quad + \langle\phi_0|G_{TS}^{(\pi)}(\hat{\rho})(\bar{\psi}\vec{\tau}\psi)^2|\phi_0\rangle + \langle\phi_0|G_{TV}^{(\pi)}(\hat{\rho})(\bar{\psi}\gamma_\mu\vec{\tau}\psi)^2|\phi_0\rangle \\
&\quad - \langle\phi_0|D_S^{(\pi)}[\boldsymbol{\nabla}(\bar{\psi}\psi)]^2|\phi_0\rangle\Big\} \, , & (7)\\
E_{coul}[\hat{\rho}] &= \frac{1}{2}\int d^3r \, \langle\phi_0|A^\mu e\bar{\psi}\frac{1+\tau_3}{2}\gamma_\mu\psi|\phi_0\rangle \quad , & (8)
\end{aligned}
$$

where $|\phi_0\rangle$ denotes the nuclear ground state. Here $E_{\rm free}$ is the energy of the free (relativistic) nucleons including their rest mass. $E_{\rm H}$ is a Hartree-type contribution representing strong scalar and vector mean fields, in connection with the leading terms of the corresponding nucleon self-energies deduced from in-medium QCD sum rules. Furthermore, E_π is the part of the energy generated by chiral $\pi N\Delta$-dynamics, including a derivative (surface) term. For more details we refer the reader to Ref. [3].

Minimization of the ground-state energy, represented in terms of a set of auxiliary Dirac spinors ψ_k, leads to the relativistic analogue of the Kohn-Sham equations in atomic physics. These single-nucleon Dirac equations are solved self-consistently in the "no-sea" approximation.

In the mean-field approximation, the single-nucleon Dirac equation is found by minimization with respect to $\bar{\psi}_k$:

$$[-i\boldsymbol{\gamma}\cdot\boldsymbol{\nabla} + M_N + \gamma_0\Sigma_V + \gamma_0\tau_3\Sigma_{TV} + \gamma_0\Sigma_R + \Sigma_S + \tau_3\Sigma_{TS}]\psi_k = \epsilon_k\psi_k \, , \qquad (9)$$

where Σ_i are the self-energies in the mean field approximation (see Ref. [3])

$$
\begin{aligned}
\Sigma_V &= [G_V^{(0)} + G_V^{(\pi)}(\rho)]\rho + eA^0\frac{1+\tau_3}{2} \, , & (10)\\
\Sigma_{TV} &= G_{TV}^{(\pi)}(\rho)\,\rho_3 \, , & (11)\\
\Sigma_S &= [G_S^{(0)} + G_S^{(\pi)}(\rho)]\rho_S + D_S^{(\pi)}\boldsymbol{\nabla}^2\rho_S \, , & (12)\\
\Sigma_{TS} &= G_{TS}^{(\pi)}(\rho)\,\rho_{S3} \, , & (13)\\
\Sigma_R &= \frac{1}{2}\left\{\frac{\partial G_V^{(\pi)}(\rho)}{\partial\rho}\rho^2 + \frac{\partial G_S^{(\pi)}(\rho)}{\partial\rho}\rho_S^2 + \frac{\partial G_{TV}^{(\pi)}(\rho)}{\partial\rho}\rho_3^2 + \frac{\partial G_{TS}^{(\pi)}(\rho)}{\partial\rho}\rho_{S3}^2\right\} \, , & (14)
\end{aligned}
$$

$A^0(\mathbf{r}) = \frac{e}{4\pi}\int d^3r' \, \frac{\rho_{ch}(\mathbf{r}')}{|\mathbf{r}-\mathbf{r}'|}$ is the Coulomb potential and ϵ_k denotes the single-nucleon energy for the k particle. Σ_R is the rearrangement term. As usual, densities are defined as (all spatial components are vanishing for spin saturated systems)

$$
\begin{aligned}
\rho &= \langle\phi_0|\bar{\psi}\gamma^0\psi|\phi_0\rangle = \rho^p + \rho^n \, , & (15)\\
\rho_3 &= \langle\phi_0|\bar{\psi}\tau_3\gamma^0\psi|\phi_0\rangle = \rho^p - \rho^n \, , & (16)\\
\rho_S &= \langle\phi_0|\bar{\psi}\psi|\phi_0\rangle = \rho_S^p + \rho_S^n \, , & (17)\\
\rho_{S3} &= \langle\phi_0|\bar{\psi}\tau_3\psi|\phi_0\rangle = \rho_S^p - \rho_S^n \, , & (18)\\
\rho_{ch} &= \langle\phi_0|\bar{\psi}\gamma^0\frac{1+\tau_3}{2}\psi|\phi_0\rangle = \rho^p \, . & (19)
\end{aligned}
$$

The density dependent couplings $G_i(\hat{\rho})$ $(i = S, V, TS, TV)$ can be written as a sum of two distinct terms:

$$\begin{aligned} G_i(\hat{\rho}) &= G_i^{(0)} + G_i^{(\pi)}(\hat{\rho}) \quad (\text{for } i = S, V) \\ \text{and} \quad G_i(\hat{\rho}) &= G_i^{(\pi)}(\hat{\rho}) \qquad\qquad (\text{for } i = TS, TV) \,, \end{aligned} \tag{20}$$

i.e. density-independent parts $G_i^{(0)}$, which arise from strong isoscalar scalar and vector background fields, and density-dependent parts $G_i^{(\pi)}(\hat{\rho})$ generated by (regularized) one- and two-pion exchange dynamics (see Ref. [3] for a complete explanation). It is assumed that only pionic processes contribute to the isovector channels. There are 7 free parameters ($G_S^{(0)}$,$G_V^{(0)}$, two isoscalar and two isovector contact terms in the contact couplings $G_i^{(\pi)}(\hat{\rho})$ and the surface term $D_S^{(\pi)}$) that have to be adjusted in order to reproduce ground-state properties of closed shell nuclei [3].

5. Results

To show the quality of present approach and the capability to make predictions about the structure of finite nuclei, we will show some (unpublished) results for the Isovector Giant Dipole Resonances (IVGDR) in closed shell nuclei. In order to perform calculations of these collective states, we worked in the Random Phase Approximation (RPA), i.e. we solved the following set of coupled equations [12]

$$\left[\omega \begin{pmatrix} 1 & 0 \\ 0 & -1 \end{pmatrix} - \begin{pmatrix} A & B \\ B^* & A^* \end{pmatrix}\right] \begin{pmatrix} X \\ Y \end{pmatrix} = 0 \,. \tag{21}$$

The A and B matrices are defined as

$$A = \begin{pmatrix} (\epsilon_p - \epsilon_h)\delta_{pp'}\delta_{hh'} & \\ & (\epsilon_a - \epsilon_h)\delta_{aa'}\delta_{hh'} \end{pmatrix} + \begin{pmatrix} V_{ph'hp'} & V_{ph'ha'} \\ V_{ah'hp'} & V_{ah'ha'} \end{pmatrix} \tag{22}$$

$$B = \begin{pmatrix} V_{pp'hh'} & V_{pa'hh'} \\ V_{ap'hh'} & V_{aa'hh'} \end{pmatrix} , \tag{23}$$

where p labels particle states, a antiparticles and h holes. X and Y amplitudes are defined as follows

$$X = \begin{pmatrix} \delta\rho_{ph} \\ \delta\rho_{ah} \end{pmatrix} \quad \text{and} \quad Y = \begin{pmatrix} \delta\rho_{hp} \\ \delta\rho_{ha} \end{pmatrix} , \tag{24}$$

in terms of the transition densities. For a derivation of the RPA equations and a complete discussion about their properties we refer the reader to the well known literature [13]. Multipole transitions are described by the reduced transition probability

$$B^T(EJ, J_i \to J_f) = \frac{1}{2J_i + 1} |\langle f || \hat{Q}_J^T || i \rangle^2 \,, \tag{25}$$

where $\hat{Q}_J^T$ corresponds to the electric multipole transition. In case of IVGDR we have

$$\hat{Q}_{1M}^{T=1} = e \sum_{i=1}^{A} \left(\tau_z^{(1)} - \frac{N-Z}{2A} \right) r_i Y_{1M}(\hat{r}_i) \,, \tag{26}$$

where the center-of-mass motion has been conveniently subtracted. $B(E)$ is usually employed to calculate

$$R(E) = \sum_i B(E_i) \frac{\Gamma^2}{4(E - E_i)^2 - \Gamma^2} \,, \tag{27}$$

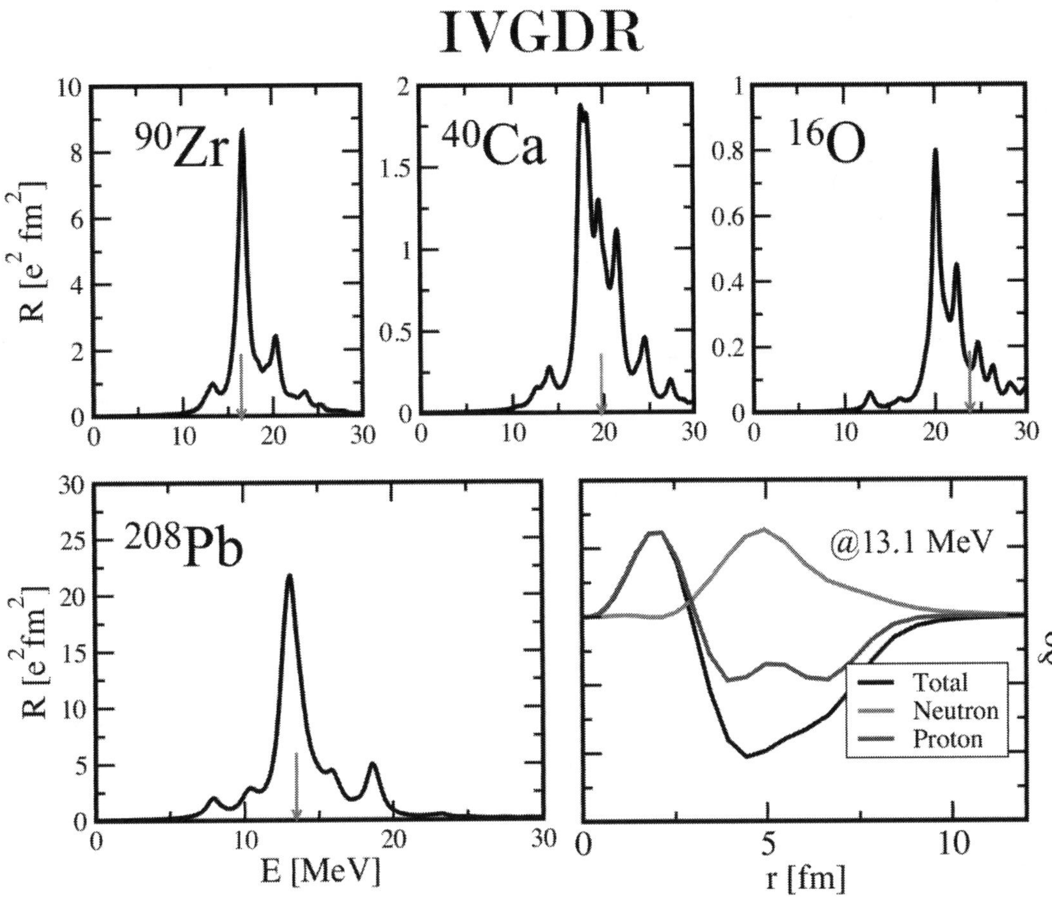

	^{16}O	^{40}Ca	^{90}Zr	^{208}Pb
Exp.	23.8±0.5	19.8±0.5	16.5±0.2	13.5±0.2
FKVW	20.1	17.7	16.7	13.1

Figure 3. Isovector giant dipole resonances for ^{16}O, ^{40}Ca, ^{90}Zr and ^{208}Pb. Experimental values [14] are denoted with red arrows. In case of lead we also show the (total, neutron and proton) transition densities at the peak energy.

where Γ is a phenomenological width (= 1 MeV). In figure 3 we show calculations for some closed-shell nuclei. The agreement with experimental data [14] is fairly good, in particular for heavy systems. Phenomenological approaches give very similar results [12].

6. Conclusions

This short contribution is able to demonstrate that chiral effective field theory provides a consistent microscopic framework in which both the isoscalar and isovector channels of a universal nuclear energy density functional can be formulated. We believe that the present approach to nuclear DFT establishes a fundamental link between low-energy QCD and ground-state properties of finite nuclei.

Acknowledgements
I would like to thank MIUR and INFN for financial support and my collaborators: S. Fritsch, N. Kaiser and W. Weise of the Technical University of Munich (Germany) and D. Vretenar, T. Nikšić and Nils Paar of the University of Zagreb (Croatia).

References
[1] Finelli P, Kaiser N, Vretenar D and Weise W, 2002 *Eur. Phys. J.* A **17** 573
[2] Finelli P, Kaiser N, Vretenar D and Weise W, 2004 *Nucl. Phys.* A **735** 449
[3] Finelli P, Kaiser N, Vretenar D and Weise W, 2006 *Nucl. Phys.* A **770** 1
[4] Finelli P, Kaiser N, Vretenar D and Weise W, 2007 *Nucl. Phys.* A **791** 57
[5] Cohen T D, Furnstahl R J and Griegel D K 1991 *Phys. Rev. Lett.* **67** 961; Furnstahl R J, Griegel D K and Cohen T D 1992 *Phys. Rev.* C **46** 1507; Cohen T D, Furnstahl R J and Griegel D K 1995 *Prog. Part. Nucl. Phys.* **35** 221; Drukarev E G and Levin E M 1990 *Nucl. Phys.* A **511** 679; Drukarev E G and Levin E M 1991 *Prog. Part. Nucl. Phys.* **27** 77
[6] Kaiser N, Fritsch S and Weise W 2002 *Nucl. Phys.* A **697** 255; Kaiser N, Fritsch S and Weise W 2002 *Nucl. Phys.* A **700** 343; Fritsch S, Kaiser N and Weise W 2002 *Phys. Lett.* B **545** 73; Fritsch S and Kaiser N 2003 *Eur. Phys. J.* A **17** 11; Fritsch S and Kaiser N 2004 *Eur. Phys. J.* A **21** 117; Fritsch S, Kaiser N and Weise W 2005 *Nucl. Phys.* A **750** 259; Kaiser N, Nikšić T and Vretenar D 2005 *Eur. Phys. J.* A **25** 257; Kaiser N 2006 *Nucl. Phys.* A **768** 99; Kaiser N, de Homont P and Weise W 2008 *Phys. Rev.* C **77** 025204; Kaiser N and Weise W 2009 *Phys. Lett.* B **671** 25;
[7] Kaiser N, Fritsch S and Weise W 2003 *Nucl. Phys.* A **724** 47; Kaiser N 2003 *Phys. Rev.* C **68** 014323;
[8] Kaiser N 2002 *Nucl. Phys.* A **709** 251; Kaiser N 2003 *Nucl. Phys.* A **720** 157; Kaiser N 2003 *Phys. Rev.* C **68** 054001; Kaiser N 2004 *Phys. Rev.* C **70** 034307;
[9] Kaiser N and Weise W 2005 *Phys. Rev.* C **71** 015203; Finelli P, Kaiser N, Vretenar D and Weise W 2007 *Phys. Lett.* B **658** 90; Finelli P, Kaiser N, Vretenar D and Weise W 2009 submitted to *Nucl. Phys.* A
[10] Kaiser N and Weise W 2008 *Nucl. Phys.* A **804** 60;
[11] Ioffe B L 2003 *Phys. At. Nucl. (Yad. Fiz.)* **66** 30 and references therein
[12] Ring P, Ma Z Y , Van Giai N, Vretenar D, Wandelt A and Cao L G (2001) *Nucl. Phys.* A **694** 249; Ma Z Y, Wandelt A, Van Giai N, Vretenar D, Ring P and Cao L G (2002) *Nucl. Phys.* A **703** 222; Paar N, Ring P, Nikšić T and Vretenar D (2003) *Phys. Rev.* C **67** 034312;
[13] Ring P and Schuck P 1980 *The Nuclear Many Body Problem* (Springer-Verlag)
[14] Gleissl P, Brack M, Meyer J and Quentin P 1990 *Ann. Phys. (N.Y.)* **197** 205

12th Conference on 'Theoretical Nuclear Physics in Italy' IOP Publishing
Journal of Physics: Conference Series **168** (2009) 012011 doi:10.1088/1742-6596/168/1/012011

Remarks on shape phase transitions in nuclei

L Fortunato[1,2], **A Vitturi**[1,2], **C E Alonso**[3] **and J M Arias**[3]

[1] Dipartimento di Fisica "G.Galilei", via Marzolo 8, I-35131 Padova, Italy
[2] I.N.F.N., via Marzolo 8, I-35131 Padova, Italy
[3] Departamento de Física Atómica, Molecular y Nuclear, Facultad de Física, Universidad de Sevilla, Apartado 1065, E-41080 Sevilla, Spain

Abstract. Various aspects of shape phase transitions in even as well as odd systems are reviewed. Firstly, the case of an odd $j = 3/2$ particle coupled to an even-even boson core that undergoes a transition from spherical limit (U(5)) to the γ-unstable limit (O(6)) is illustrated. Energy spectrum and electromagnetic transitions, in correspondence of the critical point, display behaviors qualitatively similar to those of the even core and they agree qualitatively with the model based on the E(5/4) boson-fermion symmetry. Secondly, we describe a study on two-particle transfer reactions: the evolution of the transfer spectroscopic intensities within the interacting boson model is analyzed as a possible signature of shape phase transitions. In correspondence to the critical points characterizing the phase transitions, the two-particle transfer matrix elements to both ground and excited 0^+ states display a rapid discontinuity that might help validating the experimental search for the critical point.

1. Introduction

We will report in the following on various results obtained by our collaboration during recent years [1, 2, 3] on the theme of shape phase transitions in even and odd systems. In particular we will concentrate on i) a brief introduction to shape phase transitions in the collective model and in the IBM , ii) the analysis of the γ-unstable case in odd systems and iii) two-particle transfer intensities and other signatures of the shape phase transition.

The collective model of Bohr and Mottelson treats quantistically the collective properties of nucleons in a nucleus as vibrations and rotations of an ellipsoidal surface (namely the quadrupole d.o.f.) within a single formalism that is based on a Hamiltonian operator (Bohr hamiltonian) expressed in the deformation coordinates β and γ. The Bohr hamiltonian can be solved analytically in a few cases (reviewed in Ref. [5]) depending on the potential term $V(\beta, \gamma)$ and analytic solutions correspond to algebraic structures that have their common origin in the U(6) underlying symmetry of the collective model. In particular the γ-independent harmonic oscillator potential generates a solution that has U(5) has its underlying dynamical symmetry, while the γ-unstable rotor has O(6). A third case, the axial symmetric rotor is associated with SU(3). Each of the three mentioned cases is also associated with certain spectral properties (energy spectra, electromagnetic transitions, etc.) and with a given ellipsoidal shape that characterize a given phase. Although good examples of each paradigm can be found, most nuclei do not strictly match these ideal limits, but sit somewhere in the middle between them, therefore people has studied transitions from one phase to another. Critical point for this shape phase transitions can be found.

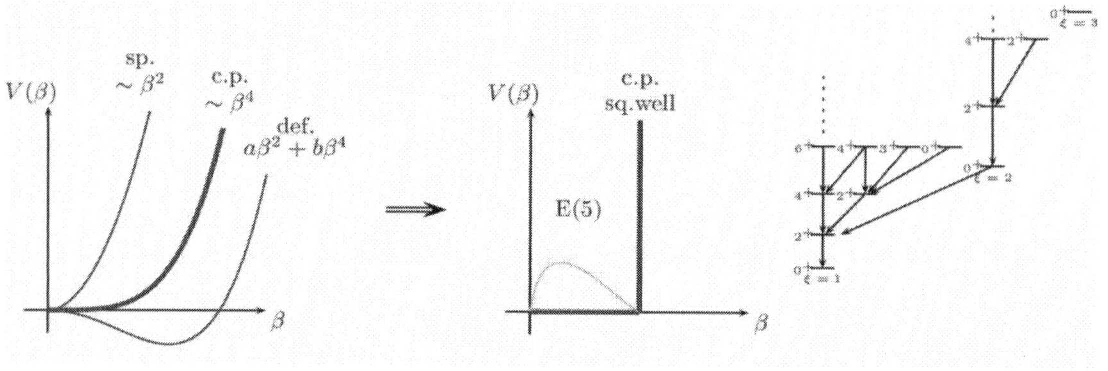

Figure 1. Evolution of the potential $V(\beta)$ along the spherical to γ-unstable shape phase transition, passing through the critical point β^4 potential. This is approximated with a square well that is analytically solvable and gives a novel dynamical symmetry, E(5) that is associated with an energy spectrum with precise rules.

Recently three papers of Iachello introduced the so-called critical point symmetries in the framework of the collective model for even nuclei. Namely he has identified the path from one phase to another with a change in the parameters of the potential and he has approximated the potential at the critical point with a solvable one that can be associated with a new algebraic structure. As an example we discuss the E(5) case (see figure 1): in the transition from the U(5) spherical limit (harmonic oscillator) to the O(6) $\gamma-$unstable limit (champagne bottle or mexican hat potential) one sees that the evolution of the potential goes from a pure β^2 to a combination of β^2 and β^4 that has a deformed minimum. At the critical point of this second order transition the potential is a pure β^4 that is not analytically solvable. This potential is approximated with a square well that is analytically solvable and is associated with a E(5) structure, the Euclidean group in five dimensions.

2. Odd-even phase transitions

Until now most of the work has been carried out for even-even nuclei, using either the Bohr Hamiltonian and the surface collective variables or algebraic approaches like the interacting boson model. Review articles with references to the original works can be found in Refs. [6, 7, 8]. In the case of odd-even nuclei, with an odd particle coupled to an even core undergoing a phase transition, attention has been put on the shape transition from spherical to deformed gamma-unstable shapes. In correspondence to the critical point in the even core two new boson-fermion critical point symmetries have been proposed, in the case of an odd particle moving in a single $j = 3/2$ shell (E(5/4) symmetry [9]) or in the $j = 1/2, 3/2, 5/2$ shells (E(5/12) symmetry [10]). Characteristic sequences of levels and ratios of electromagnetic transitions are predicted in both cases.

We consider here, within the Interacting Boson Fermion Model, the particular case of an odd $j = 3/2$ particle coupled to an even-even boson core that undergoes a transition from spherical U(5) to γ-unstable O(6) situation. The choice of the $j = 3/2$ orbital preserves in the odd case the condition of gamma-instability of the system.

Within the IBM this can be obtained, for example, from the Hamiltonian of the even-even part, as

$$H_B = x\hat{n}_d - \frac{1-x}{N}\hat{Q}_B \cdot \hat{Q}_B \; , \tag{1}$$

which produces, varying the parameter x from 1 to 0, a transition between the two extreme situations characteristic of U(5) and O(6) symmetries with a second order phase transition. The operators appearing in the Hamiltonian above are given by

$$\hat{n}_d = \sum_\mu d_\mu^\dagger d_\mu \ , \tag{2}$$

$$\hat{Q}_B = (s^\dagger \times \tilde{d} + d^\dagger \times \tilde{s})^{(2)} \ , \tag{3}$$

and N is the total number of bosons. For any value of x this Hamiltonian maintains the typical degeneracies of the O(5) symmetry. At the critical point, the energy surface acquires a β^4 behaviour, which is approximated by an infinite square well in the E(5) critical point symmetry [4] within the framework of the collective Bohr Hamiltonian.

The odd-even system is described by

$$H = H_B + H_F + V_{BF} \ , \tag{4}$$

where the term V_{BF} couples the bosonic and fermionic parts, namely

$$V_{BF} = -2\frac{1-x}{N}\hat{Q}_B \cdot \hat{q}_F \ , \tag{5}$$

where $\hat{Q}_B$ (taken of the form given in Eq.(3)) and $\hat{q}_F = (a_{3/2}^\dagger \times \tilde{a}_{3/2})^{(2)}$ are the boson and fermion quadrupole operators, respectively.

The part V_{BF}, has to be diagonalized and its eigenvalues, doubly degenerate, turn out to be γ-independent,

$$E_\pm(\beta,\gamma) = E_\pm(\beta) = \pm 2\frac{(1-x)\beta}{1+\beta^2} \ . \tag{6}$$

In other words, the addition of the odd particle does not destroy the γ-instability of the system, giving rise to energy surfaces for the different odd intrinsic states that are still γ-independent.

The resulting energy spectra in the odd system are shown in the right panel of figure 3 as a function of the control parameter $1-x$. The total number of bosons, N, has been assumed to be equal to 7. For a better comparison, we also show in the left panel of the figure the corresponding evolution of the spectrum in the even core.

The level evolution in the odd case shows a behaviour qualitatively similar to that of the even case. The group structure of $Spin^{BF}(5)$ with respect to O(5) simply leads to a richer pattern for the fermion case and slightly different ratios for the energy levels.

The position of the 3/2 state (or better the $(\tau_1 = 1/2, \tau_2 = 1/2)j = 3/2$ state) is the key element to characterize the particular situation and its position along the transitional path. The position of this state plays the same role as the key position of the first excited 0^+ state in even nuclei.

Transition probabilities, state by state, for the odd nucleus at the critical point situation are shown in figure 2. It can be observed that E2 transitions are stronger between states with the same ξ value and $\Delta\tau_1 = 1$. Transitions between states pertaining to families with different ξ are one or two orders of magnitude smaller. Transitions between states with the same ξ and τ_1 values, but different spins, corresponding to the same multiplets are also one or two orders of magnitude smaller than the ones between different multiplets in the same band.

To summarize, we have considered, within the Interacting Boson Fermion Model, the coupling of an odd $j = 3/2$ particle to a boson core that undergoes a transition from spherical U(5) to γ-unstable O(6) character. The particular choice of the Hamiltonian and of the $j = 3/2$ orbital preserves in the odd case the condition of $\gamma-$instability of the system, and it is reflected in the preservation of the degeneracies associated with the $Spin^{BF}(5)$ symmetry. As a consequence, the energy spectrum and the electromagnetic transitions for the odd nucleus with a critical core

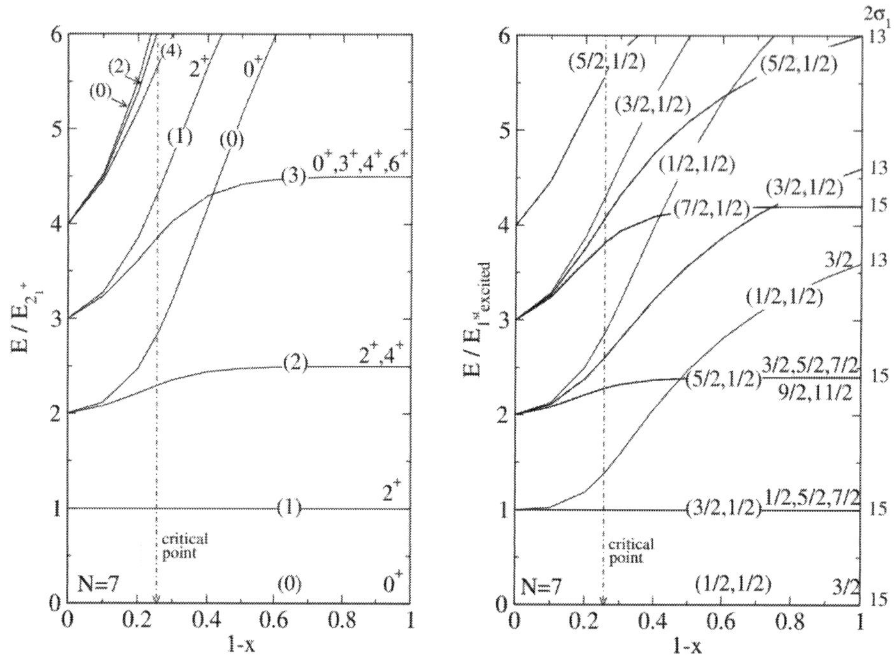

Figure 2. Energy levels (normalized to the energy of the first excited state) for the even and odd systems are displayed as a function of the parameter $(1 - x)$ for the boson (1) and boson-fermion (4) Hamiltonians. A number $N = 7$ of bosons has been assumed in both cases, while the odd particle has been taken in the $j = 3/2$ orbital. In the left panel (even case) we indicate for each level the τ quantum number (in parenthesis), spin and parity. In the right panel (odd case) we quote the (τ_1, τ_2) quantum numbers (in parenthesis) and spin. In the extreme $x = 0$ case we also indicate the σ_1 quantum number. The position of the even critical point is marked.

display behaviours qualitatively similar to those characterizing the phase transition in the even core. We have compared our results with the recently proposed E(5/4) approach, based on the Bohr hamiltonian. Both approaches display similar qualitative pictures, although we evidence a number of quantitative differences, that can be traced back to the different nature of the two schemes.

3. Two-neutron pair transfer in even-even nuclei

Pair-transfer reactions are also useful in the study of nuclear shape phase transitions, i.e., the rapid evolution of nuclear structure with mass number, such as from sphericity to axial-symmetric deformed or from sphericity to deformed γ-unstable nuclei. In our study, we have used the original version of the interacting boson model, the IBM-1 [9], which describes nuclei by assuming nucleon pairs as basic building blocks and treats them as bosons (without distinction between protons and neutrons). The transfer of one boson to a nucleus thus corresponds to the transfer of a nucleon pair, which makes the IBM model exceptionally well suited to describing two-nucleon transfer reactions.

Arima and Iachello defined the most general boson equivalent of the $L = 0$ pair-transfer operator $P^{(0)}_{+,0}$, taking into account up to cubic terms, that can be put in the form

$$P^{(0)}_{+,0} = a_1 s^\dagger + a_2[\hat{N} \times s^\dagger]^{(0)} + a_3[\hat{Q} \times d^\dagger]^{(0)} \quad (7)$$

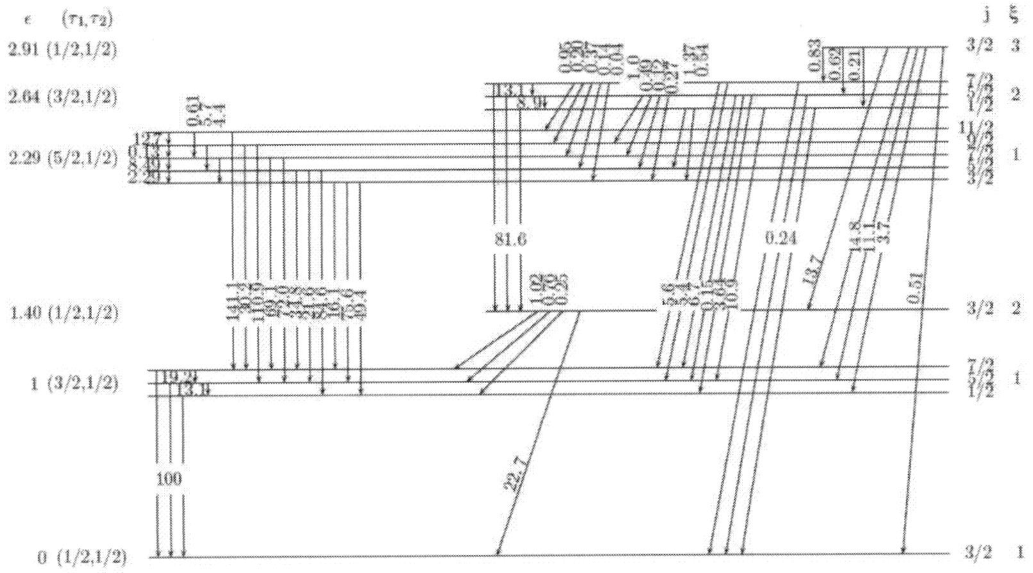

Figure 3. Energy levels and quadrupole transition rates B(E2↓) for the odd system at the core critical point. For illustration purposes the various multiplets, labeled by the (τ_1, τ_2) quantum numbers in parenthesis, have been arbitrarily split according to their j quantum number. The label on the extreme left is the energy in relative units. B(E2↓)'s have been normalized to the value 100 for the transitions (with equal strengths) between the states of the first multiplet and the ground state.

where $\hat{Q}$ is the quadrupole operator. We calculate transfer matrix elements and intensities (see [2] for details) in the Boson Coherent State formalism, obtaining exact results in terms of simple algebraic formulas that only need as input the quadrupole deformation β and β' of the nuclei between which the pair transfer takes place. We first carried out our study in a general way between each of the three dynamic limits of the IBM symmetry triangle: U(5), O(6), and SU(3). Next, we applied our results to some specific series of isotopes that are known to display phase transitions of first and second order (the ^{60}Nd, ^{62}Sm, ^{64}Gd and ^{66}Dy series, and the ^{44}Ru series, respectively). Some of our results are shown in figure 4 for the $Z \sim 60-66$ region. In this region, nuclear shapes evolve from spherical to axial-symmetric deformed Both transfer intensities from ground-to-ground and from ground-to-quasi-β bandheads are plotted. Solid lines give the exact IBM calculationswhile dashed lines give the results produced by the boson coherent-state framework (mean field). In all the isotope chains studied, a drop down is observed in ground-to-ground pair-transfer intensities at the mean-field level for a given value of N, reflecting that the structures of the isotopes involved are changing rapidly. A complementary increase in the ground-to-quasi-β pair-transfer intensity at the mean-field level is also observed. Abrupt drops in ground-to-ground transfer intensity with the complementary increments in the ground-to-β intensities clearly mark the phase transition in the case of the Nd, Sm, Gd, and Dy isotope series.

In conclusion, the intensity for pair transfer between two ground states is in general much larger than the intensity for transfer between the ground state and the β-vibrational or higher phonon 0^+ excited states. At the critical point, however, for U(5) to O(6) and U(5) to SU(3) (second-order and first-order transitions, respectively) both the IBM and the boson coherent-state framework predict sudden changes in the evolution of the transfer intensity: the gs to gs

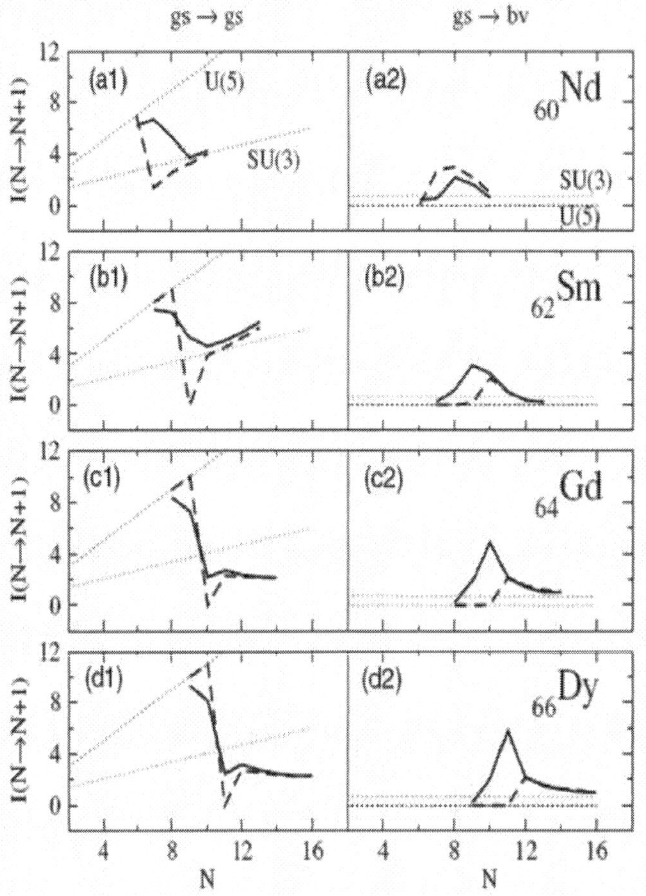

Figure 4. Two-particle transfer intensities for the $^{144-154}$Nd (with 5 proton bosons and 1 to 6 neutron bosons), $^{146-160}$Sm (p = 6 and n = 1-8), $^{148-162}$Gd (p = 7 and n= 1-8), and $^{150-166}$Dy (p = 8 and n = 1-9) isotope chains vs the total number of bosons N, for the gs to gs and gs to bv transfers.

transfer loses in strength, whereas the intensity for gs to bv and gs to dbv transfer show a peak (gs means ground state, bv means beta vibrational state and dbv means double beta vibrational state). This feature is especially present in the first-order phase transition, where in some cases the population of the ground states becomes even smaller than those to the β-vibrational 0^+ state.

References

[1] Alonso CE, Arias JM, Fortunato L and Vitturi A 2005 *Phys. Rev.* C **72** 061302
[2] Fossion R, Alonso CE, Arias JM, Fortunato L and Vitturi A 2007 *Phys. Rev.* C **76** 014316
[3] Alonso CE, Arias JM, Pietralla N, Fortunato L and Vitturi A 2008 *Phys. Rev.* C **78** 017301
[4] Iachello F 2000 *Phys. Rev. Lett.* **85** 3580
[5] Fortunato L 2005 *Eur. Phys. J.* A **26** s01 501
[6] Casten RF 2006 *Nature Physics* **2** 811
[7] Casten RF and McCutchan EA 2007 *J. Phys. G: Nucl. Part. Phys.* **34** R243
[8] Cejnar P and Jolie J 2009 *Prog. Part. Nucl. Phys.* **62** 210
[9] Iachello F 2005 *Phys. Rev. Lett.* **95** 052503
[10] Alonso CE, Arias JM and Vitturi A 2007 *Phys. Rev. Lett.* **98** 052501

Collective excitations in metallic clusters within the second random phase approximation

D Gambacurta[1,2] **and F Catara**[2,3]

[1]Consortium COMETA,
[2] Dipartimento di Fisica e Astronomia dell'Universitá di Catania
[3] Istituto Nazionale di Fisica Nucleare, Sezione di Catania, Via S.Sofia 64, I-95123 Catania, Italy

E-mail: Danilo.Gambacurta@ct.infn.it

Abstract. The second random-phase approximation (SRPA) is a natural extension of the RPA. It essentially consists in enlarging the space of the elementary modes introduced to describe the collective states by adding 2 particle - 2 hole excitations to the 1 particle - 1 hole ones of the RPA. However SRPA calculations are really few and almost all are done by resorting to strong approximations. We apply the SRPA to the study of collective states in metallic clusters. All kinds of couplings among all 1 particle - 1 hole and 2 particle - 2 hole configurations are taken into account. The coupling of 1 particle - 1 hole configurations with the 2 particle - 2 hole ones strongly pushes down the multipole strength distribution, especially for the dipole case, with respect to RPA. The inclusion of the coupling of the 2 particle - 2 hole among themselves enhances this behaviour also for the collective states whose description within RPA is widely accepted as satisfactory. Some possible origins of these unpleasant results are discussed. In particular, the need for a better treatment of short range correlations is underlined.

1. Introduction

Collective excitations are one of the most common and studied features of many-body systems. Of particular interest are the collective modes which can be interpreted in terms of vibrations. In nuclei such vibrations, both low-lying and high-lying (Giant Resonances, GR), have been known for a long time [1]. Metallic clusters show the dipole Plasmon resonance [2] which is the analogue of the nuclear dipole GR and is interpreted as due to the collective vibration of the electrons against the positive ions. The Random Phase Approximation (RPA) [3] has been extensively used as a microscopic theory to study the basic properties of these collective excitations. In both systems, RPA is able to describe the basic properties of these vibrations. However, some limitations are well known. On one hand, in the evaluation of the matrix elements appearing in the RPA equations of motion one has to resort to the Quasi Boson Approximation (QBA) introducing a clear loss of self-consistency and violations of the Pauli principle. Another limitation is that RPA predicts a perfectly harmonic spectrum with regularly spaced multiphonon states. Anharmonicities are a well established phenomenon in nuclei, both experimentally and theoretically [1, 4]. In metallic clusters no experimental evidence has been found until now for the existence of states corresponding to the double excitation of the dipole Plasmon. Theoretically such states have been predicted at energies deviating by about 10 % from the double of the Plasmon energy [5].

A natural extension of RPA, aimed to study double excitations, is the so called Second RPA (SRPA) [6, 7] where two particle-two hole ($2p-2h$) excitations, in addition to the one particle-one hole ($1p-1h$) ones, are introduced. The derivation of SRPA is also based on QBA and it has been argued [8, 9] that in this case its use is even more problematic than in RPA. This point has been recently analyzed in [10], where an approach to go beyond QBA was presented and applied to a 3 level Lipkin model. However, such analysis is limited by the fact that the model is very schematic and contains several parameters. Aim of the present paper is to analyze merits and limitations of SRPA for the study of the vibrational spectrum of a realistic many-body system. As a test laboratory we have chosen the case of metallic clusters, within the uniform jellium approximation [2] with bare Coulomb interaction both for the electrons with the jellium and for the electrons among themselves. The reason for choosing such an admittedly simplified model is that it contains many characteristics of a generic realistic many body system and the interaction does not contain any adjustable parameter.

The Paper is organized as follows. In section 2 the SRPA framework is briefly presented. In section 3 we apply it to the study of the excitation spectrum of Na metallic clusters and a comparison with RPA results is carried out. Finally, in section 4, we draw the main conclusions.

2. Formalism

In SRPA, the excited states $\mid \nu\rangle$ of the system are described as superpositions of $1p-1h$ and $2p-2h$ configurations built on top of the ground state $\mid 0\rangle$. One thus introduces the operators

$$Q_\nu^\dagger = \sum_{pi}(X^\nu_{pi} a^\dagger_p a_i - Y^\nu_{pi} a^\dagger_i a_p) + \sum_{p<m,i<j} (X^\nu_{pimj} a^\dagger_p a_i a^\dagger_m a_j - Y^\nu_{pimj} a^\dagger_i a_p a^\dagger_j a_m). \tag{1}$$

such that

$$Q_\nu^\dagger \mid 0\rangle = \mid \nu\rangle, \quad Q_\nu \mid 0\rangle = 0. \tag{2}$$

The $X's$ and $Y's$ are solutions of the equations

$$\begin{pmatrix} \mathcal{A} & \mathcal{B} \\ -\mathcal{B}^* & -\mathcal{A}^* \end{pmatrix} \begin{pmatrix} \mathcal{X}^\nu \\ \mathcal{Y}^\nu \end{pmatrix} = \omega_\nu \begin{pmatrix} \mathcal{X}^\nu \\ \mathcal{Y}^\nu \end{pmatrix}, \tag{3}$$

where

$$\mathcal{A} = \begin{pmatrix} A_{mi,pk} & A_{mi,pqkl} \\ A_{pqkl,mi} & A_{mnij,pqkl} \end{pmatrix}, \mathcal{B} = \begin{pmatrix} B_{mi,pk} & B_{mi,pqkl} \\ B_{pqkl,mi} & B_{mnij,pqkl} \end{pmatrix}, \mathcal{X}^\nu = \begin{pmatrix} X^\nu_{mi} \\ X^\nu_{mnij} \end{pmatrix}, \quad \mathcal{Y}^\nu = \begin{pmatrix} Y^\nu_{mi} \\ Y^\nu_{mnij} \end{pmatrix}.$$

The elements $A_{mi,pk}$ and $B_{mi,pk}$ of $\mathcal{A}$ and $\mathcal{B}$ are the standard RPA matrices

$$A_{pi,qj} = \langle HF \mid [a^\dagger_i a_p, [H, a^\dagger_q a_j,]] \mid HF\rangle \quad B_{pi,qj} = -\langle HF \mid [a^\dagger_i a_p, [H, a^\dagger_j a_q]] \mid HF\rangle. \tag{4}$$

while the others are

$$A_{mi,pqkl} = \langle HF|[a^\dagger_i a_m, [H, a^\dagger_p a^\dagger_q a_l a_k]]|HF\rangle, \quad A_{pqkl,mi} = A^*_{mi,pqkl}, \tag{5}$$

$$A_{mnij,pqkl} = \langle HF|[a^\dagger_i a^\dagger_j a_n a_m, [H, a^\dagger_p a^\dagger_q a_l a_k]]|HF\rangle, \tag{6}$$

$$B_{mi,pqkl} = -\langle HF|[a^\dagger_i a_m, [H, a^\dagger_k a^\dagger_l a_q a_p]]|HF\rangle, \quad B_{pqkl,mi} = B^*_{mi,pqkl} \tag{7}$$

$$B_{mnij,pqkl} = -\langle HF|[a^\dagger_i a^\dagger_j a_n a_m, [H, a^\dagger_k a^\dagger_l a_q a_p]]|HF\rangle, \tag{8}$$

Both in RPA and SRPA, since the exact ground state $\mid 0\rangle$ is not known, the HF ground state $\mid HF\rangle$ is used in the evaluation of the matrix elements appearing in the equations of motion.

This replacement is also called Quasi Boson Approximation (QBA). As a consequence of the use of the $| HF\rangle$ state in the evaluation of the SRPA matrices we obtain, in particular

$$B_{mi,pqkl} = B_{pqkl,mi} = B_{mnij,pqkl} = 0. \tag{9}$$

The matrix (5) describes the coupling of $1p-1h$ states to $2p-2h$ states, while matrix (6) takes into account the coupling between $2p-2h$ states themselves. The dimension of these matrices, especially of the latter, can be very large. If we neglect the residual interaction among the $2p-2h$ states, the matrix (6) acquires a simple form,

$$A_{mnij,pqkl} = U(ij)U(mn)\delta_{ik}\delta_{jl}\delta_{mp}\delta_{nq}(\epsilon_m + \epsilon_n - \epsilon_i - \epsilon_j) \tag{10}$$

where $U(ij)$ is the antisymmetrizer for the indices i, j and the ϵ quantities are the HF single particle energies. In this case, the SRPA problem can be reduced to an RPA eigenvalue problem, (whose dimensions are determined by the $1p-1h$ space), but where the A matrix (4) depends now on the excitation energies ω [7].

3. Results

In the following, we discuss some results obtained by applying the SRPA to the study of collective excitations in metallic clusters. The ionic background is described within the jellium approximation [2]. The reason for this choice is that the interaction of the electrons among themselves and with the ionic background is completely fixed as the bare Coulomb interaction and the model does not contain any adjustable parameter. This allows to compare in a clear way different degrees of approximation.

As a first step, we have fixed, by solving the HF equations, the single particle basis in which all the subsequent calculations are carried out. After that, we have solved the RPA and SRPA equations for natural parity states, with multipolarities ranging from $L=0$ to $L=3$. The single particle space has been truncated so that the Thouless theorem, and thus the preservation of EWSR, is satisfied in the RPA calculations better than 1 % for all the multipolarities. We focus our attention only on spin $S=0$ states. However we remark that, for what concerns the SRPA calculations, all the $1p-1h$ states with $L=0-3$, both spin $S=0$ and 1, are considered in the construction of the $2p-2h$ configurations with good total spin and angular momentum. We show the results for two sodium clusters, namely the Na_9^+ and Na_{21}^+. In the latter case an energy cutoff E_{cut} has been used and only the $2p-2h$ configurations with unperturbed energy lower than $E_{cut} = 15eV$ are included. In order to check that the configuration space is large enough, we have verified that the EWSR are numerically satisfied, (better than 1 % in the present calculations), for all the considered multipolarities. In figure 1 we plot the strength distributions for the multipole operator $F^{(\lambda)}(r) = r^\lambda Y_{\lambda 0}$ for the Na_9^+ metallic cluster. In order to make clearer the comparison, the discrete lines of RPA and SRPA spectra are folded with a Lorentzian function. In both cases an artificial width $\Gamma = 0.1eV$ has been used. In the upper and lower panel of figure 1, we show, respectively, the $L=0,1$ and $L=2,3$ multipole strength distributions with spin $S=0$ for Na_9^+ metallic cluster. Solid (red) lines refer to SRPA results, while the dashed (black) ones refer to those obtained within RPA. With the dotted (blue) lines we indicate the SRPA results when the diagonal approximation (10) is used. In the following figures we use the label "SRPA-A22D" for the results obtained when this diagonal approximation is used. We see that going from RPA to SRPA the strength distributions are shifted to lower energies, especially in the dipole case. The results shown in figure 1 refer to the lower part of the excitation spectrum, which is mainly composed by $1p-1h$ configurations. However, even for such states the inclusion of the coupling among $2p-2h$ configurations leads to a further shift down. Let us look in some detail to the collective dipole $S=0$ state, which in RPA is located at 2.98 eV, and that experimentally is found at about 2.60 eV [11]. Within the approximated

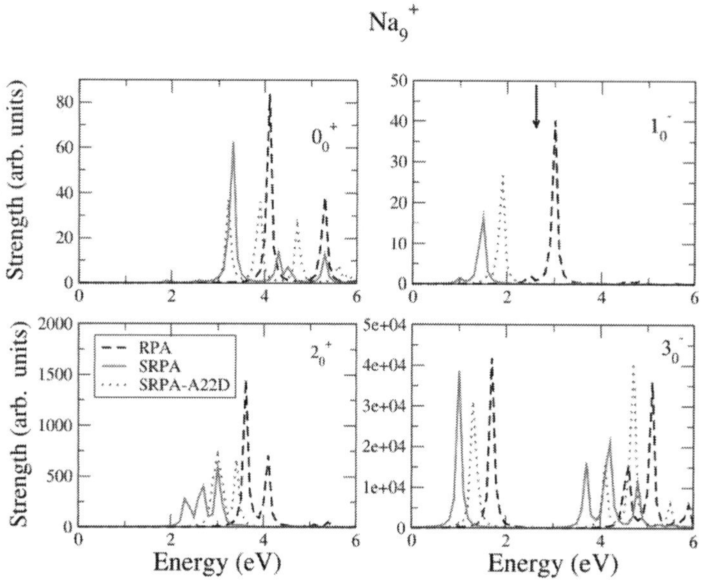

Figure 1. (Color online) Natural parity $L = 0, 1$ (upper panel) and $L = 2, 3$ (lower panel) spin=0 multipole strength distributions for Na_9^+ metallic cluster are shown. Solid (red) lines refer to SRPA results, while the dashed (black) ones refer to RPA calculations. SRPA-A22D dotted (blue) lines show the SRPA results when the approximation (10) is used. The arrow roughly indicates the positions of the experimental dipole plasmon peak.

SRPA, that is when only the coupling of $1p-1h$ to $2p-2h$ configurations is taken into account, the energy moves to 1.88 eV, while in the full SRPA one gets 1.47 eV. As shown in figure 2, qualitatively similar results have been found for Na_{21}^+. Also in this case we have a strong shift down, especially for the dipole excitations, the plasmon energy being lowered from the RPA value of 2.90 eV to the full SRPA result of 1.00 eV very far from the experimental peak at about 2.65 eV [11]. Also in the nuclear [12] case one finds that the SRPA excitations spectrum is very much different from the RPA one, even for excitations like the Giant Resonances whose description within RPA is widely accepted as satisfactory. Probably in all cases, the problems arise from the fact that the self-energy-like terms present in SRPA take contributions from very high energy "incoherent" particle -hole configurations. In addition to that, we stress again that the derivation of the SRPA equations of motion is based on QBA and this might be more doubtful than in RPA. Indeed, a better treatment of ground state correlations has been found to improve the results, within a 3 level Lipkin model [10]. The replacement of the correlated ground state with the HF one is justified only if ground state correlations are not too strong. In RPA the Y_{ph}^{ν} amplitudes are a measure of these correlations [3] and thus, looking at their behaviour going from RPA to SRPA could be useful in order to have more information about the adequacy of QBA. In table 1 we report, in the case of Na_9^+ and for different multipolarities, the sum of the squares of the Y_{ph}^{ν} amplitudes in RPA (first column), in SRPA (third column) and in SRPA when the diagonal approximation (10) is used (second column). We see that going from RPA to SRPA an increase of these quantities is observed and the inclusion of the residual interaction coupling the $2p-2h$ configurations among themselves amplifies this effect. Similar results have been found in the case of Na_{21}^+.

Journal of Physics: Conference Series **168** (2009) 012012 doi:10.1088/1742-6596/168/1/012012

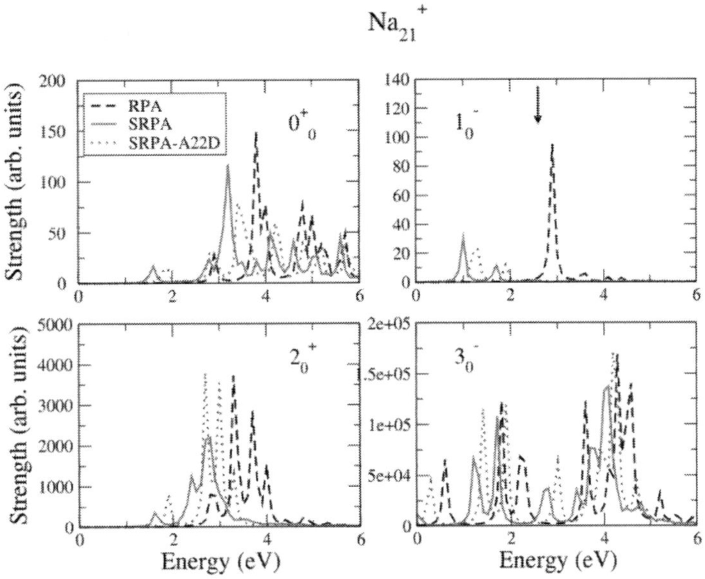

Figure 2. (Color online) As in figure 1 but for Na_{21}^{+} metallic cluster.

Table 1. The sum of the squares of the Y_{ph}^{ν} amplitudes in the case of Na_{9}^{+} and for different multipolarities, in RPA (first column), in SRPA (third column) and in SRPA when the diagonal approximation (10) is used (second column).

	$\sum_{ph,\nu} \mid Y_{ph}^{\nu} \mid^2$		
	RPA	SRPA-A22D	SRPA
L=0,S=0	0.0458	0.0512	0.0532
L=1,S=0	0.1601	0.2962	0.4170
L=2,S=0	0.0256	0.0316	0.0347
L=3,S=0	0.0198	0.0282	0.0388

The RPA occupation numbers can be calculated by using, for example, the number operator method [13] and one gets:

$$n_h = 1 - \frac{1}{2}\sum_{p,\nu} \mid Y_{ph}^{\nu} \mid^2, \qquad n_p = \frac{1}{2}\sum_{h,\nu} \mid Y_{ph}^{\nu} \mid^2 . \tag{11}$$

To our knowledge, the explicit form of the SRPA occupation numbers. Nevertheless, it could be instructive to compare the values of quantities (11) when the Y_{ph}^{ν} obtained in RPA and SRPA are used. In this sense we will refer in the following to SRPA occupation numbers. In figure 3 we show, in the case of the Na_{21}^{+} metallic cluster, the occupation numbers n_p for particle states and the opposite of the depletion numbers $n_h - 1$ for hole states obtained in SRPA and in RPA,

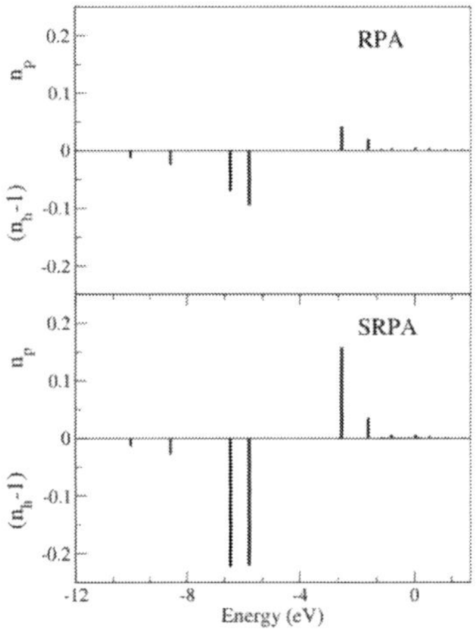

Figure 3. Occupation numbers n_p for particle states and the opposite of depletion numbers $n_h - 1$ for hole states for Na_{21}^+ metallic cluster. RPA and SRPA results are reported in the upper and lower panel, respectively. Only Spin S=0 states are included in the calculations. In the abscissa the single-particle energies in eV are indicated.

in the lower and upper panel, respectively. We note that the deviations from the HF limit, i.e. $n_h = 1$ and $n_p = 0$, are greater in SRPA than in RPA. Since, as mentioned above, the use of QBA is justified only when the HF state does not differ very much from the correlated one, the larger deviations found in SRPA suggest that this approximation could be more severe than in RPA.

As discussed at end of section 2, the SRPA problem can be reduced to an energy-dependent RPA problem. In this case, the new RPA $\tilde{A}$ matrix is

$$\tilde{A}_{1,1'}(\omega) = A_{1,1'} + \Sigma_{1,1'}(\omega) \tag{12}$$

where A is the usual RPA matrix, Σ is the energy dependent term

$$\Sigma_{1,1'}(\omega) = \sum_{2,2'} A_{1,2}(\omega + i\eta - A_{2,2'})^{-1} A_{2',1'} \tag{13}$$

and the indices 1 and 2 are a short-hand notation for the $1p - 1h$ and $2p - 2h$ configurations, respectively. The energy dependent term is connected to the self-energy of the $1p - 1h$ excitation and it is due to the coupling with the $2p - 2h$ configurations. It has been shown that its real part gives a shift of the RPA resonance energies while the imaginary part takes into account spreading width effects [14, 15]. Several SRPA calculations have been done by neglecting the real part of the particle-hole self-energy (see for example [7, 16]) and considering only the spreading width. So it is interesting to study the energy shift due to the real part of the self energy in this case. In order to have a quantitative, though approximate, evaluation of this effect we consider the self-energy acquired by a RPA phonon as a consequence of the coupling with the $2p - 2h$ configurations [15]

$$\Sigma^{RPA}(\omega_\nu) = \sum_{2p2h} \frac{|<\nu|V|2p2h>|^2}{\omega_\nu - \epsilon_{2p2h} + i\eta} \tag{14}$$

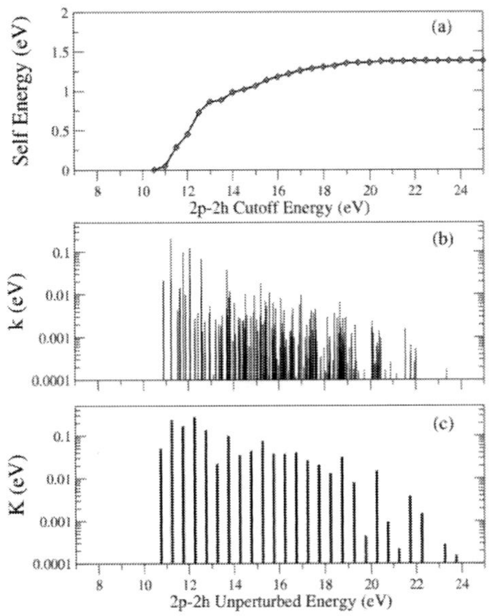

Figure 4. Panel (a): the absolute value of the real part of the RPA self-energy defined in Eq. (14) as a function of the cutoff energy of the $2p-2h$ configurations. Panel (b) and (c): the absolute value of the real part of the k and K quantities defined in Eqs. (15) and (16), respectively. The abscissa of both the figures shows the unperturbed energies of the $2p-2h$ configurations.

where ω_ν is the excitation energy of the RPA phonon $|\nu>$, V is the residual interaction and ϵ_{2p2h} is the unperturbed energy of the $2p-2h$ configuration $|2p2h>= a^\dagger_{p_2}a_{h_2}a^\dagger_{p_1}a_{h_1} \mid HF\rangle$. We have calculated the real part of this quantity for the dipole plasmon in the Na_9^+ case. The η parameter used in the calculation is $0.2eV$. As mentioned above, the RPA dipole plasmon ω_ν is located at 2.98 eV. Since the unperturbed $2p-2h$ energies ϵ_{2p2h} are larger than this value, the denominator in Eq. (14) is always negative and thus the resonance peak is shifted downwards with respect to ω_ν. In the upper panel of figure 4 we plot the absolute value of the real part of the RPA self-energy as a function of a cutoff energy E_c on the $2p-2h$ configurations, i.e. only the $2p-2h$ configurations with unperturbed energy lower than E_c are considered in Eq. (14). We see that increasing the cutoff E_c the energy shift increases until it reaches a saturation value of about 1.5 eV. We remark that this shift is consistent with the one obtained in the SRPA calculation (see figure 1). In order to have a deeper insight, we show in the medium panel of figure 4 the absolute value of the real part of the quantities

$$k(\epsilon_{2p2h}) = \frac{|<\nu|V|2p2h>|^2}{\omega_\nu - \epsilon_{2p2h} + i\eta} \tag{15}$$

i.e. the contribution to the self-energy of each $2p-2h$ configuration identified by its unperturbed energy ϵ_{2p2h}. We see that, although the largest contributions are given by the lower $2p-2h$ configurations, also the ones with higher energy contribute and the overall effect is quite large. In order to put in evidence in a clearer way this fact, we have divided the unperturbed $2p-2h$ energy range in bins of 0.5 eV and we have summed all the terms defined in Eq. (15) lying in each interval identified by the index I. In the lower panel of figure 4, where we plot the the absolute value of the real part of the quantities

$$K(I) = \sum_{\epsilon_{2p2h} \in I} k(\epsilon_{2p2h}) \qquad (16)$$

we see that the total contributions of high energy configurations, arising from a coherent sum of many small terms, become comparable with the ones of the lowest energies. Similar results are obtained in the case of Na_{21}^{+}. By the above shown analysis of the Y amplitudes and of the occupation numbers, we can conclude that a better treatment of these correlations should be necessary in SRPA.

4. Conclusions

In the present paper we have applied the SRPA to the study of collective excitations in metallic clusters. We have found strong modifications of the RPA excitation spectrum, the energy being pushed down. When the coupling of $2p-2h$ configurations among themselves is included, a further lowering is obtained even for those collective excitations whose description within RPA is widely assumed as satisfactory. Such strong modifications could be traced back to the use of QBA in calculating the expectation values appearing in the equations of motion of SRPA. although the same approximation is used in RPA, it seems more severe in SRPA.

These results suggest that a better treatment of ground state correlations along the path of extended RPA and SRPA approaches, as done for example in [10, 17], is necessary. Work in this direction is in progress. It is also interesting to analyze how much such strong modifications with respect to RPA description depend on the studied system. With this respect, in the near future, we plan to apply SRPA in nuclear systems by using effective Skyrme-type interactions.

5. Acknowledgments

This work makes use of results produced by the PI2S2 Project managed by the Consorzio COMETA, a project co-funded by the Italian Ministry of University and Research (MIUR) within the Piano Operativo Nazionale Ricerca Scientifica, Sviluppo Tecnologico, Alta Formazione (PON 2000-2006). More information is available at http://www.pi2s2.it and http://www.consorzio-cometa.it.

References

[1] Bohr A and Mottelson B R 1975 *Nuclear Structure* vol II (Benjamin: New York)
[2] Brack M 1993 *Rev. Mod. Phys.* **65** 677
[3] Ring P and Schuck P 1980 *The Nuclear Many-Body Problem* (Springer-Verlag: Berlin)
[4] Chomaz P and Frascaria N 1995 *Phys. Rep.* **252** 275
[5] Gerchikov L G *et al.* 2002 *Phys. Rev.* A **66** 053202; Catara F *et al.* Phys. Lett. 2006 A **349** 345
[6] Yannouleas C 1987 *Phys. Rev.* C **35** 1159
[7] Drozdz G, Nishizaki S, Speth J and Wambach J 1990 *Phys. Rep.* **197** 1
[8] Takayanagi K, Shimizu K and Arima A 1988 *Nucl. Phys.* A **477** 205
[9] Lauritsch G and Reinhard P G 1990 *Nucl. Phys.* A **509** 287
[10] Gambacurta D, Grasso M, Catara F and Sambataro M 2006 *Phys. Rev.* C **73** 024319
[11] Borggreen J *et al.* 1993 *Phys. Rev.* B **48** 17507
[12] Papakonstantinou P and Roth R *Preprint* arXiv:0805.4086 [nucl-th]
[13] Rowe D J 1968 *Phys. Rev.* **175** 1283
[14] Wambach J 1988 *Rep. Prog. Phys.* **51** 989
[15] Lacroix D, Ayik S and Chomaz P 2004 *Prog. Part. Nucl. Phys.* **52** 497
[16] Ait-Tahar S and Brink D M 1993 *Nucl. Phys.* A **560** 765
[17] Gambacurta D and Catara F 2008 *Phys. Rev.* B **77** 205434 and 2009 *Phys. Rev.* B **79**, 085403

Journal of Physics: Conference Series **168** (2009) 012013 doi:10.1088/1742-6596/168/1/012013

Shell-model calculations in ^{132}Sn and ^{208}Pb regions with low-momentum interactions

A Gargano[1], L Coraggio[1], A Covello[1,2] and N Itaco[1,2]

[1]Istituto Nazionale di Fisica Nucleare,
Complesso Universitario di Monte S. Angelo, I-80126 Napoli
[2]Dipartimento di Scienze Fisiche, Università di Napoli Federico II,
Complesso Universitario di Monte S. Angelo, I-80126 Napoli

E-mail: gargano@na.infn.it

Abstract.
We discuss shell-model calculations based on the use of low-momentum interactions derived from the free-space nucleon-nucleon potential. A main feature of this approach is the construction of a smooth potential, $V_{\rm low-k}$, defined within a given momentum cutoff. As a practical application of the theoretical framework, we present some selected results of our current study of nuclei around doubly magic ^{132}Sn and ^{208}Pb which have been obtained starting from the CD-Bonn potential. Focusing attention on the similarity between the spectroscopy of these two regions, we show that it emerges quite naturally from our effective interactions without use of any adjustable parameter.

1. Introduction

In the last decade, shell-model calculations employing realistic effective interactions derived from modern nucleon-nucleon (NN) potentials have entered the main stream of nuclear structure theory [1]. As is well known, the first problem one is confronted with in this kind of calculations is the strong short-range repulsion contained in the bare NN potential V_{NN}, which prevents its direct use in the derivation of the shell-model effective interaction $V_{\rm eff}$. The most popular way to overcome this difficulty has long been the Brueckner G-matrix method. However, a few years ago a new approach [2] was proposed which consists in deriving from V_{NN} a low-momentum potential, $V_{\rm low-k}$, that preserves the deuteron binding energy and scattering phase shifts of V_{NN} up to a certain cutoff momentum Λ. This is a smooth potential which can be used directly to derive $V_{\rm eff}$, and it has been shown [2, 3] that it provides an advantageous alternative to the use of the G matrix. In this connection, it should be mentioned that $V_{\rm low-k}$ potentials are currently being used in various nuclear theory contexts, such as the study of few-body systems and no-core shell-model calculations [4, 5].

Making use of the $V_{\rm low-k}$ approach, we have recently studied [6, 7, 8, 9] several nuclei beyond doubly magic ^{132}Sn, showing that their properties are well accounted for by a unique shell-model Hamiltonian with single-particle energies taken from experiment and two-body effective interaction derived from the CD-Bonn NN potential [10].

Motivated by the very good results obtained in the ^{132}Sn region and by the existence of a close resemblance [11, 12, 13, 14] between the spectroscopy of this region and that of nuclei around stable ^{208}Pb, we have found it challenging to perform a comparative study of these two regions [15]. In this paper we present some results from this study, focusing attention on the proton-proton, neutron-neutron and proton-neutron multiplets in the three far-from-stability nuclei ^{134}Te, ^{134}Sn, ^{134}Sb and in their counterparts in the ^{208}Pb region, ^{210}Po, ^{210}Pb, and ^{210}Bi.

We start by giving an outline of the theoretical framework in which our shell-model calculations are performed and then present and discuss our results. A short summary is given in the last section.

2. Theoretical framework

In the framework of the shell model an auxiliary one-body potential U is introduced in order to break up the nuclear Hamiltonian, written as the sum of the kinetic term and the NN potential, into a one-body component H_0, which describes the independent motion of the nucleons, and a residual interaction H_1. Namely,

$$H = \sum_{i=1}^{A} \frac{p_i^2}{2m} + \sum_{i<j} V_{ij} = T + V_{NN} = (T + U) + (V_{NN} - U) = H_0 + H_1. \tag{1}$$

A reduced model space is then defined in terms of the eigenvectors of H_0 and the diagonalization of the original Hamiltonian H in an infinite Hilbert space is reduced to the solution of an eigenvalue problem for an effective Hamiltonian $H_{\rm eff}$ in a finite space.

The Hamiltonian $H_{\rm eff}$ can be derived by way of the $\hat{Q}$-box folded-diagram expansion (see Ref. [1]). This implies as first step the calculation of the so-called $\hat{Q}$-box, which is made up of an infinite collection of irreducible and valence-linked Goldstone diagrams. Once the $\hat{Q}$-box has been calculated at a given order, the infinite series of the folded diagrams has to be summed up. From this procedure an effective Hamiltonian is obtained containing both one- and two-body components. Usually, only the two-body term $V_{\rm eff}$ is retained, while the one-body contributions, representing the theoretical single-particle energies, are subtracted and replaced with single-particle energies taken from experiment [16].

However, as mentioned in the Introduction no modern NN potential can be used in a perturbative nuclear structure calculation, unless its strong repulsive core is firstly "smoothed out". Here, we do not embark on any discussion of how the effective Hamiltonian is derived, and refer to [1], where a detailed description of the whole procedure can be found. Rather, in the following we focus on the $V_{\rm low-k}$ approach. We first outline the essential steps for the derivation of the $V_{\rm low-k}$ based on the Lee-Suzuki similarity transformation method [17], and then describe its main features.

Let us consider the similarity transformation on the Hamiltonian (1)

$$\mathcal{H} = X^{-1} H X, \tag{2}$$

where the operator X is defined in the whole Hilbert space. We now introduce a cutoff momentum Λ that separates fast and slow modes to the end of deriving from the original V_{NN} a low-momentum potential satisfying a decoupling condition between the low- and high-momentum spaces.

The low-momentum space is specified by

$$P = \int d\boldsymbol{p} \mid \boldsymbol{p}\rangle\langle\boldsymbol{p}\mid \;, \quad p \leq \Lambda \tag{3}$$

where $\boldsymbol{p}$ is the two-nucleon relative momentum, and the decoupling equation reads

$$Q\mathcal{H}P = 0, \tag{4}$$

with $Q = 1 - P$ being the complementary fast-mode space. The low-momentum Hamiltonian is then given by

$$H_{\rm low-k} = P\mathcal{H}P, \tag{5}$$

and it can be easily proved that its eigenvalues are a subset of the eigenvalues of the original Hamiltonian.

There are, of course, different choices for the transformation operator X. We take

$$X = e^{\omega}, \tag{6}$$

where the wave operator ω satisfies the conditions:

$$\omega = Q\omega P, \tag{7}$$

$$P\omega P = Q\omega Q = P\omega Q = 0, \tag{8}$$

the former implying that

$$X = 1 + \omega. \tag{9}$$

From Eq. (5) the low-momentum potential $V_{\rm low-k}$ can be defined as

$$V_{\rm low-k} = H_{\rm low-k} - PTP. \tag{10}$$

Employing transformation (9), this equation is written as

$$V_{\rm low-k} = PV_{NN}P + PV_{NN}Q\omega, \tag{11}$$

while Eq. (4) becomes

$$QV_{NN}P + QHQ\omega - \omega PHP - \omega PV_{NN}Q\omega = 0. \tag{12}$$

The solution of Eq. (12) gives the value of ω needed to obtain $V_{\rm low-k}$.

This decoupling equation can be solved by means of the iterative technique for non-degenerate model spaces proposed in [18], which is now sketched. We define the operators:

$$\begin{aligned} p(\omega) &= PHP + PHQ\omega, & (13)\\ q(\omega) &= QHQ - \omega PHQ, & (14) \end{aligned}$$

in terms of which one can write

$$\begin{aligned} x_0 &= -(QHQ)^{-1}QHP \ , \\ x_1 &= q(x_0)^{-1}x_0 p(x_0) \ , \\ &\ldots \\ x_n &= q(x_0 + x_1 + ... + x_{n-1})^{-1}x_{n-1}p(x_0 + x_1 + ... + x_{n-1}) \ . \end{aligned} \tag{15}$$

Once the iterative procedure has converged, $x_n \to 0$, the operator ω is given by

$$\omega_n = \sum_{i=0}^{n} x_i, \tag{16}$$

In applying this method, we have employed a momentum-space discretization procedure making use of an adequate number of Gaussian mesh points [19].

It is worth mentioning that the equation for $V_{\rm low-k}$ obtained using the similarity transformation of Lee and Suzuki is the same as that one can derive from the T-matrix equivalence approach [1, 2]. This means that the obtained low-momentum potential not only preserves the deuteron binding energy given by the original NN potential, but also its low-momentum ($\leq \Lambda$) half-on-shell T matrix.

The above $V_{\rm low-k}$ is however not Hermitian, which is not convenient for various applications, as for instance its use in the derivation of shell-model effective interactions. This $V_{\rm low-k}$ may be transformed by means of the familiar Schmidt orthogonalization procedure, which leads to a Hermitian new $V_{\rm low-k}$. As suggested in [18], another transformation, based on the Cholesky decomposition of a symmetric and positive definite matrix, can be used to this end. In fact, the matrix $P(1+\omega^+\omega)P$, being symmetric and positive definite, admits this decomposition,

$$P(1+\omega^+\omega)P = PLL^TP, \tag{17}$$

where L is a lower triangular matrix and L^T its transpose. Since L is real matrix defined within the P-space, we may write our transformation as

$$Z = L^T, \tag{18}$$

and the corresponding Hermitian $V_{\rm low-k}$ is

$$V^{\rm chol}_{\rm low-k} = PL^TP(H_0 + V_{\rm low-k})P(L^{-1})^TP - PH_0P. \tag{19}$$

In Refs. [20, 1], it has been shown that this Hermitian interaction, as well all the family of Hermitian interactions which can derived from $V_{\rm low-k}$ by means of different transformations, preserve the full-on-shell T matrix, and consequently the phase shifts of the original V_{NN}.

The so-obtained $V_{\rm low-k}$ is a smooth potential that can be used directly within the $\hat{Q}$-box folded diagram theory to derive the shell-model effective interaction. Actually, it represents an advantageous alternative to the G-matrix approach owing to the fact that it does not depend either on the energy or on the model space. This is at variance with the G matrix, which is defined in the nuclear medium. In this connection, it is worth mentioning that the merit of the $V_{\rm low-k}$ within the context of realistic shell-model calculations has been assessed by several studies evidencing that $V_{\rm low-k}$ results are as good, or even slightly better than, the G-matrix ones [2, 3].

Finally, it is a remarkable feature of the $V_{\rm low-k}$ approach that different NN potentials lead to low-momentum potentials, which are quite similar to each other [1, 21].

3. Two-valence particle nuclei in ^{132}Sn and ^{208}Pb regions

3.1. Outline of calculations

In this paper, we present some results of our current shell-model study of nuclei with two valence nucleons in the ^{132}Sn and ^{208}Pb regions, which have been obtained starting from the CD-Bonn potential renormalized by use of a $V_{\rm low-k}$ with a cutoff momentum of $\Lambda = 2.2$ fm^{-1}.

In our calculations for ^{134}Te, ^{134}Sn, and ^{134}Sb we assume that the the valence protons occupy the five levels $0g_{7/2}$, $1d_{5/2}$, $1d_{3/2}$, $2s_{1/2}$, and $0h_{11/2}$ of the $50-82$ shell, while for the neutrons the model space includes the six levels $0h_{9/2}$, $1f_{7/2}$, $1f_{5/2}$, $2p_{3/2}$, $2p_{1/2}$, and $0i_{13/2}$ of the $82-126$ shell. Similarly, for ^{210}Po, ^{210}Pb, and ^{210}Bi we take as model space for the valence protons the six levels of the $82-126$ shell and let the valence neutrons occupy the seven levels $1g_{9/2}$, $0i_{11/2}$, $0j_{15/2}$, $2d_{5/2}$, $3s_{1/2}$, $1g_{7/2}$ and $2d_{3/2}$ of the $126-184$ shell. As regards the adopted single-particle neutron and proton energies, they can be found in Refs. [6] and [22] for ^{132}Sn and ^{208}Pb, respectively.

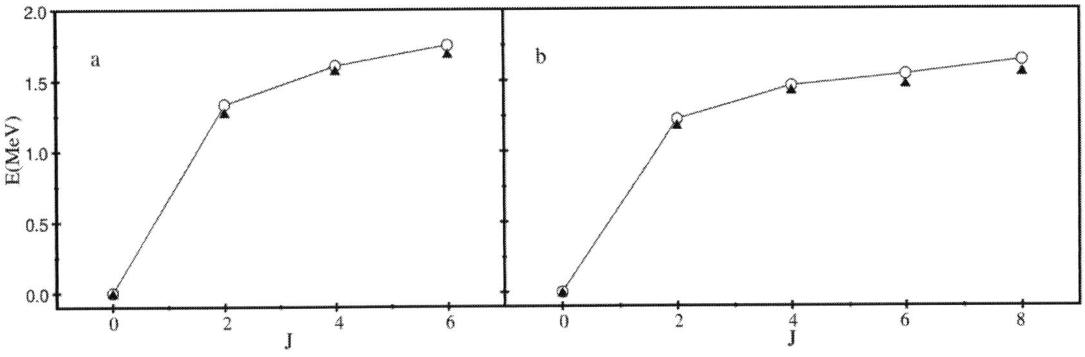

Figure 1. (a) Proton-proton $(\nu f_{7/2})^2$ multiplet in ^{134}Te.(b) Proton-proton $(\nu g_{9/2})^2$ multiplet in ^{210}Po. The theoretical results are represented by open circles while the experimental data by solid triangles.

As mentioned in section 2, the two-body matrix elements of the effective interaction are derived within the framework of the $\hat{Q}$-box folded-diagram expansion. We include in the $\hat{Q}$-box all diagrams up to second order in the interaction, given by the $V_{\text{low-k}}$ potential plus the Coulomb force for protons. These diagrams are computed within the harmonic-oscillator basis using intermediate states composed of all possible hole states and particle states restricted to the five proton and neutron shells above the Fermi surface. The oscillator parameter is 7.88 MeV for $A = 132$ region and 6.88 MeV for the $A = 208$ region, as obtained from the expression $\hbar\omega = 45A^{-1/3} - 25A^{-2/3}$. The calculations have been performed by using the NUSHELLX code [23].

3.2. Results

In figures 1, 2, and 3, we present the experimental [24, 25] and calculated excitation energies of the lowest states in the nuclei with two-proton, two-neutron, and one proton-one neutron beyond doubly magic ^{132}Sn and ^{208}Pb. We see that the agreement between theory and experiment is very good for all six nuclei considered, the discrepancies being well below 100 keV for most of the states.

All the calculated states reported in these figures are dominated by a single configuration, whose percentage ranges from 80% to 100%. In particular, they correspond to the proton-proton multiplets $(\pi g_{7/2})^2$ and $(\pi h_{9/2})^2$ in ^{134}Te and ^{210}Po, to the neutron-neutron multiplets $(\nu f_{7/2})^2$ and $(\nu g_{9/2})^2$ in ^{134}Sn and ^{210}Pb, and to the proton-neutron multiplets $\pi g_{7/2}\nu f_{7/2}$ and $\pi h_{9/2}\nu g_{9/2}$ in ^{134}Sb and ^{210}Bi.

These figures evidence the striking resemblance between the behavior of the multiplets in the three pairs of counterpart nuclei, ^{134}Te and ^{210}Po, ^{134}Sn and ^{210}Pb, ^{134}Sb and ^{210}Bi. We may only note that the curves relative to the ^{208}Pb neighbors are generally located slightly below those for the counterpart nuclei in ^{132}Sn region. In particular, from figures 3a and 3b we see that the two proton-neutron multiplets show a sizable energy gap between the 2^- state and the nearly degenerate 0^- and 1^- states, as well as a distinctive staggering, with the same magnitude and phase, between the odd and the even members starting from the 3^- state. As regards the two-identical-particle multiplets, figures 1 and 2 show four curves having all the same shape. It is worth noting, however, that the curves for the two-valence-proton nuclei are located in an energy interval larger than that pertaining to the two-valence-neutron nuclei. This means that in both ^{132}Sn and ^{208}Pb regions a weakening of the pairing gap exists for nuclei with two-valence neutrons with respect to those with two-valence protons.

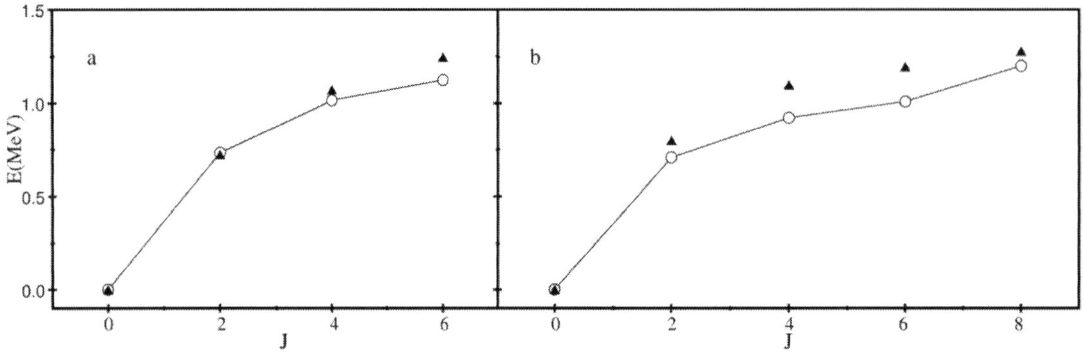

Figure 2. (a) Neutron-neutron $(\pi g_{7/2})^2$ multiplet in ^{134}Sn.(b) Neutron-neutron $(\pi h_{9/2})^2$ multiplet in ^{210}Pb. The theoretical results are represented by open circles while the experimental data by solid triangles.

It is worth mentioning that the resemblance between ^{132}Sn and ^{208}Pb regions was first pointed out by Blomqvist[11], who noticed that every ^{132}Sn single-proton and -neutron level, characterized by quantum numbers (nlj), has its counterpart around ^{208}Pb with quantum numbers $(nl+1j+1)$. However, until recent years the scarcity of information for nuclei around ^{132}Sn, which lies well away from the stability line, has prevented a detailed comparative study of the two regions. Nowadays, new data have become available which support the similarity between their spectroscopies. In our calculation, this similarity emerges quite naturally from our effective interaction which we have derived from a realistic NN potential.

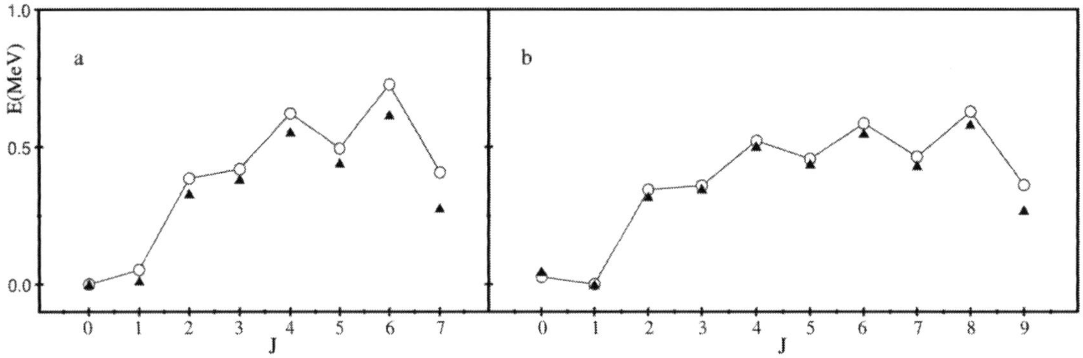

Figure 3. (a) Proton-neutron $\pi g_{7/2}\nu f_{7/2}$ multiplet in ^{134}Sb.(b) Proton-neutron $\pi h_{9/2}\nu g_{9/2}$ multiplet in ^{210}Bi. The theoretical results are represented by open circles while the experimental data by solid triangles.

4. Summary

We have briefly discussed here the theoretical framework for realistic shell-model calculations wherein use is made of low-momentum interactions derived from the free NN potential. We have shown how a smooth low-momentum potential $V_{\rm low-k}$ can be constructed, which preserves the deuteron binding energy and scattering phase shifts of the original V_{NN} up to a given momentum cutoff. We have then presented the results of a shell-model study of nuclei around doubly magic

^{132}Sn and ^{208}Pb, focusing attention on proton-proton, neutron-neutron and proton-neutron multiplets. The results obtained for the three nuclei ^{134}Te, ^{134}Sn and ^{134}Sb have been compared with those for ^{210}Po, ^{210}Pb and ^{210}Bi, which are their counterparts in the region of ^{208}Pb. In both cases, a low-momentum effective interaction derived from the CD-Bonn NN potential has been employed. It should be stressed that no adjustable parameter appears in our calculations.

Our results for all six nuclei are in very good agreement with the experimental data and account for the striking resemblance between the behavior of the multiplets in the ^{132}Sn and ^{208}Pb regions. This stimulates further studies to find out whether this resemblance extends beyond the two-valence-particle nuclei.

References

[1] Coraggio L, Covello A, Gargano A, Itaco N and Kuo T T S 2009 *Prog. Part. Nucl. Phys.* **62** 135, and references therein
[2] Bogner S, Kuo T T S, Coraggio L, Covello A and Itaco N 2002 *Phys. Rev.* C **65** 051301(R)
[3] Covello A 2003 *Proc. Int. School of Physics "E. Fermi", Course CLIII*, ed A Molinari *et al* (Amsterdam:IOS Press) pp. 79–91
[4] Deltuva A, Fonseca A C and Bogner S K 2008 *Phys. Rev.* C **77** 024002, and references therein
[5] Bogner S K, Furnstahl R J, Maris P, Perry R J, Schwenk A and Vary J P 2008 *Nucl. Phys.* A **801** 21
[6] Coraggio L, Covello A, Gargano A and Itaco N 2005 *Phys. Rev.* C **72** 057302
[7] Coraggio L, Covello A, Gargano A and Itaco N 2006 *Phys. Rev.* C **73** 031302(R)
[8] Covello A, Coraggio L, Gargano A and Itaco N 2007 *Eur. Phys. J.* ST **150** 93
[9] Simpson G S, Angelique J C, Genevey J, Pinston J A, Covello A, Gargano A, Köster U, Orlandi R and Scherillo A 2007 *Phys. Rev.* C **76** 041303(R)
[10] Machleidt R 2001 *Phys. Rev.* C **63** 024001
[11] Blomqvist J 1981 *Proc. of the 4th International Conference on Nuclei Far from Stability* CERN Report 81-09 (Geneva: CERN) p. 536.
[12] Fornal B *et al* 2001 *Phys. Rev.* C **63** 024322
[13] Isakov V I *et al* 2006 *Phys. At. Nucl.* **70** 818
[14] Korgul A *et al* 2007 *Eur. Phys. J.* A **32** 25
[15] Covello A, Coraggio L, Gargano A and Itaco N 2009 *Acta Phys. Pol.* **40** 401
[16] Shurpin J, Kuo T T S and Strottman D 1983 *Nucl. Phys.* A **408** 310
[17] Suzuki K and Lee S Y 1980 *Prog. Theor. Phys.* **64** 2091
[18] Andreozzi F 1996 *Phys. Rev.* C **54** 684
[19] Krenciglowa E M, Kung C I and Kuo T T S 1976 *Ann. Phys.* **101** 154
[20] Holt J D, Kuo T T S and Brown G E 2004 *Phys. Rev.* C **69** 034329
[21] Bogner S K, Kuo T T S and Schwenk A 2003 *Phys. Rep* **386** 1
[22] Coraggio L, Covello A, Gargano A and Itaco N 2006 *Phys. Rev.* C **76** 061303(R)
[23] NushellX for Linux, produced by William D M Rae, Garsington, Oxford, 2007/08, http://knollhouse.org
[24] Data extracted using the NNDC On-line Data Service from the ENSDF database, version of March 4, 2009
[25] Shergur J *et al* 2005 *Phys. Rev.* C **71** 064321

Pygmy resonances in Sn isotopes within a microscopic multiphonon approach

E G Lanza[1], M V Andrés[2], F Catara[1], Ph Chomaz[3] and D Gambacurta[1]

[1] I.N.F.N.-Catania and Dipartimento di Fisica e Astronomia, Universitá di Catania, Via S. Sofia 67, I-95123 Catania, Italy

[2] Departamento de Física Atomica, Molecular y Nuclear, Universidad de Sevilla, Apdo 1065, E-41080 Sevilla, Spain

[3] GANIL (DSM-CEA/IN2P3-CNRS), B.P. 55027, F-14076 Caen Cédex 5, France

E-mail: edoardo.lanza@ct.infn.it

Abstract. We study the pygmy resonances in Sn isotopes within a microscopic multiphonon approach which has been successfully applied to heavy ion reactions in recent years. In the energy region of the pygmy resonances there are a few low lying multiphonon states. The question is whether they may contribute to the observed peak. Calculations show that the inelastic cross sections in the relevant energy region have a conisderable increase depending on the isotope and on the kind of Skyrme force used.

1. Introduction

Studies of exotic modes of excitation in nuclei far from the β-stability line are very rich and in the last years have been investigated deeply due to the increase of experimental facilities now running around the world (see Ref. [1] and reference quoted therein). The dipole response has revealed a spreading of strength into the low energy region which is particularly relevant for nuclei with a pronounced neutron skin. This concentration of strength has been attributed to a resonant oscillation of the neutron in excess against an inert core and it is known as Pygmy Dipole Resonance (PDR). Whether this strength is due to a collective mode is still under discussion.

In recent past years we have acquired a skill on the study of multiple giant resonances excitation in heavy ion collisions by means of a microscopic description of the structure of the excited nucleus which includes anharmonicity [2, 3, 4, 5]. Calculations of the inelastic cross section, within a semiclassical coupled channel approach, have revealed the possibilities to excite, in an appreciable way, low lying two-phonon states which lay in the same energy range of the PDR. The motivation of this paper is to try to answer to the question whether these states may contribute to the observed PDR peak. At the same time we would like to investigate the nature of the these low lying dipole states in order to establish whether they can be considered as collective ones.

2. Multiphonon excitations

In this section we will describe, very shortly, the framework within which the calculations for the multiple excitations of giant resonance have been done and whose details can be found in

Refs. [2, 3, 4, 5].

2.1. The model

Great successes have been achieved in the study of excitation processes of heavy nuclei by using semiclassical methods techniques. These methods are based on the assumption that nuclei move on classical trajectories, while the internal degrees of freedom are treated quantum mechanically. The standard approach to study the double excitation of giant resonances implies the use of an internal Hamiltonian which is of harmonic type and an external excitation field which is linear in the phonon creation operators. In our model we have included anharmonicities in the internal Hamiltonian and non-linearities in the external field. These corrections have been included in order to overcome some anomaly in the experimental results.

One of the best microscopic theory to describe collective vibrations in nuclei is the Random Phase Approximation (RPA) whose hamiltonian, at the lowest order of a boson expansion[6], is harmonic in the boson creation operators. The introduction of anharmonicities in the internal Hamiltonian can be achieved by adding to the RPA Hamiltonian the *pppp*, *hhhh*, *ppph* and *hhhp* terms which come out by applying a boson mapping procedure to the fermionic Hamiltonian up to second order. So, the boson Hamiltonian now contains terms that mix states with the same number (≥ 2) of phonons and terms which mix states whose phonon numbers differ by one[2, 3, 4, 5].

Our one-phonon basis are chosen from the states resulting from the self-consistent Hartree Fock plus RPA calculations with angular momentum less or equal to 3 and with an appreciable percentage of energy weighting sum rule (EWSR). Then we construct all possible two- and three-phonon states out of them and by diagonalizing the residual interaction in the space of one-, two- and three-phonons states, we obtain the eigenstates of the Hamiltonian, which are mixed states

$$|\Phi_\alpha> = \sum_\nu c_\nu^\alpha|\nu> + \sum_{\nu_1\nu_2} d_{\nu_1\nu_2}^\alpha|\nu_1\nu_2> + \sum_{\nu_1\nu_2\nu_3} e_{\nu_1\nu_2\nu_3}^\alpha|\nu_1\nu_2\nu_3> \qquad (1)$$

whose corresponding eigenvalues are not harmonic.

In these semiclassical models the excitation of one of the partners of the collision is due to the mean field of the other. In standard models the excitation operator is of one-body type and it is assumed to be linear in the phonon operators because only terms of *ph* type are taken into account. Applying the same boson expansion quoted above and taking also the contributions from *pp* and *hh* terms, we obtain a non-linear excitation field which depends on time through the relative distance between the two nuclei. Two new terms appear now in the excitation field expression[2, 3, 4]: One connects states with the same number of phonons, while the other allows transitions between states whose number of phonons differs by two.

Both novel aspects may increase the excitation probability of the considered states. By expressing the state of the system in the presence of the excitation operator as a superposition of the $|\Phi_\alpha>$'s, the solution of the time dependent Schrödinger equation is cast into a set of coupled differential equations for the amplitudes which are integrated along the classical relative motion trajectories. For each $|\Phi_\alpha>$ the associated cross section is obtained by integrating the excitation probability over the whole impact parameters range modulated by the transmission coefficient [2, 3, 4].

2.2. Cross sections calculations

Calculations within this framework have been done for various Sn isotopes and with two Skyrme interactions: SGII[7] and SLY4[8]. In all the cases we run a self-consistent HF+RPA code for both interaction, we include all one-phonon states with angular momentum less or equal to 3 and with an EWSR more than 5%. Following the same prescription used previously[2, 4], we

Journal of Physics: Conference Series **168** (2009) 012014 doi:10.1088/1742-6596/168/1/012014

Table 1. RPA one-phonon basis for the nucleus ^{132}Sn and for both SGII and SLY4 interactions. For each state, spin, parity, energy and percentage of the EWSR (isovector for the GDR and the IVGQR and isoscalar for all the other states) are reported.

		SGII		SLY4	
State	J^π	E_{harm} (MeV)	$EWSR$ (%)	E_{harm} (MeV)	$EWSR$ (%)
0^+_{ll}	0^+	11.73	2	11.36	2
GMR	0^+	16.32	85	16.87	86
1^-_{ll}	1^-	9.30	1.1	9.60	1.4
GDR	1^-	13.92	56	13.81	55
1^-_{hl}	1^-	18.34	25	17.75	15
2^+_{ll}	2^+	5.03	11	5.35	11
ISGQR	2^+	13.50	77	14.11	70
IVGQR	2^+	24.76	28	24.03	36
3^-_{ll}	3^-	5.93	27	6.48	26
HEOR	3^-	24.44	32	25.81	30

put together the states in a given small energy interval in such a way to construct a single state at an energy equal to the average value obtained by weighting the energy of the single states by their B(Eλ) and whose components are obtained with the constraint that it exhausts the fraction of EWSR given by the sum of the EWSR of the states considered. In this way we obtain a reduced set of states which constitute our one-phonon basis. In tables 1 and 2 we show the RPA one-phonon basis for the two isotopes ^{132}Sn and ^{140}Sn. Then we construct all possible two- and three-phonon states out of them and in this space we diagonalize the bosonic Hamiltonian. All the details of the calculations can be found in Ref. [4].

The calculated inelastic cross sections for the Coulomb excitation of ^{132}Sn and ^{140}Sn in the collision with ^{208}Pb at 500 MeV/A are reported in figures 1 and 2. The bars are the cross section for every single state. The full curves are generated by a smoothing procedure using a Lorentzian with a 2 MeV width. In the left (right) part of the figures the results obtained with the SGII (SLY4) interaction are reported. Although we include all the states up to 30 MeV excitation energy, we show the cross sections only in the PDR energy region. We show, for each reaction, the calculations corresponding to two different approximations. The dashed (black) lines are the cross section obtained by taking the harmonic and linear limit. In this case only the basis states (shown in tables 1 and 2) are excited. In the second limit approximation we take into account the anharmonicity and the nonlinearity and the results are shown in the figures with full (red) lines. Several more states are now excited, all of them are not visible in the figure because their individual cross section is below 1 mb. Most of them are states which in the harmonic limit are pure two-phonon components. When the anharmonicities are switched on, they are pushed down and acquire a relatively strong one-phonon component. Therefore the linear term of the field can excite them through such component. In addition there is a contribution also from the other terms of the external field. This behaviour is similar for the two interactions considered here. However, the SLY4 interaction show a difference as far as the anharmonicity is concerned. Such different response will be investigated in a future work.

In order to have a quantitative measure of the effects of the low lying two-phonon states on

Table 2. Same as table 1 but for ^{140}Sn.

		SGII		SLY4	
State	J^π	E_{harm} (MeV)	$EWSR$ (%)	E_{harm} (MeV)	$EWSR$ (%)
GMR	0^+	16.22	82	16.88	79
1^-_{ll}	1^-	7.50	3	7.26	5
GDR	1^-	13.64	46	13.31	54
1^-_{hl}	1^-	17.56	22	17.63	18
2^+_{ll1}	2^+	1.70	3	1.92	2
2^+_{ll2}	2^+	5.78	6	6.37	6
ISGQR	2^+	13.01	76	13.63	74
IVGQR	2^+	24.00	32	24.08	38
3^-_{ll1}	3^-	2.97	9	3.84	9
3^-_{ll2}	3^-	6.24	17	6.82	18
HEOR	3^-	23.93	29	25.29	38

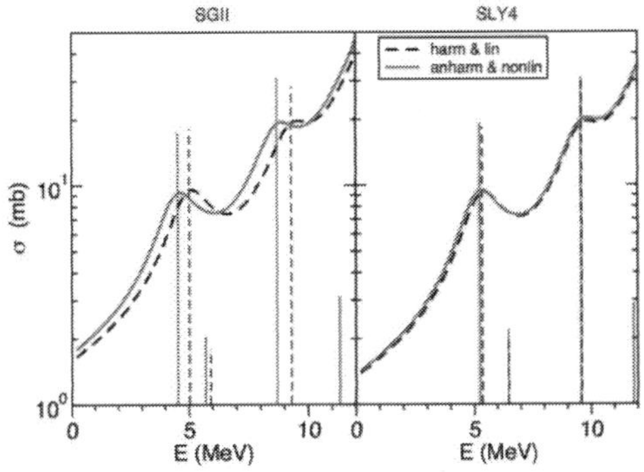

Figure 1. (Color online) Relativistic Coulomb inelastic cross section for ^{132}Sn + ^{280}Pb at 500 MeV/A with the SGII (left) and SLY4 (right) interactions. The dashed black lines are the cross section obtained in the harmonic and linear approximation. The solid red lines correspond to the anharmonic and nonlinear case.

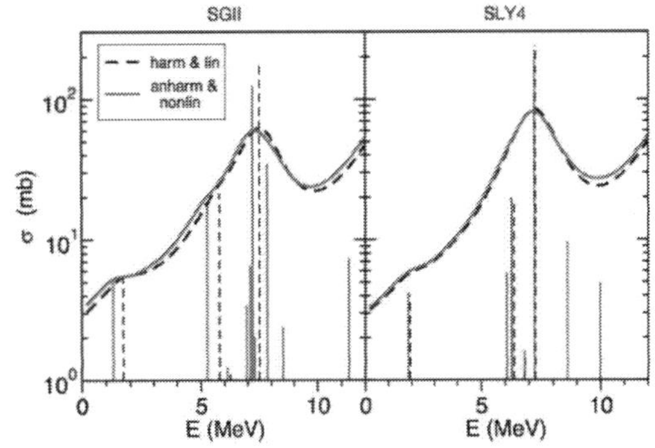

Figure 2. (Color online) Relativistic Coulomb inelastic cross section for ^{140}Sn + ^{280}Pb at 500 MeV/A with the SGII (left) and SLY4 (right) interactions. The dashed black lines are the cross section obtained in the harmonic and linear approximation. The solid red lines correspond to the anharmonic and nonlinear case.

the inelastic cross section, we have summed the cross section over the energy interval shown in the figures. In table 3 we show the summed cross section for the two reactions considered here: ^{132}Sn, ^{140}Sn + ^{208}Pb at 500 MeV/A (for both SGII and SLY4 interactions) over the interval 0-12 MeV. The values for the anharmonic and nonlinear case (AH-NL) are sensibly higher than the ones for the harmonic and linear (HL) approximations. The absolute values of the cross section is directly related to the B(Eλ) values of the low lying 1^- states. The increase is higher for the SGII case: we have 14% increase for the ^{132}Sn isotope and only 5% for ^{140}Sn; for the

Table 3. Summed inelastic cross sections (in mb) for the systems ^{132}Sn, ^{140}Sn + ^{208}Pb at 500 MeV/A (for both SGII and SLY4 interactions) over the interval 0-12 MeV. Values for the harmonic and linear case (H-L, second and fourth columns) should be compared with the ones obtained in the anharmonic and non linear calculations (AH-NL, third and fifth columns).

	SGII		SLY4	
	H-L	AH-NL	H-L	AH-NL
^{132}Sn	48.3	55.0	50.2	55.8
^{140}Sn	203.3	213.6	264.5	269.4

SLY4 interaction these increases are 11% for ^{132}Sn and only 2% for ^{140}Sn. We can certainly state that the inclusion of multiphonon states in the study of the PDR excitation has revealed to be important at least when the SGII interaction is used. The minor contribution obtained with the SLY4 interaction should be related to the weaker anharmonicity found in this case: An aspect which should be investigated in a future work.

3. The low lying 1^- states

In this section we would like to study the nature of the low lying 1^-states, in particular we would like to address the question whether or not the dipole strength at low energy corresponds to a collective mode. Previous investigations within the quasiparticle RPA have found that the peak associated to the PDR is due to a neutron strong single p-h configuration[9]. On the other hand, calculations done within the relativistic RPA [10] have got results which are at variance with those quoted above. The RPA dipole strength distribution for some Sn isotopes are reported in figure 3. The bars correspond to the RPA calculations while the full lines are generated by a smoothing procedure using a Lorentzian with a 1 MeV width. The figure shows clearly that when the neutron excess increases more strength develops or moves at low energy acquiring for the ^{140}Sn a considerable strength below 10 MeV. This result is similar for both interactions.

Another effect already underlined in previous work[11] and related to the increase of the neutron skin is a rather strong isospin mixing which develops in states of higher multipolarities. As an example we show in figure 4 the isoscalar (upper part) and isovector (lower part) RPA quadrupole strength distribution for ^{100}Sn where we can clearly distinguish two pure isoscalar states and an isovector strength concentrated around 27 MeV which can be assigned to the isovector GQR. As far as the neutron skin increases some strength of the isovector part moves to lower energy generating states with a strong isospin mixing. This general behaviour has been found also for octupole states and it is independent of the interaction used as far as our calculations are concerned.

The important question is how much collective are the dipole states whose strength lay at low energy. One measure of the collectivity is the number of p-h component entering in the RPA wave function with an appreciable weight [9, 10]. In microscopic calculations this can be analyzed in more details by defining the quantity

$$A_{ph} = |X^{\nu}_{ph}|^2 - |Y^{\nu}_{ph}|^2 \tag{2}$$

in terms of the X and Y RPA's amplitudes, with the normalization condition

$$\sum_{ph} A_{ph} = 1 \tag{3}$$

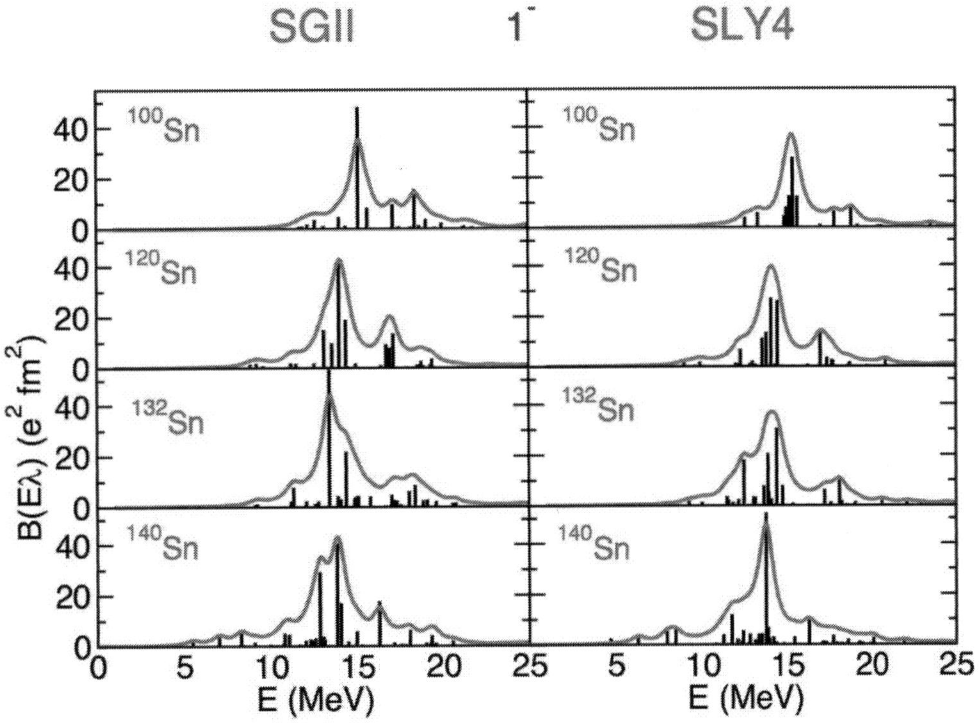

Figure 3. (Color online) Strength distributions for dipole states for tin isotopes calculated with the SGII and SLY4 interactions, left and right columns, respectively. The solid curves are obtained by adopting a smoothing procedure (see text).

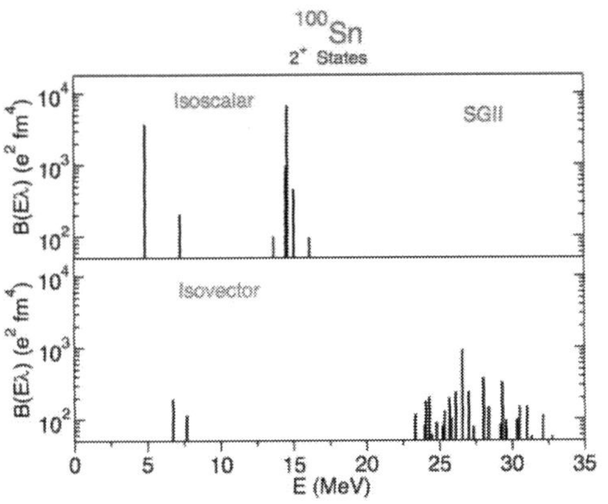

Figure 4. Isoscalar (upper part) and isovector (lower part) RPA strength distributions for quadrupole states for ^{100}Sn isotope calculated with the SGII interaction.

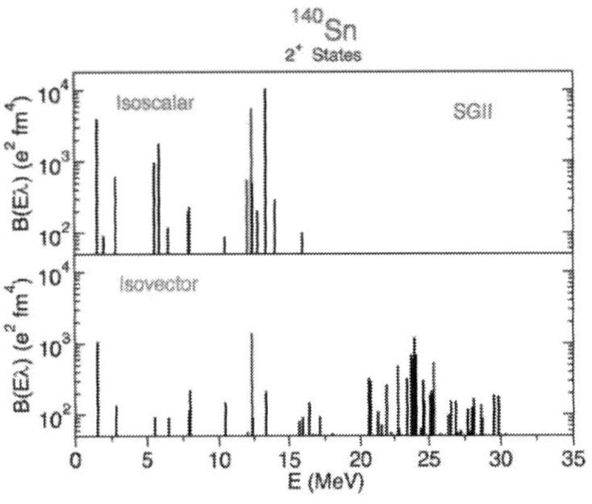

Figure 5. Isoscalar (upper part) and isovector (lower part) RPA strength distributions for quadrupole states for ^{140}Sn isotope calculated with the SGII interaction.

In table 4 we show for each p-h configuration its contribution, in percentage, to the formation of the low lying 1^- states shown in table 1. Only the configurations contributing more than 1% are shown. Note that the major contribution comes from the neutron skin. Indeed, the

Table 4. Particle-hole configurations which give the major contribution for the two dipole low-lying states (shown in table 1) for the nucleus ^{132}Sn and for the SGII and SLY4 interactions. For each p-h configuration the contribution to the norm of the state and the partial contribution to the reduced transition amplitude (in e fm units) are reported. The superscripts π, ν refer to the proton and neutron states, respectively.

SGII (E=9.3 MeV)		SLY4 (E=9.6 MeV)	
p-h conf.	A_{ph}	p-h conf.	A_{ph}
$(2p_{3/2}, 3s_{1/2})^{\pi}$	1.0%	$(1f_{5/2}, 1g_{7/2})^{\pi}$	1.1%
$(2p_{3/2}, 2d_{5/2})^{\pi}$	7.0%	$(2p_{3/2}, 2d_{5/2})^{\pi}$	1.8%
$(2p_{1/2}, 3s_{1/2})^{\pi}$	2.8%	$(1g_{9/2}, 1h_{11/2})^{\pi}$	5.6%
$(1g_{9/2}, 1h_{11/2})^{\pi}$	12.5%	$(1g_{7/2}, 1h_{9/2})^{\nu}$	1.9%
$(1g_{7/2}, 1h_{9/2})^{\nu}$	2.1%	$(2d_{3/2}, 3p_{1/2})^{\nu}$	30.0%
$(2d_{5/2}, 2f_{7/2})^{\nu}$	1.6%	$(2d_{3/2}, 2f_{5/2})^{\nu}$	41.8%
$(2d_{3/2}, 3p_{1/2})^{\nu}$	13.7%	$(3s_{1/2}, 3p_{3/2})^{\nu}$	6.1%
$(2d_{3/2}, 2f_{5/2})^{\nu}$	8.6%	$(1h_{11/2}, 1i_{13/2})^{\nu}$	7.1%
$(3s_{1/2}, 3p_{3/2})^{\nu}$	20.9%		
$(3s_{1/2}, 3p_{1/2})^{\nu}$	14.2%		
$(1h_{11/2}, 1i_{13/2})^{\nu}$	12.2%		

neutron p-h configurations contribute for 73% to the formation of the state, in the case of SGII interaction, and for 87% for the state obtained with the SLY4 interaction. For both cases our result is very similar to the one obtained in Ref. [10] where a relativistic RPA has been used for the study of the pygmy resonance. In fact, also in our case several p-h configurations contribute with an appreciable weight. A more complete analysis should also take into account the degree of coherence of such contributions. That is to say, one has to look also how the p-h contributions sum up to give the final reduced transition probability B(Eλ). Such analysis has been done in Ref. [12] and it shows how the criterion described above, based only on the magnitude of the A_{ph}, can be misleading. In fact, looking to the construction of the reduced transition probability we found, for the low lying dipole states, strong cancellations leading to a small value for the B(E1) compared with the GDR one. That the nature of these dipole states is qualitatively different from the high lying ones is illustrated in figure 6 where the proton, neutron, isoscalar and isovector transition densities for the two states, calculated with the SGII interaction, are reported. The behaviour for the low-lying state is similar to what has been found for other nuclei with neutron excess[11, 10, 9]: The proton and neutron densities in the interior region are not out of phase; in the interior the isoscalar transition density dominates over the isovector one; in the surface region only the neutrons give a contribution to both isoscalar and isovector transition densities. For the GDR and the 1^-_{hl} cases we find the expected radial dependence for a collective isovector state: the proton and neutron transition densities oscillate out of phase; the isovector part is much larger than the isoscalar one; the latter has a node in the interior of the nucleus which produce a zero result, when integrated over the radial distance, for isoscalar B(E1).

4. Summary

In this work we have extended the study of multiphonon excitations in heavy ion collisions to the case of the pygmy resonances. Indeed, we have found that a few multiphonon states, lying

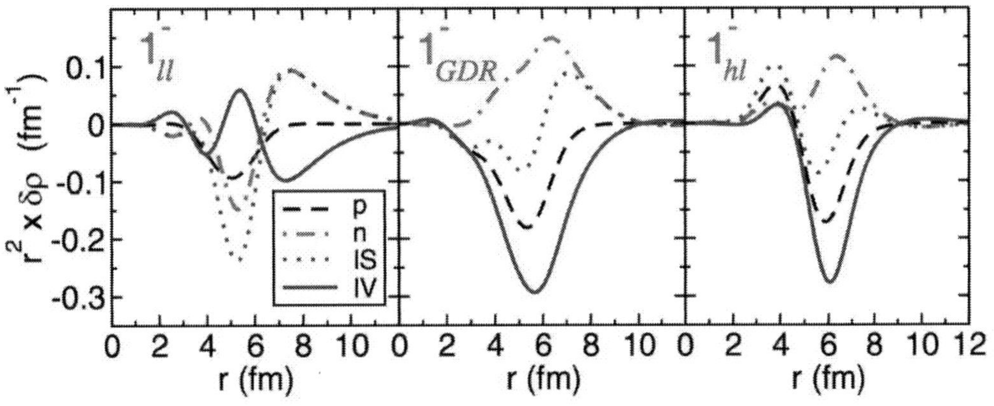

Figure 6. (Color online) Transition densities for the low-lying dipole state, the GDR and the 1^-_{hl} for the ^{140}Sn isotope calculated with the SGII interaction. We show the proton, neutron, isoscalar and isovector parts (as indicated in the legend).

in the PDR energy region, are excited in the collision process with a strong probability. Thus, they may contribute to the observed PDR peak. We start from a microscopic approach based on RPA which is extended by the introduction of anharmonicities generated by the coupling of multiphonon states. The inelastic cross sections are calculated by solving semiclassical coupled channel equations, where the external field has been implemented by taking into account non linearities. The calculations have been done for two different Skyrme interaction (SGII and SLY4) and for the reactions ^{A}Sn + ^{280}Pb at 500 MeV/A. We found an increase of the cross section in the low lying PDR energy region which is mainly due to the excitation of several states whose population is strongly suppressed by selection rules when anharmonicities and non-linearities are neglected. In general, the SLY4 interaction produces a smaller anharmonicity which is then reflected in a lower increase of the cross section in the PDR region. On the contrary, the results obtained with the SGII show the importance of anharmonicities and non linearities for the study of the PDR.

References

[1] Paar N, Vretenar D, Khan E and Colò G 2007 *Rep. Prog. Phys.* **70** 691
[2] Lanza E G, Andrés M V, Catara F, Chomaz Ph and Volpe C 1997 *Nucl. Phys.* A **613** 445; 1998 *Nucl. Phys.* A **636** 452; 1999 *Nucl. Phys.* A **654** 792c
[3] Andrés M V, Catara F, Lanza E G, Chomaz Ph, Fallot M and Scarpaci J A 2001 *Phys. Rev.* C **65** 014608
[4] Lanza E G, Andrés M V, Catara F, Chomaz Ph, Fallot M and Scarpaci J A 2006 *Phys. Rev.* C **74** 064614
[5] Fallot M, Chomaz Ph, Andrés M V, Catara F, Lanza E G and Scarpaci J A 2003 *Nucl. Phys.* A **729** 699
[6] Hage-Hassan M and Lambert M 1972 *Nucl. Phys.* A **188** 545
[7] Van Giai N and Sagawa H 1981 *Phys. Lett.* **106** B 379; Van Giai N and Sagawa H 1981 *Nucl. Phys.* A **371** 1
[8] Chabanat E, Bonche P, Haensel P, Meyer J and Schaeffer R 1998 *Nucl. Phys.* A **635** 231
[9] Coló G and Bortignon P F 2001 *Nucl. Phys.* A **696** 427; Sarchi D, Bortignon P F and Coló G 2007 *Phys. Lett.* B **601** 27
[10] Vretenar D, Paar N, Ring P and Lalazissis G A 2001 *Nucl. Phys.* A **692** 496; Litvinova E, Ring P and Vretenar D 2007 *Phys. Lett.* B **647** 111
[11] Catara F, Lanza E G, Nagarajan M A and Vitturi A 1997 *Nucl. Phys.* A **614** 86; 1997 *Nucl. Phys.* A **624** 449
[12] Lanza E G, Andrés M V, Catara F, Chomaz Ph, and Gambacurta D to be published in *Phys. Rev.* C

Journal of Physics: Conference Series **168** (2009) 012015 doi:10.1088/1742-6596/168/1/012015

Spatial dependence of the pairing field calculated with bare and induced interactions

A Pastore[1], F Barranco[2], RA Broglia[3,4,5] and E Vigezzi[3]

[1] Department of Physics, Post Office Box 35 (YFL), FI-40014 University of Jyväskylä, Finland
[2] Departamento de Fisica Aplicada III, Escuela Superior de Ingenieros, Camino de los Descubrimientos s/n, 41092 Sevilla, Spain.
[3] INFN, Sezione di Milano, via Celoria 16, 20133 Milano, Italy.
[4] Dipartimento di Fisica, Università degli Studi di Milano,via Celoria 16, 20133 Milano, Italy.
[5] The Niels Bohr Institute, University of Copenhagen, Blegdamsvej 17, 2100 Copenhagen Ø, Denmark.

Abstract. The interaction induced by the exchange of low-lying surface vibrations between pairs of orbitals close to the Fermi surface provides an important contribution to pairing correlations in superfluid nuclei. We study the spatial dependence of the pairing field obtained adding the bare and induced interaction in ^{120}Sn.

1. Introduction

The superfluid properties of nuclear systems are strongly influenced by polarization phenomena. In particular, the exchange of low-lying surface vibrations between pairs of nucleons close to the Fermi surface gives rise to an attractive induced interaction, that accounts for about one half of the value of the pairing gap derived from the experimental odd-even mass differences in superfluid nuclei [1]. If one wants to obtain a consistent calculation of the gap, one must add the effects of the nucleon-nucleon interaction, and take into account the self-energy and vertex processes that renormalize the quasiparticle strength and increase the level density around the Fermi energy. It has been shown that this procedure leads to gaps in reasonable agreement with experiment, if one starts from a single-particle spectrum associated with a $k-$effective mass $m_k \sim 0.7$, like that obtained in a Hartree-Fock calculation with the SLy4 force [2].

In this paper we study the pairing gap in coordinate space. Because our study aims at determining the basic features of the spatial dependence of the gap, rather than its precise magnitude, use is made of approximations to deal with some of these effects, in particular with self-energy effects, so as to gain in transparency in the presentation of the results. In order to make contact with other, more phenomenological approaches available in the current literature, we shall parametrize our results in terms of a density dependent, zero-range interaction.

2. The pairing induced interaction

We start by performing a Hartree-Fock calculation with the two-body interaction SLy4, obtaining a set of single-particle energy levels e_{nlj}. We then use this basis to solve the Hartree-Fock-Bogoliubov (HFB) equations in the pairing channel. The calculations are performed in a spherical box of radius $R_{box} = 15$ fm. For more details we refer to [3].

The matrix elements of the interaction induced by the exchange of a vibration will be calculated using the formalism already employed in Ref. [4]:

$$\langle \nu_1' m' \nu_2' \bar{m}' | v_{ind} | \nu_1 m \nu_2 \bar{m} \rangle =$$
$$\sum_{J^\pi M i} \frac{(f+g)^{\nu_1' m'}_{\nu_1 m; J^\pi M i} (f-g)^{\nu_2' m'}_{\nu_2 m; J^\pi M i}}{E_0 - (|e_{\nu_1'} - e_F| + |e_{\nu_2} - e_F| + \hbar\omega_{J^\pi i})} + \sum_{J^\pi M i} \frac{(f+g)^{\nu_1' m'}_{\nu_1 m; J^\pi M i} (f-g)^{\nu_2' m'}_{\nu_2 m; J^\pi M i}}{E_0 - (|e_{\nu_1} - e_F| + |e_{\nu_2'} - e_F| + \hbar\omega_{J^\pi i})}. \tag{1}$$

The index i labels the exchanged vibrational modes, having angular momentum and parity $J^\pi M$ and energy $\hbar\omega_{J^\pi i}$. The modes have been calculated in the Quasiparticle Random Phase Approximation (QRPA), using the same interaction already employed to calculate the mean-field, with the exception of the spin-orbit and of the Coulomb part [5]. E_0 is the pairing correlation energy of a Cooper pair, a quantity which is of the order of $-2\Delta_F$, where Δ_F is the average value of the gap close to the Fermi energy. In Eq. (1) f and g denote the particle-vibration coupling vertices associated with the spin-independent and spin-dependent parts of the residual interaction:

$$v_{ph}(\vec{r}, \vec{r}') = \delta(\vec{r} - \vec{r}') \times \left\{ \left[F_0(r) + F_0'(r) \vec{\tau} \cdot \vec{\tau}' \right] + \left[\left(G_0(r) + G_0'(r) \vec{\tau} \cdot \vec{\tau}' \right) \vec{\sigma} \cdot \vec{\sigma}' \right] \right\}, \tag{2}$$

where F_0, F_0' (G_0, G_0') are the generalized Landau-Migdal parameters associated with the SLy4 force controlling the isoscalar and isovector spin-independent (spin-dependent) channels. In the calculation of the particle-vibration coupling we neglected the momentum-dependent part of the interaction (this part is instead taken into account in the QRPA calculation). The vertex f is given by

$$f^{\nu' m'}_{\nu m; J^\pi M i} = i^{l-l'} \langle j' m' | (i)^J Y_{JM} | j m \rangle \times$$
$$\int dr \varphi_{\nu'} [(F_0 + F_0') \delta\rho^i_{J^\pi n} + (F_0 - F_0') \delta\rho^i_{J^\pi p}] \varphi_\nu. \tag{3}$$

The quantities $\delta\rho^i_{J^\pi n}$ and $\delta\rho^i_{J^\pi p}$ are the neutron and proton contributions to the transition densities and are given by

$$\delta\rho^i_{J^\pi}(r) = \frac{1}{\sqrt{2J+1}} \sum_{\nu_1,\nu_2} (X_{\nu_1,\nu_2}(i, J^\pi) + Y_{\nu_1,\nu_2}(i, J^\pi)) \times$$
$$(u_{\nu_1} v_{\nu_2} + u_{\nu_2} v_{\nu_1}) \langle \nu_1 || i^J Y_J || \nu_2 \rangle \varphi_{\nu_1}(r) \varphi_{\nu_2}(r), \tag{4}$$

where u_{ν_1} and v_{ν_1} are the quasiparticle amplitudes and X_{ν_1,ν_2} and Y_{ν_1,ν_2} denote the forward and backward amplitudes of the QRPA modes. The vertex g is given by a similar expression, substituting F_0, F_0' with G_0, G_0' [4].

3. Pairing gap in coordinate space

Solving the HFB equations in the pairing channel with the induced interaction (1) we obtain the matrix elements of the state-dependent pairing gap $\Delta_{nn'lj}$: the diagonal ones are shown below in figure 4(a). We can then derive the pairing field in position space $\Delta(\vec{r}, \vec{r}')$ [3]. This is shown in figure 1(a) as a function of the center of mass $R_{c.m.}$ and of the relative distance r_{12} of the pair, after averaging over the angle between $\vec{R}_{c.m.}$ and $\vec{r}_{12}$. For a given value of r_{12} the pairing field has a maximum close to the surface of the nucleus ($R_{c.m.} \sim 6$ fm), while it becomes negative in the interior, essentially due to the repulsive interaction induced by the exchange of spin modes, associated with the G_0, G_0' terms of the residual interaction (2). Taking the Fourier transform with respect to the relative distance $\vec{r}_{12}$, we can define the pairing field in momentum space

$$\Delta(\vec{R}_{c.m.}, \vec{k}) = \int d^3 r_{12} \Delta(\vec{r}, \vec{r}') \exp^{-i\vec{k}\cdot\vec{r}_{12}}, \tag{5}$$

Journal of Physics: Conference Series **168** (2009) 012015 doi:10.1088/1742-6596/168/1/012015

which is shown in figure 1(b) after averaging over the angle between $\vec{R}_{c.m.}$ and $\vec{k}$. We now consider the pairing gap obtained adding the matrix elements of the bare Argonne v_{14} nucleon-nucleon interaction v_{Arg}. We can also take into account self-energy effects in a simplified way, multiplying the total interaction $v_{Arg} + v_{ind}$ by a factor $Z = 0.7$, which is the typical value of the quasiparticle strength calculated for levels around the Fermi energy [6]:

$$v_{Arg+ind} = Z(v_{Arg} + v_{ind}). \tag{6}$$

While the pairing matrix elements are reduced by a factor Z^2, they are also enhanced by a factor $1/Z$ due to the increase in the density of levels; hence the overall multiplying factor Z [7].

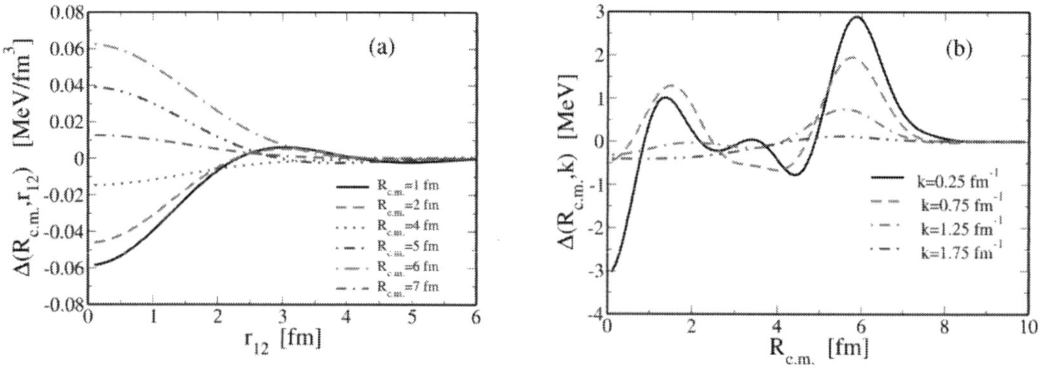

Figure 1. Pairing field in position space (a), and in momentum space (b) obtained with the induced interaction v_{ind}.

The diagonal matrix elements of the pairing gap Δ_{nnlj} obtained with the interaction $v_{Arg+ind}$ are shown below in figure 4(b) and 5(b). They correspond to a value $\Delta_F = 1.32$ MeV, obtained averaging over the single-particle levels located within an interval of ± 2 MeV around the Fermi energy (note that the value reported in [3] is about 10% larger, due to a numerical error in the calculation of some matrix elements of the induced interaction). This value is in good agreement with that derived from the empirical odd-even mass difference, but one should keep in mind that the magnitude of the gap is sensitive both to the adopted value of Z and to the effective mass of the mean field. To better assess the spatial structure of the pairing field, we prefer to introduce a local approximation [8, 9]. We define $\Delta_{loc}(R_{c.m.}) \equiv \Delta(R_{c.m.}, k_F(R_{c.m.}))$, where $k_F(R_{c.m.})$ is the local Fermi momentum. In figure 2(a) we compare the results obtained using the bare interaction and the bare plus induced interaction (6). The bare interaction leads to a peak at the surface, which is reinforced by the induced interaction. The induced interaction also produces a negative gap in the interior of the nucleus, in keeping with the results shown in figure 1.

The local approximation introduced above, based on the results obtained in the microscopic HFB calculation, leads to pairing gaps which are rather different from those obtained from the simplest Local Density Approximation, which does not take into account proximity effects associated with the nuclear surface and the fact that the nuclear radius is smaller than the coherence length in uniform matter. This can be seen in figure 2(b), where we compare the local pairing gap Δ_{loc} associated with the Argonne interaction only, with the function $\Delta_{LDA}(R_{c.m.}) = \Delta_F^{n.m.}(\rho_n(R_{c.m.}))$, where $\Delta_F^{n.m.}$ is the pairing gap calculated with the Argonne interaction at the Fermi energy in uniform neutron matter, for a density equal to the neutron density at a distance $R_{c.m.}$ from the center of the nucleus, and using the local value of the

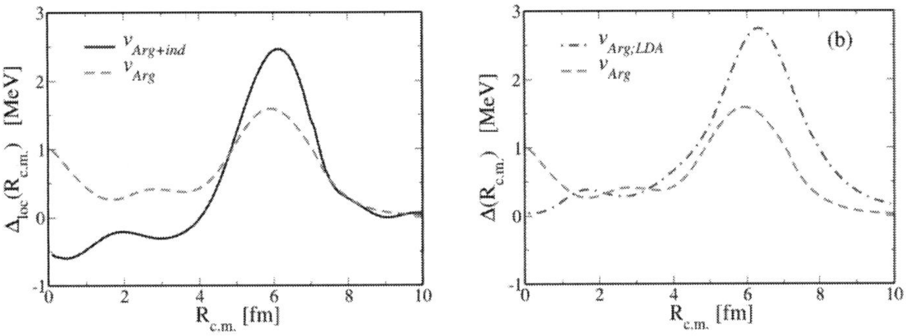

Figure 2. (a) Local pairing gaps calculated in ^{120}Sn with the interactions v_{Arg} (dashed line) and $v_{Arg+ind}$ (solid line). (b) The local pairing gap calculated with v_{Arg}, already shown in (a), is compared to the gap obtaining using the simple LDA approximation based on the pairing gap calculated with the Argonne interaction in uniform matter (solid line).

effective mass. The LDA overestimates the difference between the pairing gap on the surface and in the interior of the nucleus.

4. Parametrization of the pairing interaction

4.1. Finite range parametrization

We now introduce a parametrization of the induced interaction, v_{ind}, based on the surface or volume character of the interaction associated with the spin-independent or the spin-dependent part of the matrix elements (1). We try to determine a Gaussian function $v^G_{ind}(R_{c.m.}, r_{12})$ so as to fulfill approximately the relation

$$\Delta(R_{c.m.}, r_{12}) = -v^G_{ind}(R_{c.m.}, r_{12})\Phi^{S=0}(R_{c.m.}, r_{12}), \tag{7}$$

where $\Phi^{S=0}(R_{c.m.}, r_{12})$ is the abnormal density in the $S = 0$ channel, the dominant one. We consider separately the contributions from the attractive. spin-independent and from the repulsive, spin-dependent parts of the interaction writing $v^G_{ind}(R_{c.m.}, r_{12}) = v^G_{attr}(R_{c.m.}, r_{12}) + v^G_{rep}(R_{c.m.}, r_{12})$. We shall first fit the pairing gap obtained including only the spin-independent part of v_{ind} (that is, putting $G = G' = 0$ in Eq.(2)), using the function

$$v^G_{attr}(R_{c.m.}, r_{12}) = -b_{attr} \cdot \exp\left(-[(r_{12} - c)/a_{attr}]^2\right) \tag{8}$$

where a_{attr}, b_{attr}, c are parameters to be determined. We fix c so as to constrain the Gaussian function to be maximum when at least one of the neutrons is on the surface of the nucleus. This implies $c = |R_{nucl} - R_{c.m.}|$, where R_{nucl} is the maximum of the first derivative of the self-consistent single-particle potential. The parameter a_{attr} turns out to be close to $a_{attr} \approx 2$ fm, so in practice to simplify the fitting method we used the fixed value $a_{attr} = 2$ fm. The resulting values of the parameter b_{attr} obtained as a function of $R_{c.m.}$ are peaked on the nuclear surface and are plotted in figure 3(a).

The repulsive part of the induced interaction is active only in the interior of the nucleus, say for $R_{c.m.} \leq 4.5$ fm (cf. figure 1(a)), so we multiply the Gaussian by a Heaviside function centered at $R_0 = 4.5$ fm

$$v^G_{rep}(R_{c.m.}, r_{12}) = b_{rep}\exp\left(-[r_{12}/a_{rep}]^2\right)\Theta(R_{c.m.} - R_0). \tag{9}$$

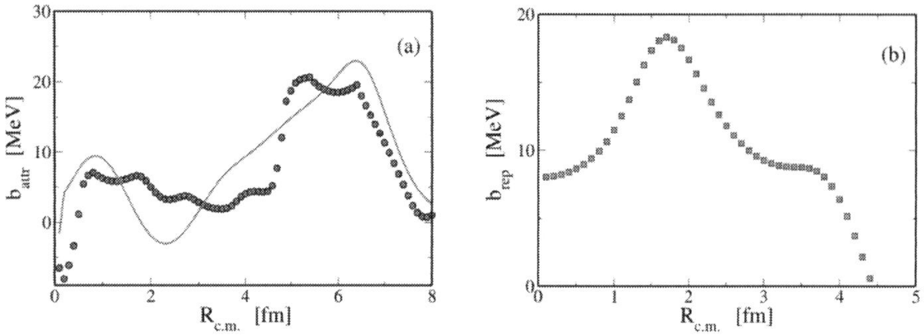

Figure 3. (a) The values of the parameter b_{attr}, obtained fitting the Gaussian interaction (cf. Eq.(8)) for ^{120}Sn, are shown as a function of the center of mass $R_{c.m.}$ (filled dots), and are compared with the function $dU(R_{c.m.})/dR_{c.m.}$ (solid line). (b) The values of the parameter b_{rep}, obtained fitting the Gaussian interaction Eq. (10), are also shown as a function of the center of mass $R_{c.m.}$ (filled squares).

Table 1. Average gaps and pairing energies (in MeV) obtained with the full expression (1) of the induced interaction and with the Gaussian parametrization v^G_{ind}.

Nucleus	Δ_F^{full}	Δ_F^G	E_{pair}^{full}	E_{pair}^G
^{108}Sn	1.02	1.07	-4.02	-5.77
^{112}Sn	1.87	1.65	-9.67	-11.20
^{116}Sn	1.34	1.18	-6.74	-7.72
^{120}Sn	0.88	0.96	-5.98	-5.85

We then determine the parameters of the repulsive Gaussian, fitting the values of a_{rep}, b_{rep} so that the resulting interaction

$$v^G_{ind}(R_{c.m.}, r_{12}) = v^G_{attr}(R_{c.m.}, r_{12}) + v^G_{rep}(R_{c.m.}, r_{12}) \tag{10}$$

satisfies Eq.(7) for values of r_{12} in the interval [0,2] fm, where we used in this case the gaps $\Delta(R_{c.m.}, r_{12})$ and the abnormal density $\Phi^{S=0}(R_{c.m.}, r_{12})$ obtained from the full calculation of the induced interaction considering both spin modes and density modes. The parameter a_{rep} turns out to be very close to $a_{rep} \approx 3.5$ fm, so in practice we used a fixed value $a_{rep} = 3.5$ fm. In figure 3(b) we show the resulting values of b_{rep} as a function of $R_{c.m.}$.

We found that it is possible to parametrize b_{attr} as $b_{attr} \sim \beta_{attr} R_{nucl} \frac{dU(R_{c.m.})}{dR_{c.m.}}$. The resulting values of β_{attr} for $A =$ 108,112,116 and 120 are $\beta_{attr}=$ 0.157, 0.172, 0.163 and 0.154 respectively.

The Gaussian parametrization gives pairing gaps and pairing energies in good agreement with the original induced interaction (1), as shown in table 1. We can then insert it in Eq. (6) in place of v_{ind}, and calculate the pairing gap associated with the bare plus induced interaction. For the case of ^{120}Sn we obtain $\Delta_F^G = 1.47$ MeV and $E^G_{pair} = -17.52$ MeV to be compared with $\Delta_F = 1.32$ MeV and $E_{pair} = -13.98$ MeV obtained with the original induced interaction (cf. table 2).

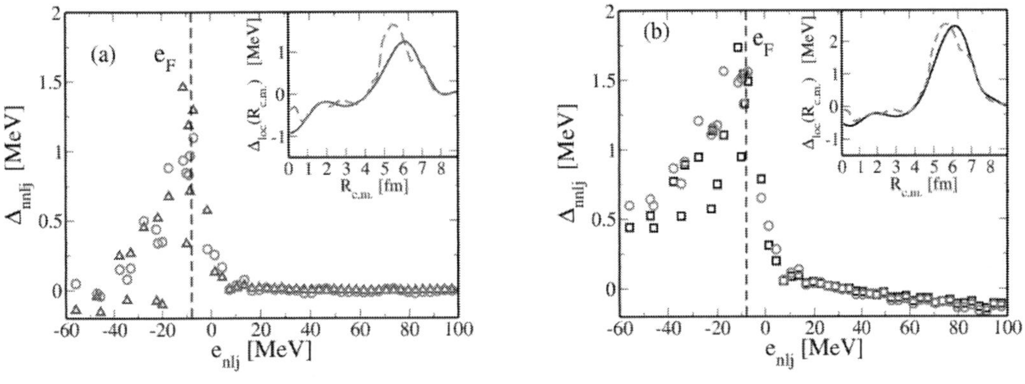

Figure 4. (a) The diagonal matrix elements of the pairing gap in ^{120}Sn, associated with the induced interaction v_{ind} (triangles), are compared with those associated with the Gaussian parametrization (circles, cf. Eq. (10)). The spatial dependence of the semiclassical pairing gap associated with the induced interaction (solid line) and with the Gaussian interaction (dashed line) are shown in the inset. (b) The same, but for the Argonne plus induced interaction.

4.2. DDDI parametrization

The local pairing fields discussed in Section 3 can be compared to those obtained by several authors, who employed a density-dependent pairing interaction (DDDI) of the form [10]-[14]:

$$v^\delta(\vec{r}_1, \vec{r}_2) = v_0 \left[1 - \eta \left(\frac{\rho\left(\frac{\vec{r}_1+\vec{r}_2}{2}\right)}{\rho_0}\right)^{\alpha}\right] \delta(\vec{r}_1 - \vec{r}_2), \tag{11}$$

where ρ_0 is the nuclear saturation density and v_0, η, α are three parameters to be determined, together with the value of a cutoff energy in the single-particle energies E_{cut}, needed to solve the HFB equations with a zero-range interaction. The parameter v_0 together with E_{cut} defines the strength of the pairing interaction, while the other two parameters determine the shape of the pairing field. For a given value of E_{cut}, the strength can be fixed at zero density so as to reproduce the neutron scattering length. We shall use the single-particle levels which lie up to 60 MeV above the Fermi energy, following Ref. [15], and we shall put $v_0 = -458.4$ MeV fm^3 accordingly.

The parameters α and η have been determined in previous works either to reproduce experimental gaps or to reproduce the pairing gap at the Fermi energy obtained with a finite range interaction like Gogny or Argonne in uniform neutron matter. In this section we want instead to determine the parameters of the DDDI from the condition that the spatial dependence of the associated gaps reproduces the spatial dependence of the local pairing fields determined in the previous section. We fit the parameters η and α, minimizing the deviation between the form of the pairing gap obtained with the DDDI of Eq. (11) and the form of the gap obtained with the local potentials. We note that if an interaction has positive values for $R_{c.m.} = 0$ the parametrization (cf. Eq.(11)) implies that $\eta > 1$. For a given value of η, the larger the value of α, the more attractive the interaction is on the surface, leading to larger values of the gap. Imposing a given value Δ_F for the average pairing gap, one obtains a relation between α and η, shown in the inset of figure 6(a) for various values of Δ_F: the spatial dependence of the pairing gaps obtained for various values of α when $\Delta_F = 1.3$ MeV is also shown in figure 6(a).

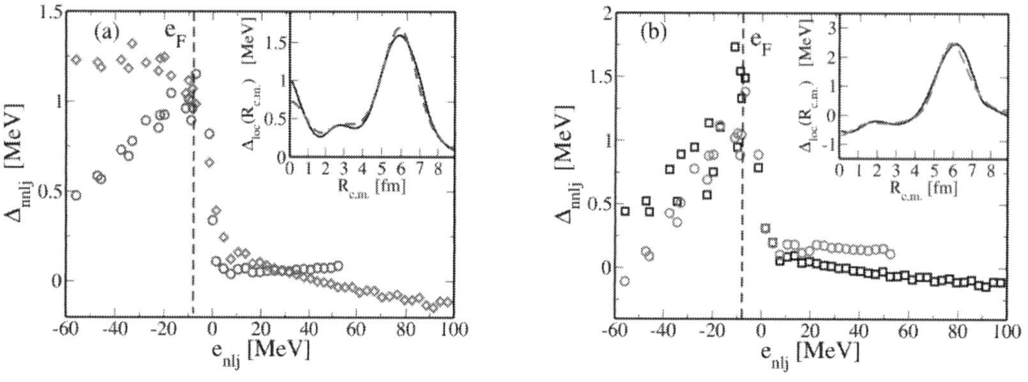

Figure 5. (a) The diagonal matrix elements of the pairing gap associated with the Argonne interaction v_{Arg} (diamonds) are compared with those associated with the DDD, zero-range interaction with the parameters $\alpha = 0.66$, $\eta = 0.84$ (circles). The local pairing gaps associated with the Argonne interaction (solid line) and with the zero-range interaction (dashed line) are shown in the inset. (b) The same, for the Argonne+induced interaction $v_{Arg+ind}$ compared with the DDDI with the parameters $\alpha = 1.5$, $\eta = 1.15$ (circles, cf. table 2).

The values of the parameters for v^δ_{Arg} and $v^\delta_{Arg+ind}$, calculated for ^{120}Sn, are reported in table 2, while the resulting fits are shown in the insets of figures 5(a) and 5(b). While the spatial dependence of the local gaps is quite well reproduced, the diagonal matrix elements of the pairing gaps, shown in figure 5 and the pairing energy (reported in table 2) are underestimated, especially in the case of the total interaction $v^\delta_{Arg+ind}$. The values of the parameters for the Argonne interaction are very close to those obtained by Matsuo for the bare interaction in uniform neutron matter [15]. We extended the analysis for the bare interaction to the nuclei $^{102-130}$Sn and $^{36-46}$Ca, finding $\bar{\alpha} = 0.67 \pm 0.04, \bar{\eta} = 0.82 \pm 0.05$, showing that the parameters given in table 2 are quite stable.
In figure 6(b) we compare the spatial dependence of the three DDD interactions listed in table 2. By construction, all interactions tend to the value v_0 = -458.4 MeV fm^{-3} for large values of $R_{c.m.}$. The bare+induced interaction $v^\delta_{ind+Arg}$ is considerably more attractive than the bare

Table 2. Parameters of the DDDI, Eq. (11), producing pairing gaps which fit the local semiclassical pairing fields obtained with the various interactions. In the last four columns we compare the pairing gap at the Fermi energy and the pairing energies (in MeV) obtained with the full calculation, Δ_F^{full} and E_{pair}^{full}, with the values obtained using the corresponding density dependent interaction, Δ_F^δ and E^δ_{pair}.

Interaction	α	η	Δ_F^{full}	Δ_F^δ	E_{pair}^{full}	E_{pair}^δ
v^δ_{Arg}	0.66	0.84	1.04	1.03	-13.2	-8.9
$v^\delta_{Arg+ind}$	1.5	1.15	1.32	1.09	-13.98	-11.78
$v^\delta_{Arg+ind}(G_0 = G_0' = 0)$	1.38	0.91	2.12	1.85	-24.45	-25.68

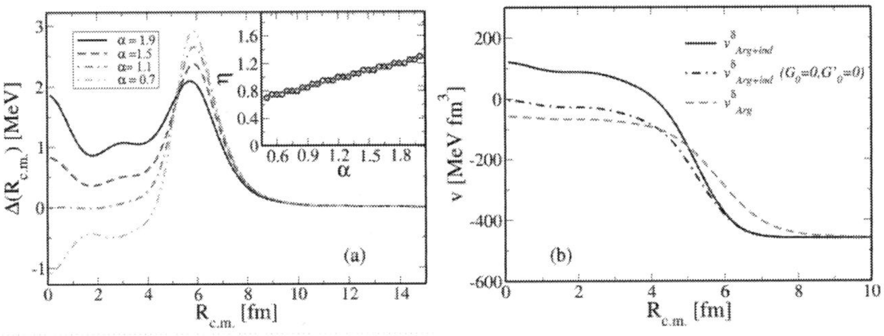

Figure 6. (a) We show in the inset the linear relation between the parameters α and η in Eq.(11) obtained requiring that $\Delta_F = 1.3$ MeV, and in the main figure the resulting pairing gaps associated with specific values of α. (b) The three pairing interactions associated with the values of the parameters α and η given in table 2.

Argonne interaction in the surface region ($R_{c.m.} \sim 6$ fm). On the other hand, $v^{\delta}_{ind+Arg}$ has a repulsive character in the interior of the nucleus, which is caused by the exchange of spin modes, as can be seen setting G_0, G_0' to zero.

5. Conclusions

We have presented a microscopic approach to the calculation of pairing correlations in superfluid nuclei, adding a bare nucleon-nucleon interaction and an induced interaction which takes into account the exchange of RPA vibrational modes, including self-energy effects in a simplified way. The resulting interaction is surface peaked, in keeping with the fact that the bare interaction is mostly attractive for low values of relative momentum, and that the exchange of collective surface modes gives the largest contribution to the induced interaction at the nuclear surface.

References

[1] Barranco F, Broglia RA, Gori G, Vigezzi E, Bortignon PF and Terasaki J 1999 *Phys. Rev. Lett.* **83** 2147
[2] Barranco F, Broglia RA, Colò G, Gori G, Vigezzi E and Bortignon PF 2004 *Eur.Phys.J.* A **21** 57
[3] Pastore A, Barranco F, Broglia RA and Vigezzi E 2008 *Phys. Rev.* C **78** 024315
[4] Gori G, Ramponi F, Barranco F, Bortignon PF, Broglia RA, Colò G and Vigezzi E 2005 Phys. Rev. C **72** 011302(R)
[5] Colò G and Bortignon PF 2001 *Nucl. Phys.* A **696** 427. Pairing correlations have been calculated in the BCS approximation with a simple seniority force, adjusted to the experimental pairing gap of ^{120}Sn.
[6] Terasaki J, Barranco F, Broglia RA and Vigezzi E 2002 *Nucl. Phys.* A **697** 127
[7] Barranco F, Bortignon PF, Broglia RA, Colò G, Schuck P, Vigezzi E and Viñas X 2005 *Phys. Rev.* C **72** 054314
[8] Ring P and Schuck P 2005 *The nuclear many-body problem* (Berlin: Springer)
[9] Barranco F, Broglia RA, Esbensen H and Vigezzi E 1998 *Phys. Rev.* C **58** 125
[10] Bertsch GF and Esbensen H 1991 *Ann. Phys. (N.Y.)* **209** 327
[11] Garrido E, Sarriguren P, Moya de Guerra E and Schuck P 1999 *Phys. Rev.* C **60** 064312
[12] Dobaczewski J, Nazarewicz W and Reinhard PG 2001 *Nucl. Phys.* A **693** 361
[13] Dobaczewski J, Nazarewicz W and Stoitsov MV 2002 *Eur. Phys. J.* A **15** 21
[14] Bertsch GF, Bertulani CA, Nazarewicz W, Schunk N and Stoitsov MV 2009 *Phys Rev.* C **79** 034306
[15] Matsuo M 2006 *Phys. Rev.* C **73** 04430.

Journal of Physics: Conference Series **168** (2009) 012016 doi:10.1088/1742-6596/168/1/012016

Nuclear physics with electroweak probes

Omar Benhar

INFN and Department of Physics, "Sapienza" Università di Roma, I-00185 Roma, Italy

E-mail: omar.benhar@roma1.infn.it

Abstract. In recent years, the italian theoretical Nuclear Physics community has played a leading role in the development of a unified approach, allowing for a consistent and fully quantitative description of the nuclear response to electromagnetic and weak probes. In this paper I review the main achievements in both fields, point out some of the open problems, and outline the most promising prospects.

1. Introduction

Over the past decades, electron scattering has provided a wealth of information on nuclear structure and dynamics. Form factors and charge distributions have been extracted from elastic scattering data, while inelastic measurements have allowed for a systematic study of the dynamic response over a broad range of momentum and energy transfer. Finally, with the advent of the last generation of continuous beam accelerators, a number of exclusive processes have been analyzed with unprecedented precision.

In electron scattering experiments the nucleus is seen as a target. Studying its interactions with the probe, whose properties are completely specified, one obtains information on the unknown features of its internal structure. In most neutrino oscillation experiments, on the other hand, nuclear interactions are exploited to detect the beam particles, whose properties are largely unknown.

Using the nucleus as a detector obviously requires that its response to neutrino interactions be quantitatively under control. Fulfillment of this prerequisite is in fact critical to keep the systematic uncertainty associated with the reconstruction of the neutrino kinematics to an acceptable level (see, e.g., Ref. [1] and references therein).

The nuclear response to neutrino interactions is also relevant to a number of problems in astrophysics. For example, the knowledge of the mean free path of low energy neutrinos in nuclear matter, over a wide range of temperature and density, is needed as an input to carry out computer simulations of both supernovæ explosions and neutron star cooling.

The community of italian theorists working on electron-nucleus scattering has a long-standing and well established record of achievements. Recently, a significant effort has been made to generalize the approaches successfully employed in electron scattering studies to the case of neutrino scattering. This paper is aimed at providing a short, and by no means exhaustive, summary of the recent results on both the electromagnetic and weak nuclear responses.

In section 2, I will briefly review the ongoing discussion on the role of short range nucleon-nucleon correlations in inclusive electron-nucleus scattering processes, while section 3 is devoted to the latest developments of the studies of two nucleon emission reactions. The results of a number of studies of the neutrino-nucleus cross section in the kinematical region relevant to the analysis of neutrino oscillation experiments are discussed in section 4. Finally, in section 5 I state the conclusions and outline my own view of the future of the field.

2. Correlation effects and final state interactions in $e + A \to e' + X$ processes

The results of electron- and hadron-induced nucleon knockout experiments have provided overwhelming evidence of the inadequacy of the independent particle model to describe the full complexity of nuclear dynamics.

While the peaks corresponding to knockout from shell model orbits can be clearly identified in the measured energy spectra, the corresponding strengths turn out to be consistently and sizably lower than expected, independent of the nuclear mass number.

This discrepancy is mainly due to dynamical correlations induced by the nucleon-nucleon (NN) force, whose effect is not taken into account in the independent particle model. Correlations give rise to scattering processes, leading to the virtual excitation of the participating nucleons to states of energy larger than the Fermi energy, thus depleting the shell model states within the Fermi sea.

The typical energy scale associated with NN correlations can be estimated considering a pair of correlated nucleons carrying momenta $\mathbf{k}_1$ and $\mathbf{k}_2$ much larger than the Fermi momentum (~ 250 MeV). In the nucleus rest frame, where the remaining $A-2$ particles have low momenta, $\mathbf{k}_1 \approx -\mathbf{k}_2 = \mathbf{k}$. Hence, knockout of a nucleon of large momentum $\mathbf{k}$ leaves the residual system with a particle in the continuum and requires an energy

$$E \approx E_{thr} + \frac{\mathbf{k}^2}{2m} , \tag{1}$$

much larger than the Fermi energy (~30 MeV). The above equation, where E_{thr} denotes the threshold for two-nucleon knockout, suggests that large nucleon removal energy and large momentum are strongly correlated. As a consequence, the spectral function $P(\mathbf{k}, E)$, yielding the probability of removing a nucleon carrying momentum $\mathbf{k}$ from the target nucleus leaving the residual system with energy E, is expected to exhibit tails extending to large $\mathbf{k}$ and E, well beyond the region corresponding to the shell model states.

A direct measurement of the spectral function of ^{12}C from the $(e, e'p)$ cross section at missing momentum and energy up to $\sim$ 800 MeV and $\sim$ 200 MeV, respectively, has been recently carried out by the JLab E97-006 Collaboration [2]. The data from the preliminary analysis appear to be consistent with the theoretical predictions of sizable high momentum and high energy components [3, 4, 5].

The search of clearcut evidence of correlation effects in the inclusive electron-nucleus cross section at high momentum transfer (for a recent review see Ref. [6]) has been pursued by a number of experimental and theoretical studies for over three decades.

As the space resolution of the electron probe is $\sim |\mathbf{q}|^{-1}$, where $\mathbf{q}$ is the momentum transfer, at large $|\mathbf{q}|$ scattering off a nuclear target reduces to the incoherent sum of elementary scattering processes involving individual nucleons. This is the basic tenet of the Impulse Approximation (IA). Under the further assumption that there are no final state interactions (FSI) between the struck nucleon and the spectator particles, the inclusive cross section can be written in the simple form

$$\left(\frac{d\sigma}{d\Omega_{e'} d\omega} \right)_{PWIA} = \int d^3k dE \left(\frac{d\sigma}{d\Omega_{e'} d\omega} \right)_{eN} P(\mathbf{k}, E) , \tag{2}$$

where $(d\sigma/d\Omega_{e'} d\omega)_{eN}$ is the cross section of the scattering process involving a *bound* nucleon *moving* with momentum $\mathbf{k}$.

It has long been recognized [7] that in the IA regime short range NN correlations move strength from the quasi-free peak, corresponding to electron energy loss $\omega \sim \omega_{QF} = Q^2/2m$, where $Q^2 = |\mathbf{q}|^2 - \omega^2$ and m is the nucleon mass, to the tails of the inclusive cross section. While the large ω region is dominated by inelastic processes, leading to the appearance of hadrons other than protons and neutrons, the y-scaling analysis [8] (to be discussed in section 4.2) clearly shows that at $\omega \ll \omega_{QF}$ the nuclear cross section is mainly due to quasi-elastic scattering off nucleons carrying high momenta.

However, the results of theoretical studies [9, 10] suggest that the interpretation of experimental data at low ω in terms of correlations may be hindered by the occurrence of FSI.

The existence of strong FSI in quasi-elastic scattering has been experimentally established by the results of $(e, e'p)$ measurements, showing that the flux of outgoing protons is strongly suppressed with respect to the predictions obtained neglecting FSI. The observed attenuation ranges from 20-40 % in Carbon to 50-70 % in Gold [11, 12, 13, 14, 15].

Being only sensitive to FSI taking place within a distance $\sim |\mathbf{q}|^{-1}$ of the electromagnetic vertex, the inclusive cross section at high momentum transfer is in general largely unaffected by FSI. However, the effects of FSI can become appreciable, indeed dominant, in the low ω region, where the cross sections become very small.

In inclusive processes FSI have two effects: i) an energy shift Δ of the cross section, due to the fact that the struck nucleon moves in the average potential generated by the spectator particles and ii) a redistribution of the strength, leading to the quenching of the quasielastic peak and an enhancement of the tails, as a consequence of NN scattering processes coupling the one particle-one hole final state to more complex n-particle n-hole configurations.

As a result, the inclusive cross section can be written in the convolution form [9]

$$\frac{d\sigma}{d\Omega_{e'} d\omega} = \int d\omega' \left(\frac{d\sigma}{d\Omega_{e'} d\omega'} \right)_{PWIA} f_{\mathbf{q}}(\omega - \omega') \ , \tag{3}$$

where $(d\sigma/d\Omega_{e'} d\omega')_{PWIA}$, given by Eq.(2), is the cross section in the absence of FSI. The folding function, embodying FSI effects, is trivially related to the spectral function of particle states. It can be obtained within the eikonal approximation, i.e. assuming that: i) the struck nucleon moves along a straight trajectory with constant speed, and ii) the spectator particles act as fixed scattering centers. The resulting $f_{\mathbf{q}}(\omega)$ can be cast in the form

$$f_{\mathbf{q}}(\omega) = \sqrt{T_{\mathbf{q}}}\ \delta(\omega - \Delta) + (1 - \sqrt{T_{\mathbf{q}}})\ F_{\mathbf{q}}(\omega - \Delta) \ , \tag{4}$$

where the δ-funcion term accounts for the probability that the outgoing nucleon does not interact with the recoiling nucleus and T_q is the nuclear transperency measured in semi-inclusive nucleon knockout experiments. The energy shift Δ, whose typical size is $\sim$ 10 MeV, is barely visible on the energy loss scale of inclusive processes at momentum transfer $\gtrsim$ 1 GeV. On the other hand, the redistribution of the strength induced by the finite width of the function $F_{\mathbf{q}}(\omega)$ may lead to a large enhancement of the cross section at $\omega \ll \omega_{QF}$. This mechanism is illustrated in figure 1, showing the inclusive cross section off uniform nuclear matter, obtained by extrapolation of the available data to $A \to \infty$ [16]. The solid line corresponds to the cross section in the absence of FSI, while the dashed line shows the folding function (displayed in linear scale and multiplied by a factor 10^3) [9]. It clearly appears that going from $\omega \sim$1.2 GeV, where the theoretical cross section is in good agreement with the data, to $\sim$ 0.8 GeV, where the experiment is severely underestimated, the measured cross section drops by more than two orders of magnitude. As a consequence, even a a tiny tail of the folding function extending to $|\omega - \omega'| \sim$ 400 MeV can produce a large enhancement of the cross section at $\omega \sim$ 0.8 GeV.

Theoretical calculations show that the FSI effects on the low energy loss tail of the cross section, corresponding to $x \gg 1$, where $x = Q^2/2m\omega$ is the Bjorken scaling variable, is large, and must be included to reproduce the data.

Journal of Physics: Conference Series **168** (2009) 012016 doi:10.1088/1742-6596/168/1/012016

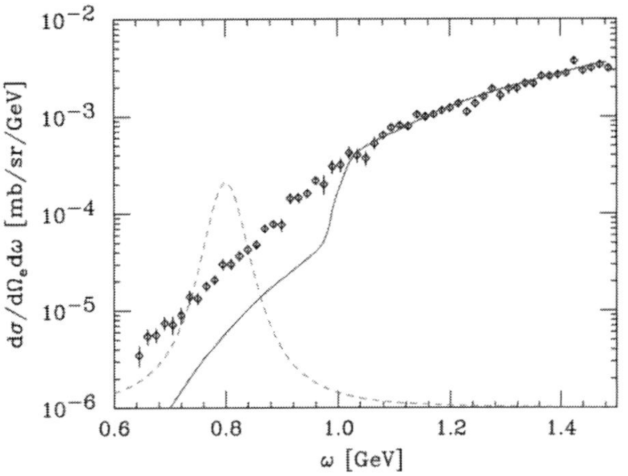

Figure 1. Inclusive cross section off uniform nuclear at beam energy 4.0 GeV and electron scattering angle 30° [16], as a function of the energy loss ω. The solid line corresponds to the cross section in the absence of FSI, while the dashed line shows the folding function (see Eq.(4)), displayed in linear scale and multiplied by a factor 10^3 [9].

As an example, in figure 2 the SLAC data corresponding to iron target, beam energy $E = 3.6$ GeV and electron scattering angle 30° and 39° [17], are compared to the results of Ref. [18]. It clearly appears that, while the dotted line, obtained including only the effect of NN correlations in the quasi elastic channel fails to explain the measured cross section at low energy loss, the inclusion of FSI and inelastic channels brings theory and experiment into agreement over the whole ω range.

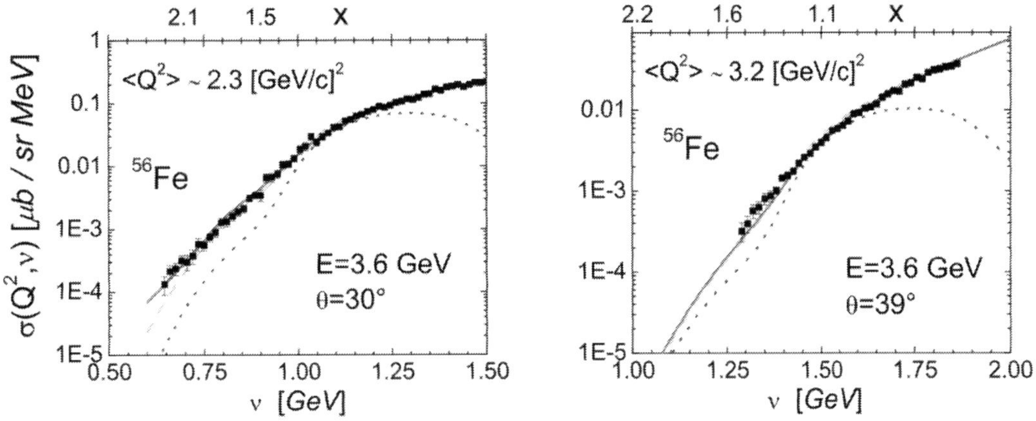

Figure 2. Comparison between the measured cross section of the process $^{56}Fe(e, e')X$ at beam energy $E = 3.6$ GeV [17] and the results of theoretical calculations including the effects of both short range correlations and FSI, represented by the solid lines. The left and right panel correspond to electron scattering angle 30° and 39°, respectively. The labels of the bottom and top horizontal axes refer to the energy transfer and the Bjorken scaling variable x, respectively. (After Ref. [18]).

Journal of Physics: Conference Series **168** (2009) 012016 doi:10.1088/1742-6596/168/1/012016

The remarkable agreement between data and the results of theoretical calculations including *both* NN correlations and FSI is also illustrated in figure 3 [20], showing the Carbon cross section at beam energy $E = 5.8$ GeV and scattering angle 32°, measured at JLab [19].

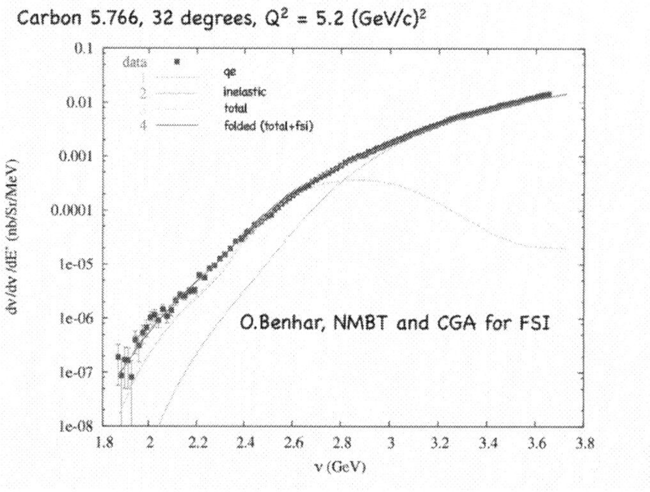

Figure 3. Comparison between the measured cross section of the process $^{12}C(e,e')X$ at beam energy $E = 5.8$ GeV and electron scattering angle 32° [19] and the results of theoretical calculations based on the approach of Ref. [9], represented by the solid line. The labels of the bottom axis refer to the energy transfer. (After Ref. [20]).

The results of figures 2 and 3 strongly suggest that a quantitative understanding of FSI is required to unambiguously identify correlation effects in the region of low energy loss.

A procedure aimed at extracting information on NN correlations, i.e. on the high momentum components of the nuclear wave function, from the inclusive cross section at $x > 1$ has been proposed in Refs. [21, 22], whose authors argue that the appearance of a plateau in the ratio of the cross sections corresponding to different targets is due to the fact that the nuclear momentum distributions have a similar behavior at large momenta, and essentially differ only by an overall factor. Hence, according to Refs. [21, 22] the plateau is a signature of the cancellation of FSI effects in the ratio.

As an example, figure 4 shows the ratios

$$R_A = \frac{3}{A} \frac{d\sigma(e + A \to e' + X)}{d\sigma(e + ^3He \to e' + X)}, \tag{5}$$

for A=4, 12 and 56, corresponging to ^{4}He, ^{12}C and ^{56}Fe, in the range $1 < x < 2.8$. It is of course tempting to interpret the strength near $x = 2$ ($x = 3$) as originating from scattering off correlated two(three)-nucleon systems, with mass $2m$ ($3m$). This interpretation, however, ignores the fact that the data exhibit clear scaling in the variable y [8], hereby proving that the electron scatters off constituents with nucleonic mass and nucleonic form factor. The interpretation of cross section ratios as ratios of correlation strengths is also hindered by the fact that x, unlike y, is not simply related to the momentum carried by the struck nucleon [23].

The assumption of cancellation of FSI effects, underlying the analysis proposed in Refs. [21, 22], is still controversial. Establishing its validity will require a systematic study of the ratios within the approaches which have proved successful in explaining the measured cross sections.

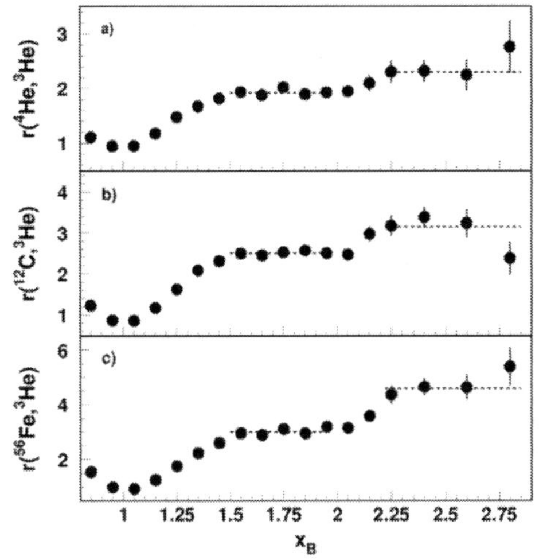

Figure 4. Cross section ratios of (a) ^{4}He, (b) ^{12}C and (c) ^{56}Fe to ^{3}He as a function of the Bjorken scaling variable for $Q^2 > 1.4$ GeV^2. (After Ref. [22]).

3. Electron-induced two-nucleon knockout

The interest in the experimental investigation of two-nucleon emission processes was triggered, at the end of the 1980s, by the prospect of the upcoming new generation of continuous beam electron accelerators [24]. In the following years a number of measurements of $(e, e'NN)$ cross sections have been carried out at NIKHEF-K [26, 27, 28] and MAMI [29].

Ideally, the two-nucleon knockout reaction can be regarded as the cleanest source of experimental information on NN correlations, as it may give access to the two-nucleon spectral function [25]

$$P(\mathbf{k}_1, \mathbf{k}_2, E) = \sum_n |\langle n|a_{\mathbf{k}_1} a_{\mathbf{k}_2}|0\rangle|^2 \delta(E - E_n + E_0) , \qquad (6)$$

yielding the probability of removing two nucleons of momenta $\mathbf{k}_1$ and $\mathbf{k}_2$ from the target ground state $|0\rangle$, leaving the residual $(A-2)$-nucleon system with excitation energy E.

However, it was soon realized that the role of correlations may be obscured by the presence of competing mechanisms, such as FSI, and that extracting information on the spectral function from the data requires i) a careful choice of the kinematical setup and ii) the development of consistent theoretical models including all the relevant effects.

The comparison between the ^{16}O$(e, e'pp)$ cross sections measured at NIKHEF-K and theoretical calculations [30, 31] has shown that correlation effects are dominant in the transition to the ground state of ^{14}C when the two protons are emitted back-to-back with small total momentum. The large cross section measured in this kinematical conditions is in fact regarded as a clean signature of correlations.

Accurate theoretical calculations must include a consistent description of both the two-nucleon overlap functions, containing the information on correlations between the pair of knocked out nucleons in the initial state, and the nuclear electromagnetic current. The effects of FSI of the outgoing nucleons, with one another and with the recoiling nucleus, must be also taken into account.

The results of recent work carried out by the Pavia Group [32, 33] suggest that the requirement that single particle bound and scattering states be orthogonal may play a significant role.

The orthogonalization procedure developed in Refs. [32, 33] takes into account the spurious contributions associated with the center of mass motion, and is therefore suitable for application to targets such as oxygen.

The relevance of the treatment of orthogonalization turns out to depend on the specific process and kinematical setup. While in many instances the effect of the spurious contributions is small, it is very large in the so called super-parallel kinematics, in which the momenta of the two knocked out nucleons are parallel and anti-parallel to the momentum transfer $\mathbf{q}$. This kinematical setup has been adopted in the measurements of the $^{16}O(e,e'pp)^{14}C$ [29] and $^{16}O(e,e'pn)^{14}N$ [34] cross sections carried out at MAMI.

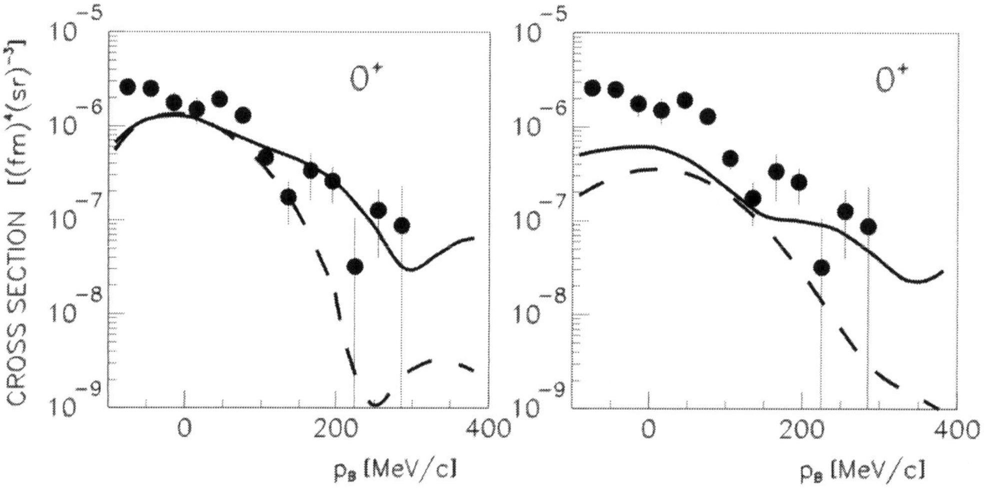

Figure 5. Differential cross section of the reaction $^{16}O(e,e'pp)^{14}C$ in super-parallel kinematics at beam energy $E = 855$ MeV, energy transfer $\omega = 215$ MeV and momentum transfer $|\mathbf{q}| = 316$ MeV, as a function of the recoil momentum $\mathbf{p}_B$. The solid lines have been obtained including the full FSI effect, while the results represented by the dashed lines do not take into account the interactions between the two ejected nucleons. The left and right panels correspond to the different orthogonalization procedures described in the text. The data is taken form Ref. [29]. (After Ref. [35]).

As an example, figure 5 shows the differential cross section of the reaction $^{16}O(e,e'pp)^{14}C$ to the 0^+ ground state in super-parallel kinematics. The data have been taken at incident electron energy $E = 855$ MeV, energy transfer $\omega = 215$ MeV and momentum transfer $|\mathbf{q}| = 316$ MeV [29]. Different values of the recoil momentum $\mathbf{p}_B$ correspond to different kinetic energies of the two outoging nucleons. Positive (negative) values are associated with $\mathbf{p}_B$ parallel (anti-parallel) to $\mathbf{q}$.

The results of the orthogonalized approach with and without removal of the the spuriosity are displayed in the left and right panels, respectively. The solid lines have been obtained including the full FSI effect, while the results represented by the dashed lines do not take into account the interactions between the two ejected nucleons. It clearly appears that FSI effects are large and a correct treatment of the center of mass motion is needed to bring theory and data into agreement at $\mathbf{p}_B > 0$.

4. Neutrino-nucleus scattering

As pointed out in the Introduction, experimental searches of neutrino oscillations exploit neutrino-nucleus interactions to detect the beam particles, whose properties are unknown. The use of nuclear targets as detectors, while allowing for a substantial increase of the event rate, entails non trivial problems, since data analysis requires a quantitative understanding of the neutrino-nucleus interactions. In view of the present experimental accuracy, the treatment of nuclear effect is in fact regarded as one of the main sources of systematic uncertainty.

The description of nuclear dynamics is even more critical to neutrino experiments aimed at obtaining *nucleon* properties from *nuclear* cross sections [36, 37].

Starting in 2001, a series of Workshops on "Neutrino-Nucleus Interactions in the Few GeV Region" (NUINT) has been devoted to the discussion of nuclear effects in neutrino interactions (see Ref.[1] and references therein). The NUINT Workshops are mainly aimed at establishing a connection between the communities of electron- and neutrino-nucleus scattering, with the purpose of generalizing the approaches successfully employed in the analysis of electron-nucleus scattering to the case of neutrino scattering. The simulation codes currently employed by many oscillation experiments are based on the Relativistic Fermi Gas Model (RFGM). According to this model, the nucleon momentum distribution $n(\mathbf{k})$ is flat up to the Fermi momentum k_F, and vanishes at $|\mathbf{k}| > k_F$, while the removal energy is fixed to a constant value ϵ.

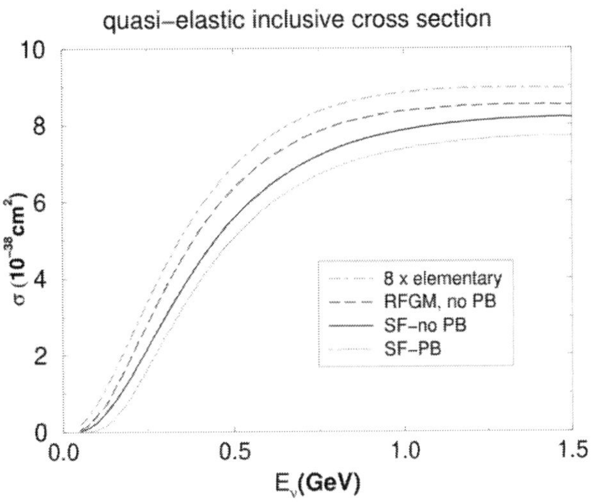

Figure 6. Total quasi-elastic cross section of the process $\nu_e + ^{16}O \to e^- + X$. The dot-dash line represents eight times the elementary cross section; the dashed line is the result of the RFGM with Fermi momentum $k_F = 225$ MeV and binding energy $\epsilon = 25$ MeV; the dotted and solid lines have been obtained using the spectral function of Ref. [5], with and without inclusion of Pauli blocking, respectively. (After Ref. [38]).

In order to gauge the magnitude of the nuclear effects under discussion, and the need of improving upon the RFGM description, figure 6 shows the energy dependence of the quasi elastic contribution to the total cross section of the process $\nu_e + ^{16}O \to e^- + X$ computed using different approximations. The dot-dash line represents the result obtained describing oxygen as a collection of noninteracting stationary nucleons, while the dashed and solid line have been obtained from the RFGM and Eq.(2), with the spectral function of Ref. [5], respectively. It is apparent that replacing the RFGM with the approach based on a realistic spectral function leads to a sizable suppression of the total cross section. Comparison between the dot-dash line and the dotted one, obtained from a generalization of Eq.(2) designed to include the effect of Pauli blocking [39], shows that the overall change due to nuclear effect is ~ 20 %.

Note that FSI between the nucleon produced at the elementary weak interaction vertex and the spectator particles have not been taken into account, as they *do not* contribute to the total cross section.

4.1. Validation of nuclear models through comparison to electron scattering data

The accuracy of the models of nuclear effects to be used in the analysis of neutrino oscillation experiments can be tested by comparing theoretical results to electron scattering data.

The relevant kinematical domain can be readily identified considering the quantum mechanical phase difference developed by two neutrino mass eigenstates over a distance L

$$\Delta\varphi_{jk} = (E_k - E_j)L \approx \frac{\Delta m^2_{jk}}{2E_\nu}L \ , \tag{7}$$

where $\Delta m^2_{12} \approx 6.9 \times 10^{-5}$ eV and $\Delta m^2_{23} \approx 2.5 \times 10^{-3}$ eV. Knowing Δm^2_{23}, the energy E_ν that maximizes the oscillation can be obtained from

$$E_\nu = 0.60 \left[\frac{\Delta m^2_{23}}{3 \times 10^{-3}\ \mathrm{eV}^2}\right] \left[\frac{L}{150\ \mathrm{Km}}\right] \mathrm{GeV} \ , \tag{8}$$

showing that for long baseline experiments E_ν ranges between several hundreds MeV to ~ 1.5 GeV. In this kinematical region quasielastic scattering and pion production, mainly trough Δ excitation, are known to provide the dominant contributions to the cross section [40].

As an example, figure 7 shows a comparison between the cross section of the process $e +^{16} O \to e' + X$ at beam energy $E = 1.2$ GeV and 880 MeV and electron scattering angle $\theta = 32°$, measured at Frascati [41], and the results of the theoretical calculations of Nakamura *et al.* [42]. The solid lines correspond to the full calculation, carried out within the approach based on a realistic spectral function, while the dashed lines have been obtained within the RFGM.

The peaks corresponding to quasi elastic scattering and Δ production are both clearly visible in the data. While the former is reproduced with an accuracy of ~ 10 %, more sizable discrepancies between theory and data occur at ω above pion production threshold. The authors of Ref. [43] argued that they may be ascribed to the lack of accurate parametrizations of the neutron structure functions in the Δ production region at the low values of Q^2 corresponding to the data of Ref. [41] ($Q^2 \sim 0.2$ GeV^2).

Independent of the description of the electron-nucleon vertex, however, the results of figure 7 clearly show that replacing the RFGM with the approach of Ref. [42] leads to a much better overall agreement between theory and data.

4.2. Superscaling analysis

Scaling is observed in a variety of scattering processes involving many-body systems [44]. For example, at large momentum transfer the response of liquid helium measured by inclusive scattering of thermal neutrons, which in general depends upon *both* $\mathbf{q}$ and the energy transfer ω, exhibits a striking scaling behavior, i.e. it becomes a function of the single variable $y = (M/|\mathbf{q}|)(\omega - |\mathbf{q}|^2/2M)$, M being the mass of the helium atom [45]. Scaling in a similar variable occurs in inclusive electron-nucleus scattering at $|\mathbf{q}| \gtrsim 500$ MeV and electron energy loss $\omega < Q^2/2M$ [8]. Another most celebrated example is scaling of the deep inelastic proton structure functions, measured by lepton scattering at large Q^2, in the Bjorken variable x [46].

Being a consequence of the kinematics of the underlying electron-nucleon scattering process, scaling provides a strong handle on the reaction mechanism. Furthermore, the observation of scaling violations reveals that the dynamics go beyond the IA picture.

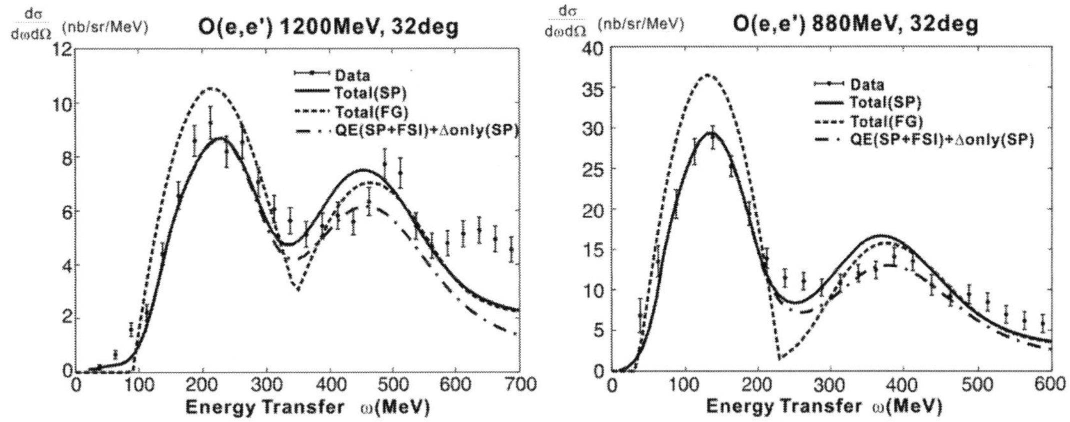

Figure 7. Comparison between the measured cross section of the process $^{16}O(e,e')X$ at electron scattering angle $\theta = 32°$ [41] and the results of theoretical calculations including the effects of both short range correlations and FSI, represented by the solid lines. The left and right panels correspond to electron beam energy $E = 1.2$ GeV and 880 MeV, respectively. For comparison, the results of the RFGM are also shown by dashed lines. (After Ref. [42]).

Figure 8 provides an illustration of y-scaling in electron nucleus scattering. It clearly appears that the inclusive cross sections off Iron at momentum transfer ranging between ~1 GeV and ~3.5 GeV, spanning more than seven orders of magnitude, when plotted as a function of the scaling variable y collapse to a single curve at $y < 0$. The scaling violations observed in the region of positive y, corresponding to large electron energy loss, have to be ascribed to the occurrence of inelastic electron-nucleon scattering. The results of figure 8 suggest that the occurrence of y-scaling can be exploited to predict the inclusive cross section at any value of $|\mathbf{q}|$ in the kinematical regime in which IA is applicable.

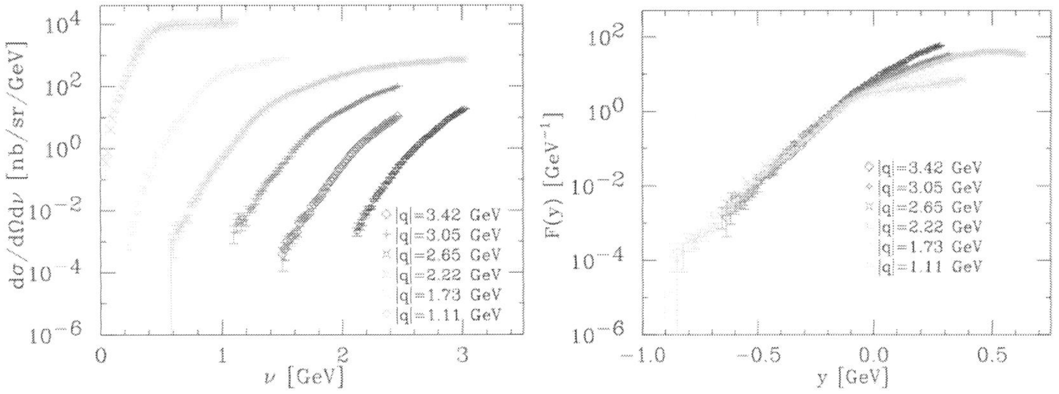

Figure 8. Left panel: Inclusive cross sections of the process $^{56}Fe(e,e')X$ measured at JLab [47]. The different curves correspond to different values of the momentum transfer at the quasi free peak. Right panel: y-scaling functions obtained from the cross sections shown in the left panel.

The scaling analysis has been recently pushed one step further [48]. Motivated by the Fermi gas model, in which all momentum distributions only differ by an overall scale factor, the Fermi momentum, the authors of Ref. [48] have investigated whether the scaling functions of different

Journal of Physics: Conference Series **168** (2009) 012016 doi:10.1088/1742-6596/168/1/012016

nuclei can be related to one other, by adjusting one overall scale factor. As it turned out, this appears to be possible for nuclei with $A \geq 12$.

Figure 9 shows an example of the scaling function $f(\psi')$, obtained from the analysis of the data of Ref. [47], plotted as a function of the new scaling variable ψ', corresponding to y scaled by a Fermi momentum.

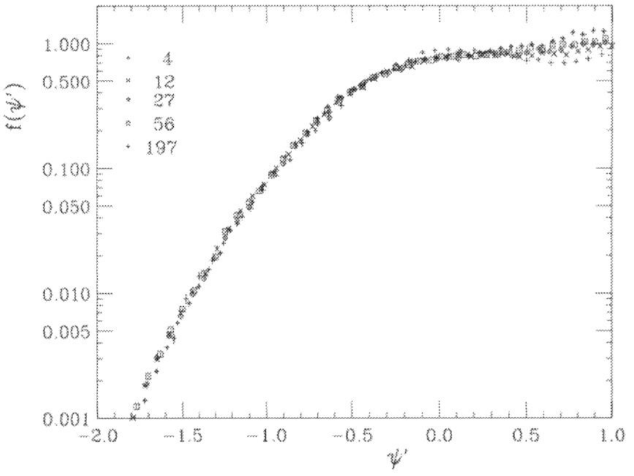

Figure 9. Scaling function of nuclei with A in the range $4 - 197$ at fixed kinematics, corresponding to $|\mathbf{q}| \sim 1$ GeV, plotted as a function of the scaling variable ψ'. (After Ref. [6]).

Scaling as a function of the nuclear mass number, called scaling of second kind, or *superscaling*, appears to be realized better than y-scaling, which is broken by the non quasi elastic contributions to the cross section. At the same kinematics, these contributions are in fact not too dissimilar for different nuclei. As the momentum used to scale y is a slowly varying and smooth function of A, superscaling is very useful to interpolate data and predict the cross sections off nuclei not experimentally investigated.

Inclusive electron scattering and charged current neutrino scattering are closely related, the underlying nuclear physics being the same. Hence, it has been suggested that superscaling may be exploited to accurately predict the cross sections of neutrino induced reactions relevant to the ongoing experimental activity (see, e.g. Ref. [50]).

As shown in figure 7, however, in the kinematical region relevant to long baseline oscillation experiments both quasi elastic scattering and Δ-production contribute to the cross section. The extension of the superscaling approach to the Δ-production region, carried out in Ref. [49], is based on the observation that the cross section exhibits a pronounced peak at $\omega = (|\mathbf{q}|^2 + m_\Delta^2)^{1/2}$, m_Δ being the Δ mass, whose width is determined by the nuclear Fermi momentum. Figure 10 shows the quasielastic and Δ-production superscaling functions obtained from the analysis of electron scattering data.

As an example of the application of the supercaling analysis to neutrino-nucleus scattering, figure 11 shows the charged current interaction cross section corresponding to beam energy $E_\nu =$ 1 GeV and muon emission angle $\theta = 45°$.

The comparison with the predictions of the RFGM, also displayed in both figures 10 and 11, clearly indicates that nuclear dynamics plays a critical role and should not be described using oversimplified approaches. Although the idea of superscaling has been inspired by the RFGM, the scaling functions turn out to be quite different from the ones obtained from this model. As pointed out in Ref. [52], this is a clear indication that some physical effects not included in the RFGM, while not affecting the appearance of scaling, are in fact non negligible.

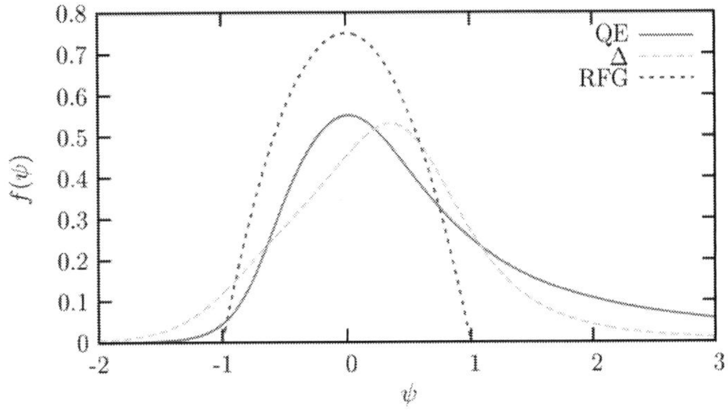

Figure 10. Superscaling function resulting from the analysis of electron scattering data. The solid and long-dashed lines correspond to the quasi elastic and the Δ-production contributions, respectively. For comparison, the short-dashed line shows the results of the RFGM. (After Ref. [51])

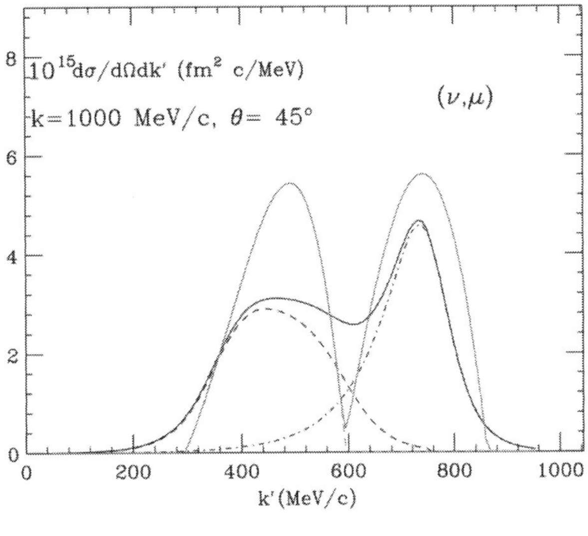

Figure 11. Charged current interaction neutrino-nucleus cross section corresponding to beam energy $E_\nu = 1$ GeV and muon emission angle $\theta = 45°$, plotted as a function of the muon energy. The dot-dash and dashed lines correspond to the quasi elastic and Δ-production contributions, respectively, while the solid line represents the total cross section. For comparison, the result of the RFGM is also shown by a solid line. (After Ref. [51]).

Figure 12 shows the results of calculations including a number of effects neglected in the RFGM: the finite size of the system, its collective excitations, NN correlations, FSI and meson exchange currents [52]. It is apparent that the inclusion of these effects produces sizable modifications of the scaling functions, with respect to the RFGM results. These modifications remarkably improve the agreement with the experimental scaling functions. On the other hand, they do not heavily affect the scaling behavior of the functions f_L and f_T.

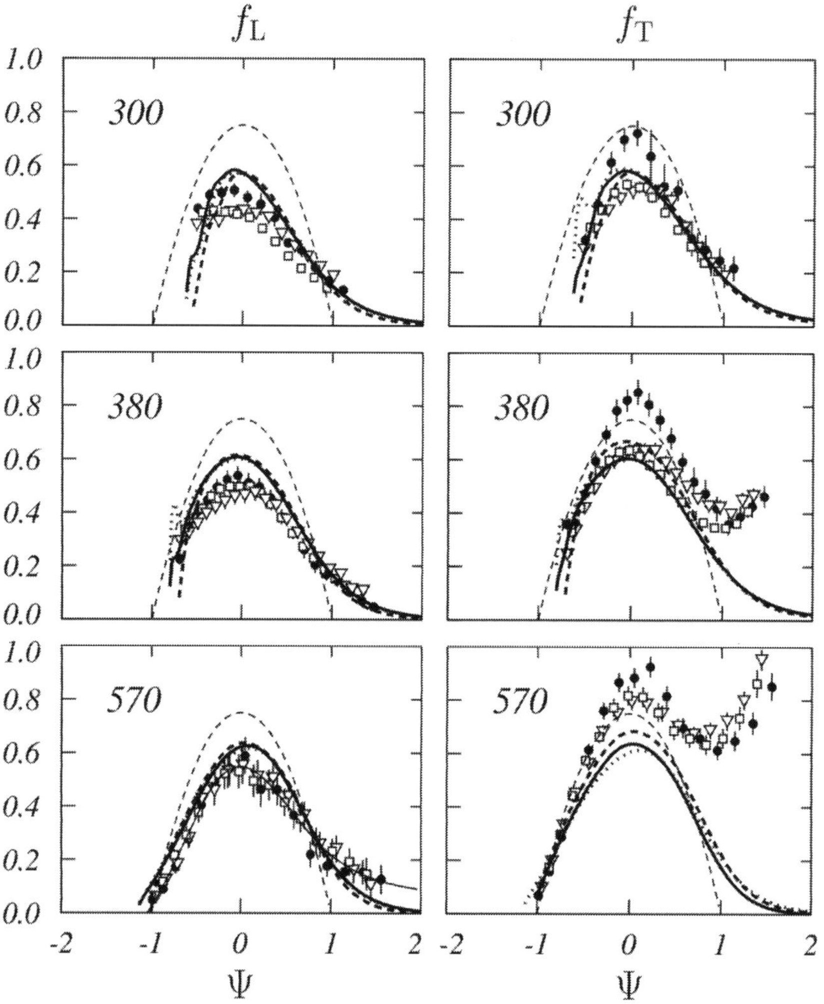

Figure 12. Longitudinal, f_L, and transverse, f_T, scaling functions obtained from the data of Ref. [53]. The panels are labelled according to the value of the momentum transfer in MeV. Full circles, white squares and white triangles refer to ^{12}C, ^{40}Ca and ^{56}Fe , respectively. The thin black line in the f_L panel at 570 MeV represents the empirical scaling function obtained by fitting the data, while the thick lines show the results including effects beyond the RFGM. The full, dotted and dashed lines correspond to ^{12}C, ^{16}O, and ^{40}Ca, respectively. For comparison, the thin dashed lines show the RFGM scaling functions, scaled by a Fermi momentum. (After Ref. [52]).

4.3. Deuteron breakup by low energy neutrinos

Unlike the long baseline neutrino oscillation experiments discussed in the previous section, the experiments aimed at detecting solar neutrino oscillations are sensitive to nuclear interactions of low energy neutrinos.

The search for oscillations of solar neutrinos carried out by the SNO collaboration [54] is based on the determination of the yield of the neutral and charged current deuteron disintegration processes

$$\begin{aligned}\nu_x + d &\longrightarrow \nu_x' + n + p\ ,\\ \nu_e + d &\longrightarrow e^- + p + p\ ,\end{aligned}$$

at neutrino energy E_ν up to $\sim$ 20 MeV. Extracting the information on the solar neutrino flux from the data requires the knowledge of the cross sections of the neutrino- and antineutrino-deuteron breakup cross section.

Accurate theoretical calculations of these processes have been carried out using currents derived from elementary hadron amplitudes, extracted in the tree approximation from the chiral Lagrangians, and nuclear wave functions generated from realistic nuclear potentials [55].

In principle, the alternative approach based on effective field theories should be regarded as more fundamental. However, it involves parameters that cannot be determined from processes involving elementary particles.

Recently, Mosconi *et al* have analized the uncertainties associated with the calculations carried out within the two different approaches. Their work was aimed at assessing the model dependence of the theoretical results, that contributes to the systematic error of the experiment.

Figure 13 shows the energy dependence of the quantity

$$\delta_i^a = 1 - \frac{\sigma_{EFT}^a}{\sigma_{pot,i}}\ , \tag{9}$$

yielding a measure of the difference between the cross sections obtained from potential models, $\sigma_{pot,i}$, and effective field theories, σ_{EFT}^a.

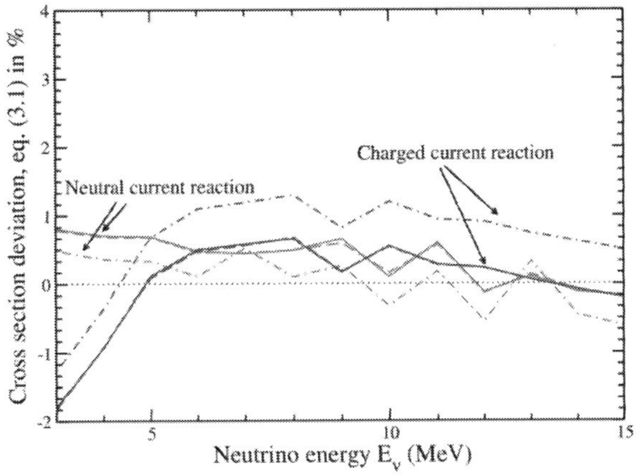

Figure 13. Energy dependence of the quantity δ_i^a, defined by Eq.(9), providing a measure of the differences between the deuteron breakup cross sections obtained from potential models and effective field theories. (After Ref. [56]).

Although the uncertainty turns out to be quite small for both the neutral current and charged current breakup, it is comparable to the full effect of the two-body currents. Based on this observation, the authors of Ref. [56] conclude that the accurate determination of the effect of the two-body currents is still an open issue.

5. Summary and prospects

The italian theoretical nuclear physics community, keeping up its long-standing tradition, is carrying out first class research in the field of electron-nucleus scattering, working in many instances in close connection with experimental collaborations.

After more than three decades, a significant effort is being devoted to the study of the effects of NN correlations, which still elude a precise experimental determination. At the Workshop on Short Range Structure of Nuclei at 12 GeV, held at Jlab in the fall of 2007, it was suggested that, in order to understand the interplay between correlation and FSI contributions to the inclusive cross section at $x \gg 1$, theorists should agree on a well defined *homework* problem, to be used as a test case to compare the results of various approaches. Such a comparison is definitely much needed, as in the past different formalisms have been mostly used to obtain different observables.

While inclusive processes are likely to help identify correlation effects, the $(e, e'p)$ cross section appears to be better suited to carry out a direct measurement of the tails of the nucleon spectral function at high momentum and high removal energy [2]. More exclusive experiments, measuring the double coincidence $(e, e'NN)$ cross sections, may even provide direct access to the internal dynamics of pair correlated nucleons.

In spite of the many difficulties associated with these experiments, as well as with the consistent and realistic theoretical description of the two-nucleon emission process, we are finally approaching the level of development required to make significant comparisons between theory and data. This will help to pin down the role of the competing reaction mechanisms, as well as the kinematical setups in which NN correlations dominate.

The italian groups have been quick to realize that the available theoretical and experimental information on electron-nucleus scattering can be very useful in the analysis of neutrino oscillation experiments. Their contributions, clearly visible in the Proceedings of the Workshops of the NUINT series (see Ref. [1] and references therein) are likely to have a large influence in shaping the field.

The approaches based on the direct calculation of the neutrino-nucleus cross section using realistic spectral functions, as well as the superscaling analysis of electron-scattering data, appear to describe the scattering process much better than the RFGM, currently employed for the data analysis of most long baseline neutrino experiments.

In spite of their success, however, it has to be realized that the potential for exploiting more advanced nuclear models in the analysis of oscillation experiments, in which the nucleus is seen as a detector rather than a target, largely depends upon the possibilty of implementing the formalism in Monte Carlo simulations. This is likely to be one of the main focuses of the research activity in the years to come.

Acknowledgements

I wish to express my gratitude to the organizers of the XII Workshop on Theoretical Nuclear Physics in Italy, for inviting me to a very interesting and productive meeting. A number of illuminating discussions with the participants in the meeting of the INFN Theory Collaboration PI31, held at ECT*, Trento, on March 27-28, 2008, are also gratefully acknowledged.

References

[1] Zeller G P, Morfin J G and Cavanna F (ed) 2007 *Proceedings of NUINT07, AIP Conf. Proc.* **967**
[2] Rohe D *et al* (JLab E97-006 Collaboration) 2004 *Phys. Rev. Lett.* **93** 182501
[3] Benhar O, Fabrocini A and Fantoni S 1989 *Nucl. Phys.* A **505** 267
[4] Ramos A, Polls A and Dickhoff W 1989 *Nucl. Phys.* A **503** 1
[5] Benhar O, Fabrocini A, Fantoni S and Sick I 1994 *Nucl. Phys.* A **579** 493
[6] Benhar O, Day D and Sick I 2008 *Rev. Mod. Phys.* **80** 189
[7] Czyz W and Gottfried K 1963 *Ann. Phys.* **21** 47

[8] Sick I, Day D and McCarthy J S 1980 *Phys. Lett.* **45** 872
[9] Benhar O *et al* 1991 *Phys. Rev.* C **44** 2328
[10] Ciofi degli Atti C and Simula S 1994 *Phys.Lett.* B **325** 276
[11] Garino G. *et al* 1992 *Phys. Rev.* C **45** 78
[12] O'Neill T G *et al* 1995 *Phys. Lett.* B **351** 87
[13] Abbott D *et al* 1998 *Phys. Rev. Lett.* **80** 5072
[14] Garrow K *et al* 2002 *Phys. Rev.* C **66** 044613
[15] Rohe D *et al* (JLab E97-006 Collaboration) 2005 *Phys. Rev.* C **72** 054602
[16] Day D *et al* 1989 *Phys. Rev.* C **40** 1011
[17] Day D *at al* 1987 *Phys. Rev. Lett.* **59** 427
[18] Alvioli M *et al* 2008 *AIP Conf. Proc.* **1056** 307
[19] Fomin N 2009 Ph. D. Thesis, University of Virginia, http://www.jlab.org/~fomin/test.pdf
[20] Day D 2007, talk at SRC07 Jefferson Lab, October 26-27, 2007, http://conferences.jlab.org/src2007/talks/Donal-Day.pdf
[21] Egiyan K S *et al* 2003 *Phys. Rev.* C **68** 14313
[22] Egiyan K S *et al* (JLab CLAS Collaboration) 2006 *Phys. Rev. Lett.* **96** 082501
[23] Liuti S 1993 *Phys. Rev.* C **47** 1854
[24] Benhar O and Fabrocini A (ed) 1990 *Two Nucleon Emission Reactions* (ETS, Pisa)
[25] Benhar O and Fabrocini A 2000 *Phys. Rev.* C **62** 034304
[26] Onderwater C J G *et al* 1997 *Phys. Rev. Lett.* **78** 4893
[27] Onderwater C J C *et al* 1998 *Phys. Rev. Lett.* **81** 2213
[28] Starink R *et al* 2000 *Phys. Lett.* B **474** 33
[29] Rosner G 2000 *Prog. Part. Nucl. Phys.* **44** 99
[30] Ryckebusch J *et al* 1997 *Nucl. Phys.* A **624** 581
[31] Giusti C *et al* 1998 *Phys. Rev.* C **57** 1691
[32] Giusti C, Pacati F D, Schwamb M and Boffi S 2007 *Eur. Phys. J.* A **31** 155
[33] Giusti C, Pacati F D, Schwamb M and Boffi S 2007 *Eur. Phys. J.* A **33** 29
[34] Middleton D G *et al* 2006 *Eur. Phys. J.* A **29** 261
[35] Giusti C *et al* 2007 talk at the XVII International School on Nuclear Physics, Neutron Physics and Applications, Varna, Bulgaria, September 24-30, 2007, *Preprint* arXiv:0801.2304
[36] Gran R *et al.* (K2K Collaboration) 2008 *Phys. Rev.* D **74** 052002
[37] Aguilar Arevalo A A *et al.* (MiniBooNE) 2008 *Phys. Rev. Lett.* **98** 231801
[38] Benhar O and Meloni D 2007 *Nucl. Phys.* A **789** 379
[39] Benhar O *et al* 2005 *Phys. Rev.* D **72** 053005
[40] Lipari P, Lusignoli M and Sartogo F 1995 *Phys. Rev. Lett.* **74** 4384
[41] Anghinolfi M *et al* 1996 *Nucl. Phys.* A **602** 405
[42] Nakamura H, Nasu T, Sakuda M and Benhar O 2007 *Phys. Rev.* C **76** 065208
[43] Benhar O and Meloni D 2006 *Phys. Rev. Lett.* **97** 192301
[44] West G B 1975 *Phys. Rep.* **18** 263
[45] Sosnick T R , Snow W M , Silver R N and Sokol P E 1991 *Phys. Rev.* B **43** 216
[46] Friedman J I and Kendall H W 1972 *Ann. Rev. Nucl. Sci.* **22** 203
[47] Arrington J *et al* 2001 *Phys. Rev.* C **64** 014602
[48] Donnelly T W and Sick I 1999 *Phys. Rev.* C **60** 065502
[49] Amaro J E *et al* 2005 *Phys. Rev.* C **71** 15501
[50] Barbaro M B 2006 *Nucl. Phys.* B *(Proc. Suppl.)* **159** 186
[51] Barbaro M B *et al* 2007 talk at the 26th International Workshop on Nuclear Theory, Rila Mountains, Bulgaria, June 25-30, 2007, *Preprint* arXiv:0710.4089
[52] Martini M, Cò G, Anguiano M and Lallena A M 2007 *Phys. Rev.* C **75** 034604
[53] Jourdan J 1996 *Nucl. Phys.* A **603** 117
[54] Ahmed S N *et al.* 2004 *Phys. Rev. Lett.* **92** 181301
[55] Mosconi B, Ricci P and Truhlík E 2006 *Nucl. Phys.* A **772** 81
[56] Mosconi B, Ricci P, Truhlík E and Vögel P 2007 *Phys. Rev.* C **75** 044610

Weak response of nuclear matter at low momentum transfer

Nicola Farina

INFN, Sezione di Roma. I-00185 Roma, Italy

E-mail: nicola.farina@roma1.infn.it

Abstract. A quantitative understanding of the weak nuclear response is a prerequisite for the computer simulations of astrophysical phenomena like supernovæ explosions and neutron star cooling. In order to reduce the systematic uncertainties associated with the simulations, a consistent framework, able to take into account dynamical correlation effects, is needed to compute neutrino-nucleon and neutrino-nucleus reaction rates. In this paper we describe the many-body theory of the weak nuclear response at low energy regime. We show how to include both short and long range correlations effects in a consistent fashion.

1. Introduction

The description of neutrino interactions at low momentum transfer ($E_\nu \sim 10$ MeV) with nuclei, and nuclear matter in general, is relevant to the study of many different problems, from supernovæ explosions [1] to neutron star cooling [2].

The systematic uncertainty associated with computer simulations depends heavily on the values of the neutrino-nucleon and neutrino-nucleus reaction rates used as inputs.

Nuclear many body theory provides a scheme allowing for a consistent treatment of neutrino-nucleus interactions. Within this approach, nuclear dynamics is described by a phenomenological hamiltonian, whose structure is completely determined by the available data on two- and three-nucleon systems, and both short and long range dynamical correlations are taken into account.

Over the past decade, the formalism based on correlated wave functions, originally proposed to describe quantum liquids [3], has been employed to carry out highly accurate calculations of the binding energies of both nuclei and nuclear matter, using either the Monte Carlo method [4, 5, 6] or the cluster expansion formalism and the Fermi Hypernetted Chain integral equations [7, 8, 9].

A different approach, recently proposed in Refs. [10, 11] exploits the correlated wave functions to construct an effective interaction suitable for use in standard perturbation theory. This scheme has been employed to obtain a variety of nuclear matter properties, including the neutrino mean free path [10] and the transport coefficients [11, 12].

In this work we describe the application of the formalism based on correlated wave functions and the effective interaction to the calculation of the weak response of uniform nuclear matter.

2. Many-body theory of the weak nuclear response

2.1. Neutrino-nucleus cross section

The differential cross section of the process

$$\nu_\ell + A \to \ell + X \ , \tag{1}$$

in which a neutrino carrying initial four-momentum $k \equiv (E_\nu, \mathbf{k})$ interacts with a nuclear target, producing a lepton in a state of four-momentum $k' \equiv (E_\ell, \mathbf{k}')$, the target final state being undetected, can be written in Born approximation as (see, e.g., Ref. [17])

$$\frac{d\sigma}{d\Omega_\ell dE_\ell} = \frac{G^2}{32\pi^2} \frac{|\mathbf{k}'|}{|\mathbf{k}|} L_{\mu\nu} W^{\mu\nu} \ , \tag{2}$$

where $G = G_F \cos\theta_C$, G_F and θ_C being Fermi's coupling constant and Cabibbo's angle. The leptonic tensor, that can be written, neglecting all lepton mass, as

$$L_{\mu\nu} = 8 \left[k_\mu k'_\nu + k_\nu k'_\mu - g_{\mu\nu}(kk') - i\epsilon_{\mu\nu\alpha\beta} k^\alpha k'^\beta \right] \ , \tag{3}$$

is completely determined by lepton kinematics, whereas the nuclear tensor $W^{\mu\nu}$ contains all the information on target structure. Its definition involves the initial and final hadronic states $|0\rangle$ and $|X\rangle$, carrying four-momenta p_0 and p_X, respectively, as well as the nuclear weak current operator J^μ:

$$W^{\mu\nu} = \sum_X \langle 0|J^\mu|X\rangle\langle X|J^\nu|0\rangle \delta^{(4)}(p_0 + q - p_X) \ , \tag{4}$$

where the sum includes all hadronic final states and $q = k - k'$.

In the low momentum transfer regime, we expect the non relativistic approximation to be applicable. Within this approach, the initial and final states can be obtained from nuclear many-body theory (NMBT), while the weak current entering the definition of the hadronic tensor [4] is expanded in powers of $|\mathbf{q}|/m$. At leading order, the resulting response can be written in the simple form

$$S(\mathbf{q}, \omega) = \frac{1}{N} \sum_n \langle 0|O^\dagger_{\mathbf{q}}|n\rangle\langle n|O_{\mathbf{q}}|0\rangle \delta(\omega + E_0 - E_n) \ . \tag{5}$$

where, in the case of charged current interactions, $O_{\mathbf{q}}$ is the operator corresponding to Fermi or Gamow-Teller transitions, whose expressions in the coordinate space are:

$$\begin{aligned} O^F_i(\mathbf{q}) &= g_V \delta(\mathbf{r}_i - \mathbf{r}'_i) \ e^{i\mathbf{q}\mathbf{r}_i} \tau^+_i \\ \mathbf{O}^{GT}_i(\mathbf{q}) &= g_A \delta(\mathbf{r}_i - \mathbf{r}'_i) \ e^{i\mathbf{q}\mathbf{r}_i} \boldsymbol{\sigma}_i \tau^+_i \ , \end{aligned} \tag{6}$$

where $\mathbf{r}_i$ specifies the position of the i–th particle.

2.2. Description of the initial and final states

Understanding the properties of matter at densities comparable to the central density of atomic nuclei is made difficult by *both* the complexity of the interactions *and* the approximations implied in any theoretical description of quantum mechanical many-particle systems.

The main problem associated with the use of the nuclear potential models in a many-body calculation lies in the strong repulsive core of the nucleon-nucleon (NN) force, which cannot be handled within standard perturbation theory.

Journal of Physics: Conference Series **168** (2009) 012017 doi:10.1088/1742-6596/168/1/012017

Within NMBT, a nuclear system is seen as a collection of point-like protons and neutrons whose dynamics are described by the hamiltonian

$$H = \sum_i t(i) + \sum_{j>i} v(ij) + \ldots \ , \tag{7}$$

where $t(i)$ and $v(ij)$ denote the kinetic energy operator and the *bare* NN potential, respectively, while the ellipses refer to the presence of additional many-body interactions.

Carrying out perturbation theory in the basis provided by the eigenstates of the noninteracting system requires a renormalization of the NN potential. This is the foundation of the widely employed approach developed by Brückner, Bethe and Goldstone, in which $v(ij)$ is replaced by the well-behaved G-matrix, describing NN scattering in the nuclear medium (see, e.g. Ref.[13]). Alternatively, the many-body Schrödinger equation, with the hamiltonian of Eq.(7), can be solved using either the variational method or stochastic techniques. These approaches have been successfully applied to the study of both light nuclei [4] and uniform neutron and nuclear matter [7, 5, 6, 18].

Our work has been carried out using the scheme of correlated basis function (CBF) theory in which nonperturbative effects due to the short-range repulsion are embodied in the basis functions.

The *correlated* states of nuclear matter are obtained from the Fermi gas (FG) states $|n_{FG}\rangle$ through the transformation [3, 19]

$$|n) = \frac{F|n_{\rm FG}\rangle}{\langle n_{\rm FG}|F^\dagger F|n_{\rm FG}\rangle^{1/2}} \ . \tag{8}$$

In the above equation, $|n_{\rm FG}\rangle$ is a determinant of single particle states describing N noninteracting nucleons. The operator F, embodying the correlation structure induced by the NN interaction, is written in the form

$$F(1,\ldots,N) = \mathcal{S} \prod_{j>i=1}^{N} f_{ij} \ , \tag{9}$$

where $\mathcal{S}$ is the symmetrization operator which takes care of the fact that, in general,

$$[f_{ij}, f_{ik}] \neq 0 \ . \tag{10}$$

The structure of the two-body *correlation functions* f_{ij} must reflect the complexity of the NN potential. The shape of the radial functions $f^n(r_{ij})$ is determined through functional minimization of the expectation value of the nuclear hamiltonian in the correlated ground state

$$E_0^V = (0|H|0) \ . \tag{11}$$

As an example, figure 1 shows the radial dependence of the potentials and correlation functions acting in the spin-isospin channels $S = 0$ and $T = 0$ and $S = 0$ and $T = 1$.

It has to be pointed out that the correlation operator of Eq.(9) is defined such that, if any subset of the particles, say $i_1, \ldots i_p$, is removed far from the remaining $i_{p+1}, \ldots i_N$, it factorizes according to

$$F(1,\ldots,N) \to F_p(i_1,\ldots i_p) F_{N-p}(i_{p+1},\ldots i_N) \ . \tag{12}$$

The above property is the basis of the cluster expansion formalism, that allows one to write the matrix element of a many-body operator between correlated states as a sum, whose terms correspond to contributions arising from isolated subsystems (*clusters*) involving an increasing number of particles.

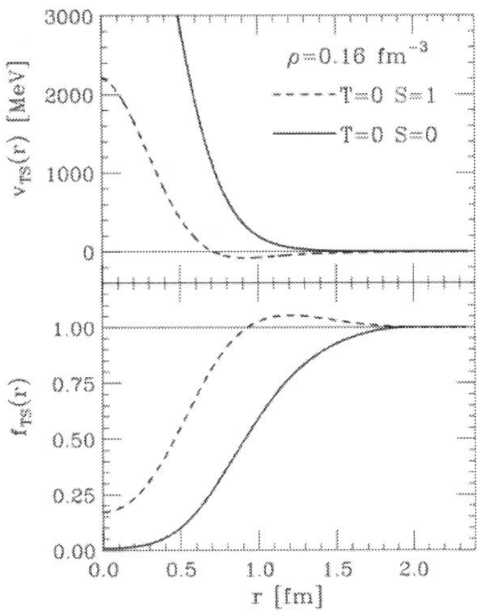

Figure 1. Interaction potentials (upper panel) and correlation functions (lower panel) acting in the spin-isospin channels $S = 0$ and $T = 0$ (solid lines) and $S = 0$ and $T = 1$ (dashed lines). The potential is the Argonne v_8' [15] and the correlation functions correspond to nuclear matter at equilibrium density.

3. Correlation effects in the transition matrix elements

Using correlated states implies severe difficulties in the explicit calculation of the weak matrix element. In the FG model, the nuclear response is non vanishing only when the final nuclear state differs from the initial state for the presence of a particle excited outside the Fermi sea and a hole in the Fermi sea. In the presence of correlations, which can induce virtual nucleon-nucleon scattering processes leading to excitation of nucleons to states outside the Fermi sea, more complex scenarios must also be considered. For example, if the initial state has a two particle-two hole component, the final state can be a three particle-three hole state or, if the probe interacts with an excited nucleon, a two particle-two hole state.

In the following we will consider only the dominant transition, between the *correlated* ground state and a *correlated* one particle-one hole (ph) state. The corresponding weak matrix element can be written

$$M_{ph} = \frac{\langle ph|F^\dagger O F|0\rangle}{\langle ph|F^\dagger F|ph\rangle^{\frac{1}{2}}\langle 0|F^\dagger F|0\rangle^{\frac{1}{2}}} , \tag{13}$$

where F is the correlation operator defined in Eq.(9). Here the kets $|0\rangle$ and $|ph\rangle$ correspond to the ground and one particle-one hole Fermi Gas states, respectively, and $O = \sum_i O_i$, O_i being the Fermi or the Gamow-Teller transition operator (see Eqs.(6)).

Note the the $g_{ij} = f_{ij} - 1$ is short ranged, and therefore its matrix elements are small. At two-body level, the cluster expansion of Eq.(13) yields [21]

$$\langle ph|F^\dagger O F|0\rangle \simeq \langle ph|(1 + \sum_{j>i} g_{ij})O(1 + \sum_{j>i} g_{ij})|0\rangle . \tag{14}$$

The above equation suggests the definition of an *effective operator* O_{12}^{eff}, acting on Fermi Gas states. From

$$\langle ph|F^\dagger O F|0\rangle = \langle ph|O^{eff}|0\rangle. \tag{15}$$

it follows that, at the two-body cluster level (compare to Eq.(14))

$$\frac{1}{N} O_{12}^{eff} = O_1 + \frac{N-1}{2}\{O_1 + O_2, g_{12}\} + \frac{N-1}{2}[g_{12}(O_1 + O_2)g_{12}] \; , \tag{16}$$

where $\{A, B\} = AB + BA$ and N denotes the number of particles. Note that the O^{eff} is a two-body operator, as it includes screening effects arising from nucleon-nucleon correlations.

We have carried out the calculation of M_{ph} on a cubic lattice, for a discrete set of N_h states $\mathbf{h_i}$, satisfying the conditions $\mathbf{h_i} < k_f$ and $\mathbf{h_i} + \mathbf{q} > k_f$, k_f being the Fermi momentum. Figure 2 shows a comparison between the correlated and the Fermi Gas matrix element for a Fermi transition. It has to be noticed that the effects of correlations is enhanced in the calculation of the response, see Eq.(5), whose definition involves the square of the correlated matrix element.

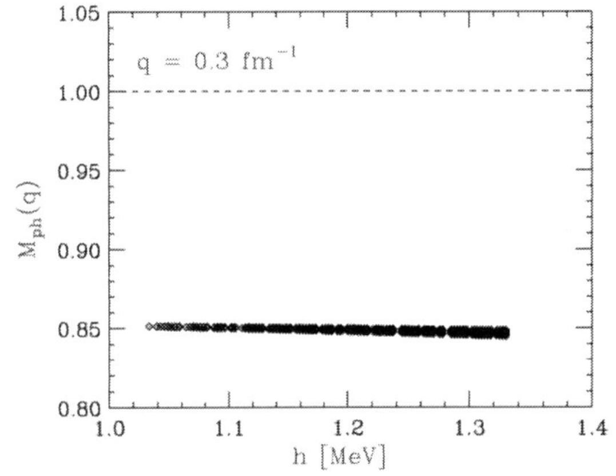

Figure 2. Fermi transition matrix element at $|\mathbf{q}| = 0.3$ fm^{-1} as a function of the magnitude of hole momentum $|\mathbf{h}|$. The dashed horizontal line corresponds to the result of the FG model.

4. The effective interaction

At lowest order of CBF, the effective interaction $V_{\rm eff}$ is *defined* by

$$\langle H \rangle = \langle 0_{FG}|F^\dagger H F|0_{FG}\rangle = \langle 0_{FG}|T_0 + V_{\rm eff}|0_{FG}\rangle \; . \tag{17}$$

As the above equation suggests, the approach based on the effective interaction allows one to obtain any nuclear matter observables using perturbation theory in the FG basis. However, as discussed in the previous Section, the calculation of the hamiltonian expectation value in the correlated ground state, needed to extract $V_{\rm eff}$ from Eq.(17), involves severe difficulties. In this work we follow the procedure developed in Refs. [14, 10], whose authors derived the expectation value of the effective interaction by carrying out a cluster expansion of the rhs of Eq.(17), and keeping only the two-body cluster contribution. The resulting expression can be written

$$\langle 0_{FG}|V_{\rm eff}|0_{FG}\rangle = \sum_{i<j}\langle ij|v_{\rm eff}(12)|ij\rangle_a = \sum_{i<j}\langle ij|f_{12}\left[-\frac{1}{m}(\nabla^2 f_{12}) + v(12)f_{12}\right]|ij\rangle_a \; , \tag{18}$$

where the laplacian operates on the relative coordinate and the suffix a denotes that the ket $|ij\rangle_a$ is antisymmetrized.

In the left panel of figure 3 the central components of the effective interaction obtained from the v_8' potential of Ref.[15] and the three-nucleon interaction (TNI) effects from Ref.[16] at

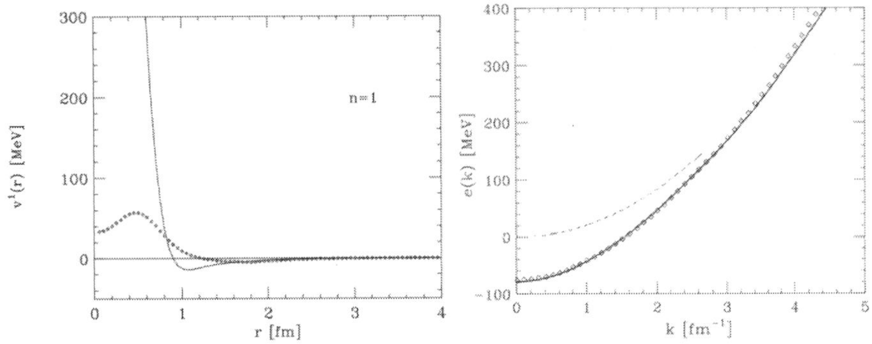

Figure 3. Left panel: Comparison between the central component of the bare Argonne v_8'[15] potential (solid line) and the effective potential defined by Eq.(17) (dashed line), calculated at nuclear matter equilibrium density. Right panel: momentum dependence of the single particle energies obtained from the CBF effective interaction in the Hartree-Fock approximation (solid line). The diamonds show the results of Ref.[20]. For reference, the energies of the Fermi Gas model are also shown (dashed line).

equilibrium density is compared to the corresponding component of the bare interaction. It clearly appears that screening effects due to NN correlations lead to a significant quenching.

Using the effective interaction described above, the single particle energies $e_{\mathbf{k}}$ in nuclear matter can be easily computed in Hartree-Fock (HF) approximation. The risulting expression is:

$$e_{\mathbf{k}} = \frac{\mathbf{k}^2}{2m} + \sum_{h<k_f} \langle hk|V_{eff}|hk\rangle_a \ , \tag{19}$$

The right panel of figure 3 shows the energy spectrum for symmetric nuclear matter at equilibrium density.

5. Correlation effects on the response.

5.1. Short range correlations.

We have calculated the nuclear response for a Fermi transition (Eqs.(5) and (6)). The inclusion of interaction leads to sizable modifications of the FG response. Correlation effects in the transition matrix elements, taken into account through the use of the effective operator, produce a quenching of ~ 15 %. This feature is apparent in figure 4, where the FG response, represented by the dot-dashed line, is compared to that obtained using the effective operator in the calculation of M_{ph} and the FG single particle spectrum, represented by the dashed line. An even larger modification is produced by interaction effects on the single particle energies. Replacing the FG single particle energies with the HF energies (Eq.(19)) leads to a sizable broadening of ω region corresponding to non-vanishing response as shown by the solid line in figure 4.

5.2. Long range correlations: the Tamm-Dancoff Approximation (TDA)

In the previous section we have discussed the nuclear response in the correlated HF approximation, in which the bare Fermi transition operator is replaced by the effective operator of Eq.(15) and the final state is assumed to be a one particle-one hole state.

It is important to realize that the FG one particle-one hole states, while being eigenstates of the HF hamiltonian, defined as

$$H_{HF} = \sum_i e_i \ , \tag{20}$$

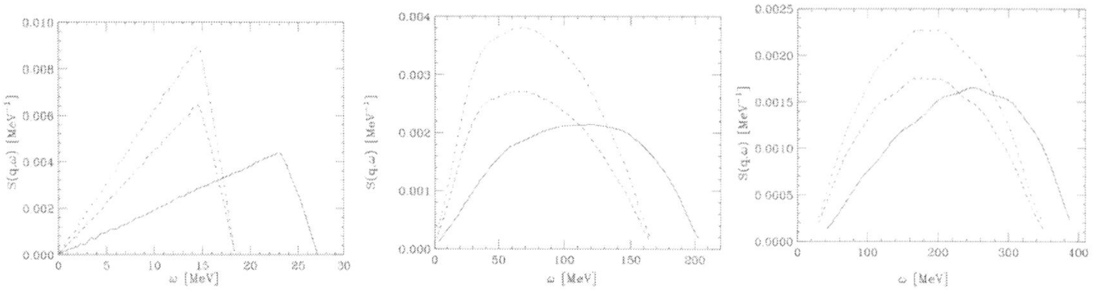

Figure 4. Nuclear response for Fermi transition, for $\mathbf{q} = .3$ fm^{-1} (left), $\mathbf{q} = 1.8$ fm^{-1} (center), and $\mathbf{q} = 3$ fm^{-1} (right). Solid line is full calculation. Confrontation with Fermi Gas (dot-dashed line) and effective operator with Fermi Gas energy spectrum (dashed line) are shown.

with e_i given by Eq.(19), are not eigenstates of the full nuclear hamiltonian. As a consequence, there is a residual interaction V_{res}, which can be identified with the effective interaction defined in Section 4, that can induce transitions between different one particle-one hole states, as long as their total momentum $\mathbf{q}$, spin and isospin are conserved.

In order to include the effects of these transitions, we use the TDA, which amounts to expanding the final state in the basis of one particle-one hole states according to[22]

$$|f\rangle = |\mathbf{q},\ TSM\rangle = \sum_i c_i^{TSM} |\mathbf{h}_i,\ \mathbf{p}_i = \mathbf{h}_i + \mathbf{q},\ TSM\rangle \ , \tag{21}$$

where S and T denote the total spin and isospin of the particle-hole pair and M is the spin projection.

At fixed $\mathbf{q}$, the excitation energy of the state (21), ω^f, as well as the coefficients c_i^{TSM}, are obtained solving the eigenvalue equation

$$H|f\rangle = (H_{HF} + V_{res})|f\rangle = \left(\sum_i e_i + \sum_{ij} v_{eff}(ij) \right) |f\rangle = (E_0 + \omega^f)|f\rangle \ , \tag{22}$$

where E_0 is the ground state energy. In the TDA, the response can be written as

$$S(\mathbf{q}, \omega) = \sum_{T,M,S} \sum_n \left| \sum_i (c_n^{TSM})_i \langle \mathbf{h}_i,\ \mathbf{p}_i,\ TSM | O_{eff}(\mathbf{q}) | 0 \rangle \right|^2 \delta(\omega - \omega_n^{TSM}) \ , \tag{23}$$

where $(c_n^{TSM})_i$ denotes the i-th component of the eigenvector belonging to the eigenvalue ω_n^{TSM}.

We have performed the diagonalization of the Hamiltonian operator over a basis of $N_h \sim 3000$ one particle-one hole states for any spin-isospin channel. The appearance of an eigenvalue lying well outside the particle-hole continuum, corresponding to a collective excitation, reminiscent to the plasmon mode of the electron gas, is clearly visible in figure 5, where the calculation of the response for a Fermi transition within TDA is compared with the response of the correlated HF approximation. The TDA response exhibits a sharp isolated peak corresponding to the collective mode, lying $\sim$ 4 MeV above the upper limit of the particle-hole continuum.

6. Conclusions

We have carried out calculations of the charged current weak response of nuclear matter in the low momentum transfer regime. The quantitative understanding of this quantity is required in

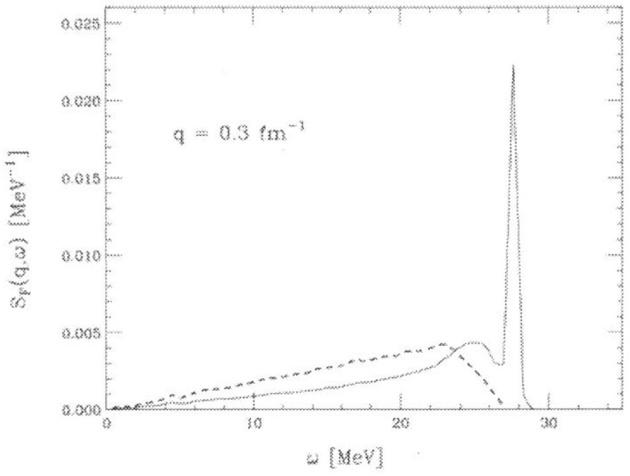

Figure 5. Nuclear matter response in the TD approximation at $|\mathbf{q}| = 0.3$ fm^{-1} (solid line). The dashed line shows the results of the correlated HF approximation.

many problems of physics, ranging from simulations of supernovæ explosions to neutron star cooling.

The calculation has been performed using a many-body approach based on a realistic nuclear hamiltonian, including two- and three-nucleon interactions, yielding a good description of the properties of both the two-nucleon systems and uniform nuclear matter. The response associated with Fermi transitions at low momentum transfer has been calculated from an effective interaction, derived using the formalism of correlated basis functions and the cluster expansion technique. Our work improves upon existing effective interaction models in that it includes the effects of many-nucleon forces, which become sizable, at high density. The responses calculated within the correlated HF approximation show that the inclusion of short range correlations leads to a significant quenching of the transition matrix elements and shifts the strength towards larger values of the energy transfer, ω, for all values of $|\mathbf{q}|$. At low momentum transfer, long range correlations have been taken into account within the TDA. The excitation of the coherent state can be clearly seen in our results at $|\mathbf{q}| = 0.3$ fm^{-1}.

References

[1] Fryer C L and Young P 2007 *Ap. J.* **659** 1438
[2] Yakovlev D G, Kaminker A D, Gnedin O Y and Haensel P 2001 *Phys. Rep.* **354** 1
[3] Feenberg E 1969 *Theory of Quantum Fluids* (New York: Academic Press)
[4] Pieper S C and Wiringa R B 2001 *Ann. Rev. Nucl. Part. Sci.* **51** 53
[5] Carlson J, Morales J Jr, Pandharipande V R and Ravenhall D G 2003 *Phys. Rev.* C **68** 025802
[6] Sarsa A, Fantoni S, Schmidt K E and Pederiva F 2003 *Phys. Rev.* C **68** 024308
[7] Akmal A and Pandharipande V R 1997 *Phys. Rev.* C **56** 2261
[8] Bisconti C, de Saavedra F A, Cò G and Fabrocini A 2006 *Phys. Rev.* C **73** 054304.
[9] Bisconti C, de Saavedra F A, Cò G and Fabrocini A 2007 *Phys. Rep.* C **450** 1
[10] Cowell S and Pandharipande V R 2004 *Phys. Rev.* C **70** 035801
[11] Benhar O and Valli M 2007 *Phys. Rev. Lett.* **99** 232501
[12] Benhar O, Farina N, Fiorilla S and Valli M 2008 *AIP Conf. Proc.* **1056** 248
[13] *Nuclear Methods and Nuclear Equation of State* 1999 Ed. Baldo M (Singapore: World Scientific)
[14] Cowell S 2004 Ph.D. Thesis, University of Illinois at Urbana-Champaign
[15] Pudliner B S, Pandharipandhe V R, Carlson J, Pieper S C and Wiringa R B 1997 *Phys. Rev.* C **56** 1720
[16] Lagaris I and Pandharipande V R 1981 *Nucl. Phys.* A **359** 349
[17] Benhar O, Farina N, Nakamura H, Sakuda M and Seki R 2005 *Phys. Rev.* D **72** 053005
[18] Gandolfi S, Pederiva F, Fantoni S and Schmidt K E 2007 *Phys. Rev. Lett.* **98** 102503
[19] Clark J W 1979 *Prog. Part. Nucl. Phys.* **2** 89
[20] Wiringa R B 1988 *Pyhs. Rev. C* **38** 2967
[21] Farina N 2008 Ph.D. Thesis *Preprint* arXiv:0901.2507v1[nucl-th]
[22] Boffi S 2004 *Da Heisenberg a Landau* (Napoli: Bibliopolis)

Journal of Physics: Conference Series **168** (2009) 012018 doi:10.1088/1742-6596/168/1/012018

Effective nucleon-nucleon interaction and low-lying nuclear magnetic states

C Maieron[1], V De Donno[1], G Co'[1], M Anguiano[2], A M Lallena[2] and M Moreno Torres[2]

[1] Dipartimento di Fisica, Università del Salento and INFN, sezione di Lecce, Lecce, I-73100, Italy

[2] Departamento de Fisíca Atómica, Molecular y Nuclear, Universidad de Granada, Granada, E-18071, Spain

E-mail: chiara.maieron@le.infn.it

Abstract. We present a calculation of low energy magnetic states of doubly-closed-shell nuclei. Our results have been obtained within the random phase approximation using different nucleon-nucleon interactions, having zero- or finite-range and including a possible contribution in the tensor channel.

1. Introduction

In the last thirty years electron-nucleus scattering experiments have produced a large amount of high precision data, which impose severe constraints on nuclear models and effective theories, such as the Random Phase Approximation (RPA). In particular the description of low energy excited states within the RPA is known to be very sensitive to the details of the effective nucleon-nucleon (NN) interaction used in the calculations.

We present here a selection of results from a systematic study of the low energy spectra of several doubly-closed-shell nuclei we have made within the RPA theory [1]. We have focused, in particular, on the unnatural parity states, which are sensitive to the spin, spin-isospin and tensor channels of the residual NN interaction [2].

In the first step of our project we have employed a purely phenomenological approach, using a single-particle mean-field basis generated by a Woods-Saxon well, and constructing phenomenological residual NN interactions which reproduce some selected nuclear states. In order to study the sensitivity of our results to the details of the residual interaction, we have used four different NN interactions, which have zero- and finite-range and may include contributions in the tensor channels. We have found some states which are very sensitive to the details of the interaction, and in general we have obtained a satisfactory description of the low energy states of the nuclei under consideration. We can thus consider our phenomenological approach as an "optimal" RPA approach in terms of comparison with the experimental data.

In our study we have then used the RPA amplitudes X and Y to calculate electron scattering response functions. As a further independent extension of this approach, we have also considered the computation of neutrino cross sections, for which an example will be shown in the following.

We have then proceeded to the second step of our study, performing RPA calculations within a self-consistent approach, where the single-particle mean-field basis is obtained by means of

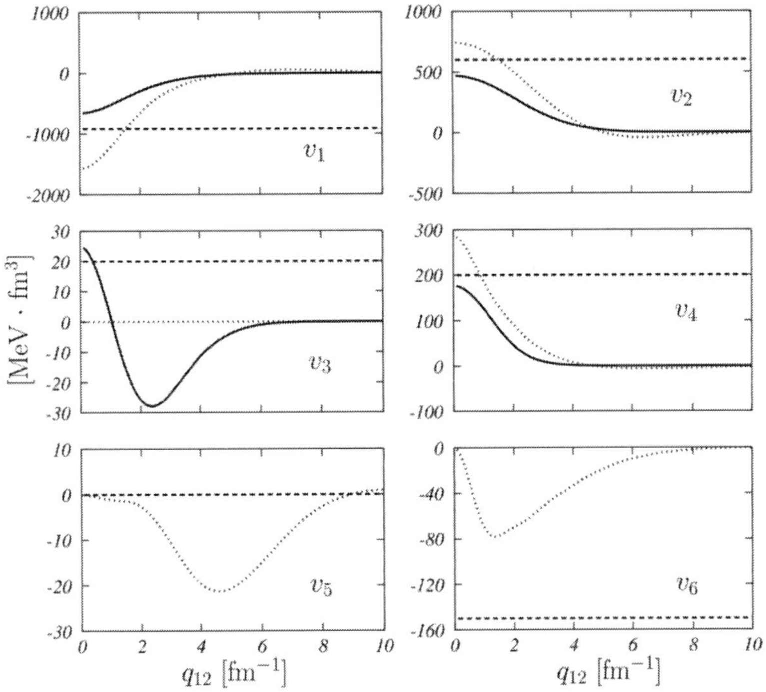

Figure 1. Effective NN interactions as functions of the relative momentum. The dashed lines represent the LMtt interaction, the dotted the FRtt and the solid the Gogny D1. The central channels v_i (i=1-4) of the LMtt and FRtt interactions are identical to those of LM and FR, respectively.

a Hartree-Fock calculation which uses the same effective NN interaction used in the RPA calculation. In particular we have used the Gogny D1 finite-range interaction [3, 4, 5], finding, in this case, remarkable disagreement with the experimental spectra.

2. Formalism

The inputs required by RPA calculations are a set of single-particle energies and wave functions and a residual NN interaction.

In the purely phenomenological calculations, for the former we have used the single-particle basis generated by a Woods-Saxon well, whose parameters are taken from the literature [6]. For the latter, we have considered a generic residual interaction written as

$$\begin{aligned} V_{\rm eff}(1,2) &= v_1(r_{12}) + v_1^\rho(r_{12})\,\rho^\alpha(r_1,r_2) \\ &+ \left[v_2(r_{12}) + v_2^\rho(r_{12})\,\rho^\alpha(r_1,r_2)\right]\,\boldsymbol{\tau_1}\cdot\boldsymbol{\tau_2} \\ &+ v_3(r_{12})\,\boldsymbol{\sigma_1}\cdot\boldsymbol{\sigma_2} + v_4(r_{12})\,\boldsymbol{\sigma_1}\cdot\boldsymbol{\sigma_2}\,\boldsymbol{\tau_1}\cdot\boldsymbol{\tau_2} \\ &+ v_5(r_{12})\,S_{12}(\hat{r}_{12}) + v_6(r_{12})\,S_{12}(\hat{r}_{12})\,\boldsymbol{\tau_1}\cdot\boldsymbol{\tau_2}\,, \end{aligned} \tag{1}$$

where $\boldsymbol{\sigma}$ and $\boldsymbol{\tau}$ are the usual spin and isospin operators and S_{12} is the tensor operator. As suggested by past phenomenological RPA studies [4, 7] we have included density dependent terms in the the central and isospin channels, with

$$\rho(r_1,r_2) = \left[\rho(r_1)\rho(r_2)\right]^{1/2}\,, \tag{2}$$

and using $\alpha = 1$ in Eq. (1).

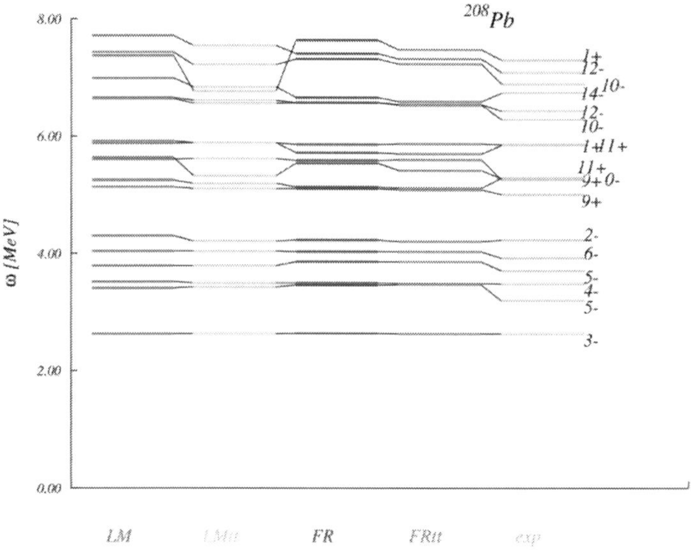

Figure 2. (Color online) Low energy spectrum of ^{208}Pb calculated with the four interactions used in this work and compared with the experimental energies [11].

We have considered four different forms of the NN interaction, parametrizing them according to the following criteria: (i) we have chosen a unique set of parameters for all the nuclei under investigation, with the exception of the density dependent terms which are different for each nucleus; (ii) the density dependent terms have been set to reproduce the first 2^+ state in ^{12}C and the first 3^- states in the other nuclei; (iii) the remaining contributions in the central and isospin channels have been chosen to get a reasonable description of the centroid energy of the isovector giant dipole resonance; (iv) the spin, spin-isospin and tensor channels have been adjusted to describe the low energy (below 8 MeV) magnetic spectrum of ^{208}Pb , with particular attention to the 12^- and 1^+ states and, in addition, taking care that the energy of the first 4^- state of ^{16}O is reproduced reasonably.

Following a Landau-Migdal approach, we have first considered zero-range interactions without and with a tensor-isospin channel contribution (LM and LMtt in the following).

We have then constructed finite-range interactions (FR and FRtt), by keeping the long-range behavior of the Argonne v_{18} potential [8] and substituting its short-range part with a sum of Gaussians. We have also used Gaussians to parametrize the density dependent terms of the interaction and, for the FRtt case, we have obtained the tensor channel terms by multiplying the corresponding terms of the v_{18} interaction by a correlation function obtained in variational calculations [6].

The behavior of the various interactions we have considered is shown in figure 1 as a function of the relative momentum of the interacting pair.

In the self-consistent approach we have obtained the single particle basis by solving Hartree-Fock equations [9, 10], using the same effective NN interaction used for the RPA calculations. We have used the Gogny D1 interaction [3, 4, 5], shown in figure 1 by the solid lines, which has finite-range components in the central, isospin, spin and spin-isospin channels and a zero-range density dependent contribution. We have not included the spin-orbit term in the RPA

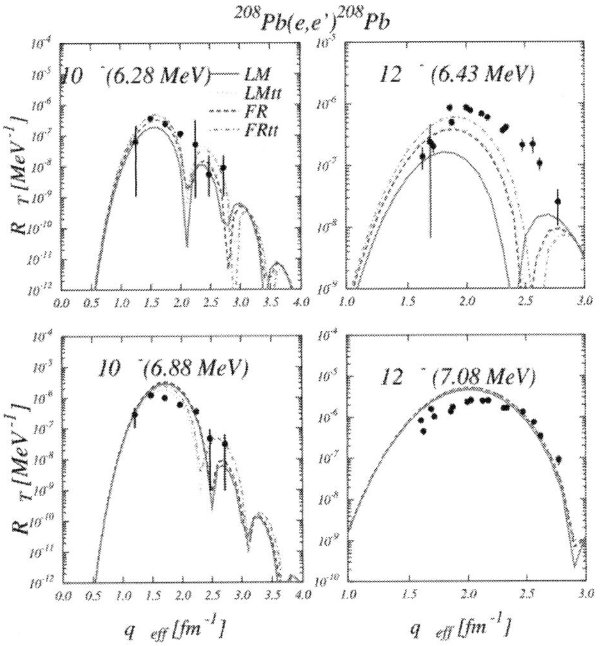

Figure 3. (Color online) Right panels: electron scattering transverse responses for the first 12^- states of ^{208}Pb , versus the effective momentum transfer. Different residual interactions are used as indicated. Left panels: the same for the first 10^- states. Experimental data from [12].

calculations.

For the self-consistent calculations we have included the contributions of both direct and exchange matrix elements of the interaction, whereas for the phenomenological approach we have considered direct terms only, assuming that the effect of the exchange terms is effectively included in the choice of the parameters characterizing the various NN interactions.

Finally we observe that our calculations have been obtained by discretizing the continuum. We have checked that our results are stable with respect to the parameters characterizing the continuum discretization and also with respect to the size of the single-particle configuration space used for each nucleus. More details on the role of the continuum discretization in self-consistent RPA calculations can be found in [2].

3. Results

We have performed systematic calculations of the low energy spectra of ^{12}C , ^{16}O , ^{40}Ca , ^{48}Ca , ^{90}Zr and ^{208}Pb . For each state we have also computed electromagnetic response functions, and we have compared them with the available experimental data.

As an example of low energy spectrum, in figure 2 we show the case of ^{208}Pb . Here we present only those experimental states we have been able to identify with those obtained in our calculations. We observe that the general agreement is quite good and that, except for some cases, the various energies have little sensitivity to the details of the interactions.

The transverse response functions of the 12^- states of ^{208}Pb are shown in the right panels of

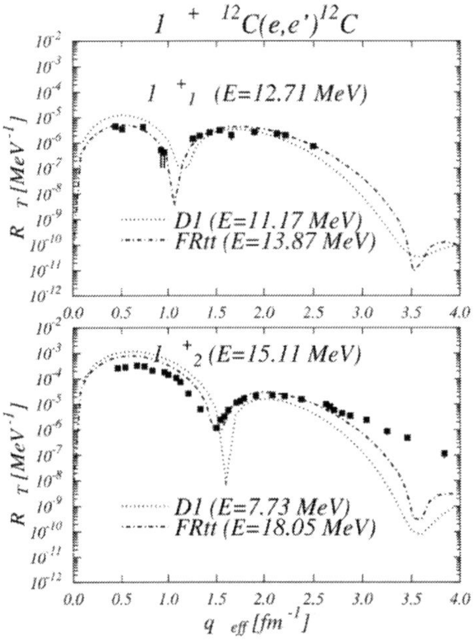

Figure 4. Electron scattering transverse responses of the first 1^+ states of ^{12}C , versus the effective momentum transfer. The self-consistent results obtained with the Gogny D1 interaction are compared with those corresponding to the FRtt phenomenological case. Experimental data from [13, 14].

figure 3. We can see that for the higher state at 7.08 MeV (lower right panel) the experimental data are rather well reproduced with all NN interactions, which do not produce significant differences in the curves. On the other hand, for the lower energy state (upper right panel) we observe a very strong dependence on the residual interaction, both when finite-range and when tensor channel contributions are included. We remark that a better description of this state

Table 1. Energies of the low-lying magnetic states of ^{12}C (in MeV). Experimental energies from [11].

^{12}C			
excitation	D1	FRtt	exp
1^+	11.17	13.87	12.71
1^+	7.73	18.05	15.11
4^-	19.16	17.75	18.27
4^-	15.63	19.49	19.15

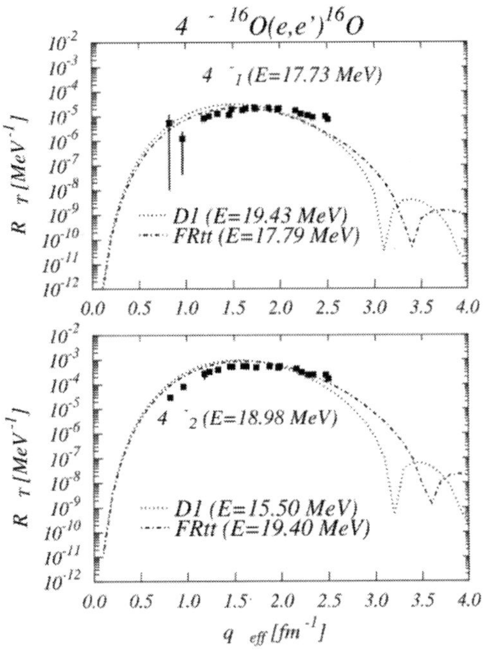

Figure 5. Same as figure 4 for the first 4^- states of ^{16}O . Experimental data from [15].

alone could be obtained with a different choice of the parameters of the residual interaction, but this would worsen the global descritpion of the various magnetic spectra we have considered.

In the left panels of the same figure the transverse responses of the 10^- states are also shown. It is interesting to notice that, in this case, only the curves which include tensor contributions are able to reproduce the second peak shown by the data, for both states.

An example of the results we have obtained within the self-consistent approach is given in table 1, where the low energy magnetic spectrum of ^{12}C is presented, and in figure 4, where we show the corresponding transverse responses of the 1^+ isospin-doublet of states. For the sake of comparison between the self-consistent results and the purely phenomenological ones, we have taken as reference the FRtt case, considered to be the most complete interaction we have obtained.

We see that the magnetic states of ^{12}C are reproduced rather badly by the self-consistent calculations: the response functions agree in shape and magnitude with the data, but the order of the states forming isospin doublets is inverted.

This happens also for ^{16}O , as illustrated, for the case of the 4^- states, in figure 5, where the values of the excitation energies obtained with the D1 and FRtt interactions are also reported.

We have systematically obtained this kind of inversion for all magnetic states of all the nuclei we have studied . This result indicates the inadequacy of the D1 interaction in isospin-dependent channels, and it may be considered the most important outcome of the self-consistent part of our study.

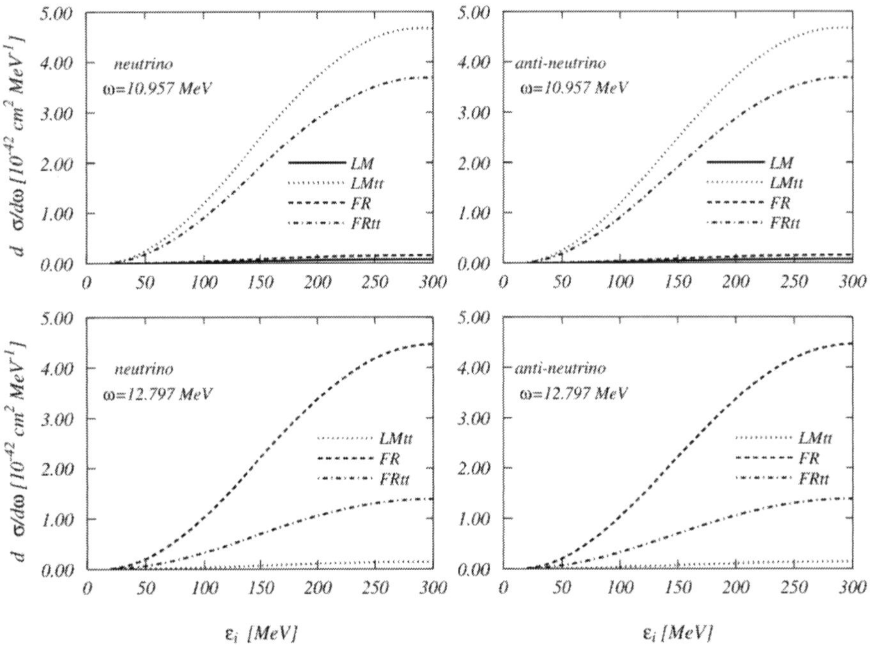

Figure 6. Cross sections for the neutral current neutrino and antineutrino excitation of 0^- states of ^{16}O at excitation energy ω, versus the incident (anti)neutrino energy.

4. Extension of the phenomenological approach to neutrino scattering

After obtaining an "optimal" phenomenological RPA approach, by tuning NN interactions to reproduce data from electron scattering at best, we have also considered the possibility of applying it to the study of low energy neutrino scattering cross-sections, with the purpose of studying their sensitivity to the tensor components of the interaction.

In figure 6 we show the cross sections for the neutral-current neutrino and antineutrino excitation of the low energy 0^- states of ^{16}O , as a function of the incident neutrino energy. The values of the excitation energies ω are indicated inside the figure.

We observe that, in this test-case, extremely large differences in the cross sections are obtained when tensor channel contributions are included in the effective NN interaction. This indicates that the role of the residual interaction in neutrino scattering cross section is a very interesting topic, for example in connection with the problem of nuclear uncertainties in the detection of supernova neutrinos [16].

5. Summary and conclusions

We have made a systematic study of the low-lying (magnetic) spectra of doubly-closed-shell nuclei. Our study indicates that a simultaneous description of all states imposes strong constraints on the residual NN interaction. Within a purely phenomenological approach it is possible to get a good description of the spectra and of the response functions of most of the states, thus obtaining an "optimal" RPA approach. Some states which are not well described exhibit a strong sensitivity to some details of the residual interaction, and a deeper investigation

Journal of Physics: Conference Series **168** (2009) 012018 doi:10.1088/1742-6596/168/1/012018

could be used to obtain further constraints on it.

On the contrary, the self-consistent calculations we have performed with the Gogny D1 interaction produce results in disagreement with the experimental spectra, indicating the presence of problems in particular in the isospin dependent parts of the interaction. A detailed study can be found in [2].

Finally we have considered an example of a potentially important application of our phenomenological approach to the calculation of neutrino scattering cross sections, to investigate the sensitivity of the latter to the tensor components of the residual interaction.

References

[1] De Donno V 2008 *Nuclear excited states within the Random Phase Approximation theory,* Ph.D. thesis Università del Salento (Lecce, Italy), unpublished
[2] De Donno V, Co' G, Maieron C, Anguiano M, Lallena A and Moreno-Torres M 2009 *Low-lying magnetic excitations of doubly-closed-shell nuclei and nucleon-nucleon effective interactions*, submitted for publication
[3] Gogny D 1975 *Nuclear selfconsistent fields* (G.Ripka and M.Porneuf, North Holland)
[4] Blaizot J P and Gogny D 1977 *Nucl. Phys.* A **284** 429
[5] Dechargè J and Gogny D 1980 *Phys. Rev.* C **21** 1568
[6] Arias de Saavedra F, Bisconti C, Co' G and Fabrocini A 2007 *Phys. Rep.* **450** 1
[7] Speth J, Werner E and Wild W 1977 *Phys. Rep.* **33** 127
[8] Wiringa R B, Stoks V G J and Schiavilla R 1995 *Phys. Rev.* C **51** 38
[9] Co' G and Lallena A M 1998 *Phys. Rev.* C **57** 145
[10] Bautista A R, Co' G and Lallena A M 1999 *Nuovo Cim.* A **112** 1117
[11] Lederer C M and Shirley V S 1978 *Table of isotopes, 7th ed.* (John Wiley and sons, New York)
[12] Lichtenstadt J *et al* 1979 *Phys. Rev.* C **20** 497
[13] Buti T N *et al* 1986 *Phys. Rev.* C **33** 755
[14] Hicks R S *et al* 1984 *Phys. Rev.* C **30** 1
[15] Hyde-Wright 1984 Ph.D. thesis Massachusetts Institute of Technology, unpublished
[16] Co' G, De Donno V and Maieron C 2008, proceedings of NOW 2008 (to be published), *Preprint* arXiv:0812.2765 [nucl–th]

Journal of Physics: Conference Series **168** (2009) 012019 doi:10.1088/1742-6596/168/1/012019

Hadron structure and hadronic matter

Elena Santopinto
Istituto Nazionale di Fisica Nucleare, Sezione di Genova
Genova, I-16146 Italy
E-mail: santopinto@ge.infn.it

Abstract. Some of the most relevant and recent results of the italian theoretical groups active in hadronic physics are briefly reviewed.

1. Introduction

The studies of hadron properties can be considered to follow roughly two main approaches. The first one, which can be called microscopic or systematic, starts from the internal hadron dynamics, that is from the quark degrees of freedom and their interactions. The widely accepted framework is of course provided by Quantum ChromoDynamics (QCD), which however can be presently treated only in the perturbative regime (pQCD). A big effort is devoted to the analysis of hadron structure in the Lattice QCD (LQCD): there are now many important results [1] and more are expected in the future, when the computer capabilities will allow to reach the required precision of the calculations in order to extract the hadron properties in a systematic way. This will presumably take still some time and in the meanwhile one can rely on models, eventually based on QCD or LQCD. Moreover chiral perturbation theory can be considered as the low-energy effective field theory of QCD [2]. The second approach, which can be denoted as "phenomenological", uses some parametrization of single hadron properties within a theoretical framework, based on general aspects of quarks and gluons dynamics. The two approaches are not sharply separated, but on the contrary they can be strongly correlated and they are introduced mainly for simplifying the discussion.

Many models have been built and applied to the description of hadron properties. An important class is provided by Constituent Quark Models (CQM), in which quarks are considered as effective internal degrees of freedom and can acquire a mass and even, in certain approaches, a finite size. The idea of quarks as constituent particles of hadrons has been introduced very soon [3]; quark confinement was not yet known and quarks were considered to have a very large mass, thereby supporting a non relativistic approach. The modern version of the non relativistic CQM has been formulated by Isgur and Karl [4] (IK), using a h.o. confinement potential and a hyperfine interaction, which is spin dependent and inspired by the One-Gluon-Exchange mechanism (OGE). The IK model has been subsequently reformulated using a relativistic kinetic energy operator and a three-body linear confinement, inspired by a flux tube approach [5] (CI). Further models have been recently proposed. In the $U(7)$-algebraic model [6], $U(7) \otimes SU_{sf}(6) \otimes SU_c(3)$ is used as spectrum generating algebra for the baryons, that is for the excitation spectrum and other properties, such as elastic form factors and helicity amplitudes. The hypercentral CQM (hCQM) is characterized by a linear confinement and a coulomb-like term [7]; the hypercentral interaction depends on the coordinates of the three quarks and can

include three-body effects. The main feature of the chiral CQM [8] is the quark interaction derived from the exchange of Golstone Bosons (GBE), like pions and kaons. In a recent CQM [9], instanton interactions are introduced in order to describe the baryon properties.
Constituent models have been considerably refined to include consistently relativity, formulating or reformulating them in Instant, Front (called also light cone) or Point Form Dynamics. Bethe-Salpeter approches have been also developed. Now the current main effort is towards a systematic and consistent inclusion of the higher Fock components in order to arrive at a new generation of CQMs.

Besides CQMs, there are various models formulated in a consistent way within chiral theory. Many papers have been written along this line of thought and the interested reader is referred to the abundant literature (see e.g. Ref. [10]). A completely different approach is provided by the instanton liquid model (for a review see [11]).

In the following, the recent work done by the italian theoretical groups in the field of hadron physics will be briefly reviewed, considering that they have used all the tools at disposal at the moment, ranging from LQCD up to phenomenological models.

2. CQMs

The nucleon has a rich spectrum, which however becomes very hard to be studied experimentally in the higher energy part because of the large widths of the states and the increasing overlapping of the various resonances. In any case, the description of the spectrum is the first task of a model builder: it serves to determine a quark interaction to be used for the calculation of other physical quantitites. The form factors (elastic and inelastic) are particularly suited for a test of the models, since they use explicitly the baryon wave functions determined by the description of the spectrum. All the various forms of CQMs are able to reproduce more or less the spectrum, even though they differ in the used potential, in particular in the form of the $SU(6)$ spin flavour breaking interaction. Regarding the elastic form factors the use of a relativistic model is essential [12, 13, 14, 15, 16, 17], while for the helicity amplitudes it seems to be less important[18, 19, 20].

A very important test of the models is provided by the helicity amplitudes for the electromagnetic excitation of the nucleon resonances. The transverse amplitudes, $A_{1/2}$ and $A_{3/2}$, are defined as the matrix elements of the transverse electromagnetic interaction, $H^t_{e.m.}$, between the nucleon, N, and the resonance, B, states:

$$\begin{aligned} A_{1/2} &= \langle B, J', J'_z = \tfrac{1}{2}\ |H_{em}| N, J = \tfrac{1}{2}, J_z = -\tfrac{1}{2}\ \rangle \\ A_{3/2} &= \langle B, J', J'_z = \tfrac{3}{2}\ |H_{em}| N, J = \tfrac{1}{2}, J_z = \tfrac{1}{2}\ \rangle \end{aligned} \tag{1}$$

The building of a CQM implies the knowledge of model wave functions and therefore the above matrix elements can be calculated. The results of the IK model were soon available [21], however they were obtained adjusting the proton size to a value of about $0.5 fm$ [22]. More recently, results from the CI in the light front approach have been obtained by the Rome group [14]. A systematic study of the helicity amplitudes for the negative parity resonances has been performed within the hCQM [20]; this study is now extended to all resonances with an observable electromagnetic transition strength and to the longitudinal excitation amplitudes as well [23]. In the case of the negative parity non strange resonances, the experimental behaviour is fairly well reproduced by the parameter-free calculations made with the hCQM [20], specially for the S_{11}, whose trend has been predicted in advance with respect to the recent Jlab data [24]. In the medium Q^2 region the theoretical amplitudes are not damped, at variance with the h.o. calculations. This softer dependence is due, in the hCQM calculation, to the presence of the Coulomb-like term in the interaction. At low Q^2, in particular for the $A_{3/2}$ amplitudes, there is a lack of strength, which is however typical of all CQMs [18]. Such discrepancy can be attributed

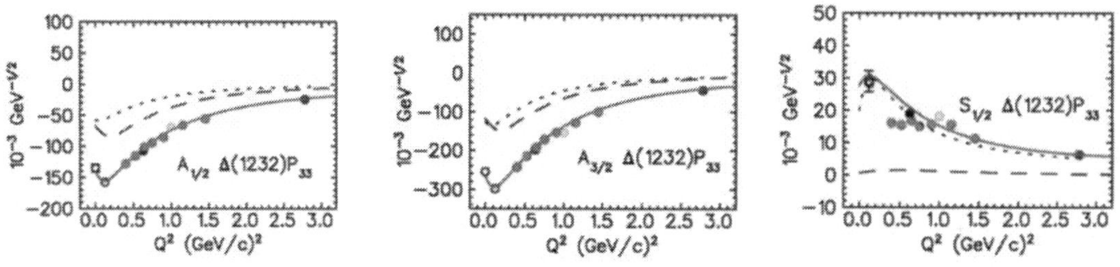

Figure 1. The Q^2 dependence of the $N \to \Delta$ helicity amplitudes. The solid and dashed curves are the results of the superglobal fit with MAID and the predictions of the hCQM, respectively. The dotted lines show the pion cloud contributions calculated with DMT [27]. The data points at finite Q^2 are the results of the Mainz single-Q^2 fits [26]. At $Q^2 = 0$ for $A_{1/2}$ and $A_{3/2}$ the photon couplings from PDG are shown [28].

to the fact that all CQMs miss some fundamental mechanism, such as the quark-antiquark pair production [18]. A confirmation of this statement is provided by a recent analysis performed by a collaboration of the Genoa and Mainz groups [25]. In figure 1 [25] the transition amplitudes for the Δ excitation, calculated with the hCQM, are reported in comparison with a global fit performed by the Mainz group [26]. The meson cloud contributions, evaluated by means of a dynamical model [27], are also reported. Their importance decreases with increasing Q^2, going rapidly to zero, as expected; however, for low Q^2, they are able to fill most of the gap between data and the hCQM results. This feature is quite general, since it happens systematically also for the excitation of higher resonances, such as $P_{11}(1440), S_{11}(1535), D_{13}(1520), F_{15}(1680)$ [25].

To conclude, it should be noted that the proton radius calculated in the hCQM is about $0.5 fm$, just the value previously fitted [21, 22] to the $D13$ transition strength. In the outer region, that is for low Q^2, there is a lack of strength, which can be at least partially accounted for by pion cloud contributions. The emerging picture is therefore that of a quark core with a dimension of about $0.5 fm$ plus a meson (or sea-quark) cloud.

3. Spin Structure of the Hadrons and Transversity effects

During the last years the interest in spin physics has considerably grown. The study of the nucleon spin structure has made important progress and new ideas and models have been developed. All that can help to learn about the spin structure of hadrons and to describe some spin effects in inclusive and semi-inclusive hadron production, even if still many aspects have to be understood. In parallel with the theoretical efforts, a strong experimental program has already started at RHIC (BRAHMS, PHENIX, STAR), HERA (HERMES), CERN (COMPASS) and KEK (Belle), or is being planned at JLab, JPARC.

The role of intrinsic transverse momentum has received a lot of attention in the context of polarization effects, in particular in transverse Single Spin Asymmetries (SSA) at moderate p_T values. These, contrary to former expectations of pQCD, show sizeable experimental values. It was originally suggested by Sivers [29] (1990) to start from pQCD including transverse momentum effects in parton distribution functions. Subsequently this idea has been developed and extended in a certain number of papers [30, 31, 32, 33, 34, 35].

In particular, a great effort has been devoted, in several steps, to a generalization of the usual collinear pQCD approach with the inclusion of spin and transverse momentum effects by the Torino-Cagliari group [36]. The main steps in the construction, under the assumption of a generalized QCD factorization scheme, and in the refinement of the so-called Generalized Parton Model (GPM), are:

1) The inclusion of spin and transverse momentum dependences in the parton distribution functions (1995) [30, 31];
2) The improved treatment of kinematics with inclusion of transverse momentum effects in the study of unpolarized cross sections and the Sivers effect in pp collisions (2004) [37];
3) The inclusion of all transverse momentum dependences of partons in hadrons and of hadrons produced by fragmenting partons and in the elementary hard interactions computed at leading order with noncollinear kinematics, thus arriving to the GPM [36].

The GPM is based on a factorization scheme, separating the soft, long distance from the hard, short distance contributions. The hard part is computable in perturbative QCD, while information on the soft one has to be extracted from the experiments or modeled. This can be considered a very reasonable starting point (although for non collinear processes factorization has been proved only in particular kinematical situations). Moreover in this scheme leading twist Transverse Momentum Dependent (TMD) distributions keep their partonic interpretation, and are expected to be universal and process-independent. All these reasonable assumptions have, however, to undergo phenomenological scrutiny, considering that this is the best one can do at the moment.

In the context of the GPM the Torino-Cagliari group has started a systematic program of analysis of the existing data, aiming at reproducing all the available data on Single Spin Asymmetries from Semi Inclusive Deep Inelastic Scattering (SIDIS), hadronic collisions and e^+e^- annihilations with a single set of universal TMD distributions. This recently led to the extraction of TMDs, with the possibility of giving very useful predictions for many forthcoming experiments. In particular they were able to obtain, for the first time, the following important results:
i) The simultaneous extraction of the transversity distribution and the Collins fragmentation function [38]. In figure 2 we report the first phenomenological extraction of the transversity distribution for the valence up and down quarks;
ii) Taking into account new higher statistics data from HERMES, COMPASS and Belle, a new phenomenological extraction of the transversity distribution for up and down quarks with a smaller error-band (see figure 3) [39], and of the Collins (see figure 4) [39] and Sivers functions (see figure 5) [40].

By using the Sivers function [41], the transversity distribution and the Collins fragmentation function [38] extracted by means of high quality global fits, within the same scheme, they predict [42] the $p^\uparrow p \to h+X$ single spin asymmetries in remarkable agreement with RHIC experimental data.

Single-spin asymmetries in processes with transversely polarized targets have been measured in proton-proton collisions, $p^\uparrow p \to h + X$ [43], and in semi-inclusive deep inelastic scattering (SIDIS), $l\ p^\uparrow \to l'\ \pi + X$ [44]. Different theoretical approaches [45, 36] have been adopted to study and interpret these asymmetries and make predictions for new possible experiments. In a recent Letter [46], different clear-cut predictions are made for a simple process accessible at RHIC, using the color-gauge QCD invariant formalism [47] and the GPM approach. A weighted asymmetry in the azimuthal distribution of photon-jet pairs, produced in the process $p^\uparrow p \to \gamma$ jet $+X$ with a transversely polarized proton, is studied from both the phenomenological and theoretical point of view. In Ref. [46] different predictions for the asymmetry are obtained. In particular the results with the GPM [36] are compared with the gluonic-pole cross sections calculable in perturbative QCD [47]. The two predictions have opposite signs. Experimental measurements of the asymmetry will therefore test the present understanding of single spin asymmetries.

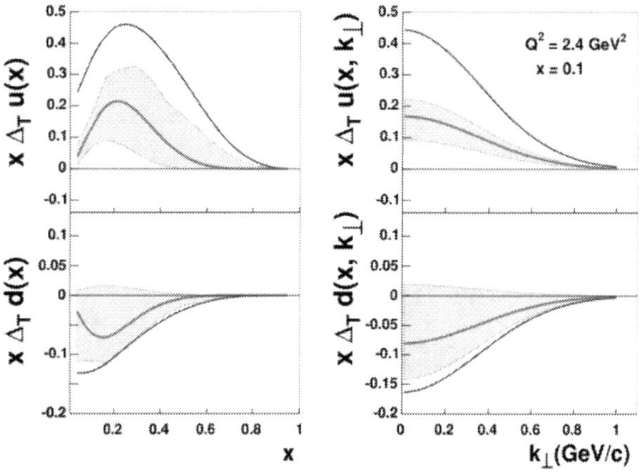

Figure 2. The first phenomenological extraction of the transversity distribution for the up and down quarks from Ref. [38] [APS copyright].

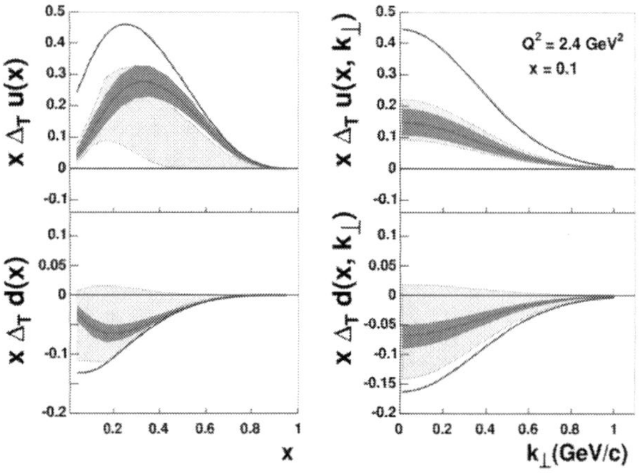

Figure 3. Transversity distribution for the up and down quarks - II extraction [39].

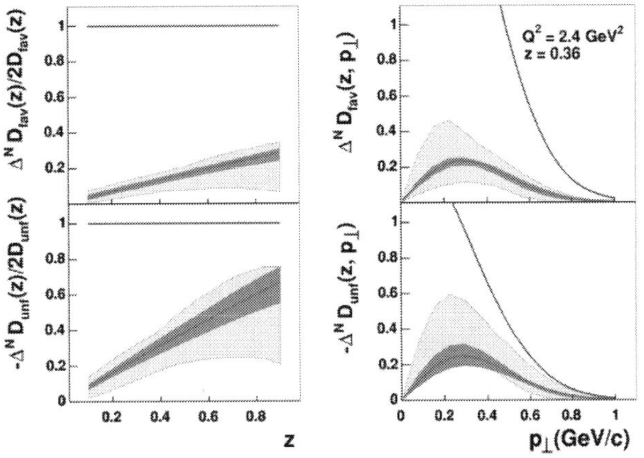

Figure 4. Phenomenological extraction of the Collins function (SIDIS) [39].

3.1. Transverse Momentum Dependent distributions and effective models

As discussed in the previous section, the study of SSA experimental data has required the introduction of a new class of Tranverse Momentum Dependent parton distributions and Fragmentation Functions (FFs). Phenomenologically the most relevant ones are at leading twist, the Sivers and the Boers Mulders distribution functions and the Collins fragmentations functions, since in a factorized pQCD approach they are responsible for most of the azymuthal spin asymmetries in SIDIS, Drell-Yan (DY) processes and e^+e^- annihilations. Factorization theorems have been proved for SIDIS and DY processes in the kinematical regime in which two well distinct energy scales are present: the large scale of the photon virtuality or the invariant mass of the leptonic pair and the low-moderate scale of the transverse momentum p_T of the observed hadron or the leptonic pair in respect to the colliding beams.

Although many models calculations of integrated PFDs are available, there are not so many for TMDs- also called unintegrated PDFs. The Transverse Momentum Dependent Distributions describe the probability of finding in a hadron a parton with longitudinal momentum fraction x and transverse mometum p_T. At leading twist there are eight TMDs [48], three of them surviving when integrated over the transverse momentum and giving rise to parton density, helicity and transversity distributions. Since TMDs are typical non perturbative quantities, they are not directly calculable in QCD, and their modelling requires assumptions on the nucleon wave function.

In Ref.[49] all the leading twist T-even functions were calculated in a light cone spectator model with scalar and axial-vector diquarks. Recently the Pavia group has performed a calculation using light cone three quark wave function [50]. The formalism and the analytic expressions in the SU(6) symmetry limit are derived for the T-even TMDs, studied in a light-cone description of the nucleon with the Fock expansion truncated to three valence quarks [50]. Numerical results for the T-even TMDs are shown in their paper [50] by using the Schlumpf [51] parametrization of the momentum wave function. It is interesting to observe figure 4 of Ref. [50] (see figure 6 of this article) where the Pavia results obtained with both the Schlumpf [51] and the hypercentral wave function [52, 7] are compared with results obtained with other

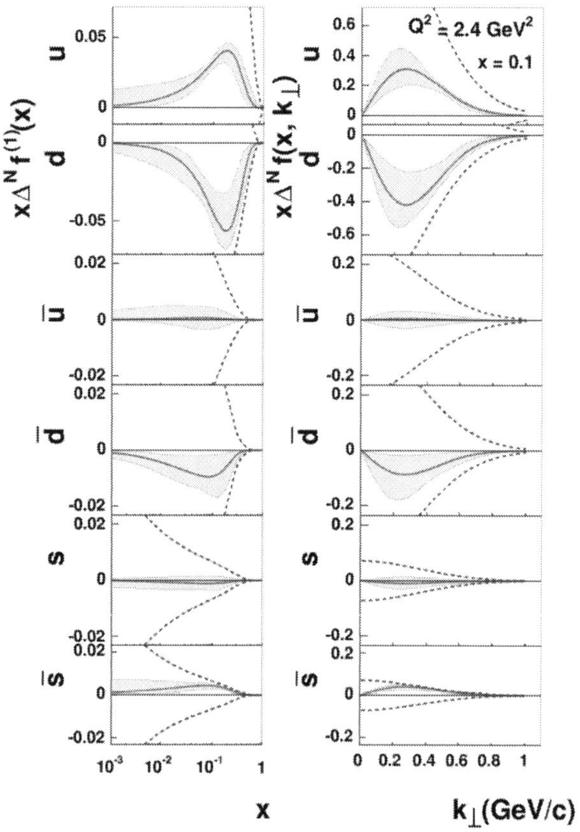

Figure 5. Quark Sivers functions from SIDIS [40].

models, in particular with a bag model and the spectator model with scalar and axial-vector diquarks of Ref. [49]. The results for the pretzelosity distribution and its transverse moments are rather different in different models, and this sensitivity to the adopted model suggests that new data could give useful insights to model the momentum dependence of the nucleon wave function. A preliminary work on the transversity distribution calculated by the Pavia group [53] using the hypercentral wave function [52, 7] shows an interesting and intriguing coincidence with a phenomenological extraction of the transversity distribution.

T-odd functions were calculated in a spectator model with scalar diquarks in [54] and scalar and vector diquarks in [55], in the MIT bag model by Yuan[56] and by Cherednikov, D'Alesio, Kokelev and Murgia [57], in the non relativistic Isgur Karl constituent quark model by the Valencia-Perugia collaboration [58] and in a spectator model for the pion [59].

A calculation of all the TMDs in a spectator model with a scalar diquark can be found in Ref. [60], while in the paper by Bacchetta, Conti and Radici [61] the authors calculate all the TMDs by using also an axial vector diquark. They consider different options for the possible diquark polarization states, as well as for the form factors that parametrize the nucleon quark diquark vertex. Their results for the transversity distributions are given in figures 7 and 8 , that correspond to figures 8 and 9 of Ref. [61]. It is interesting to observe that there is a change

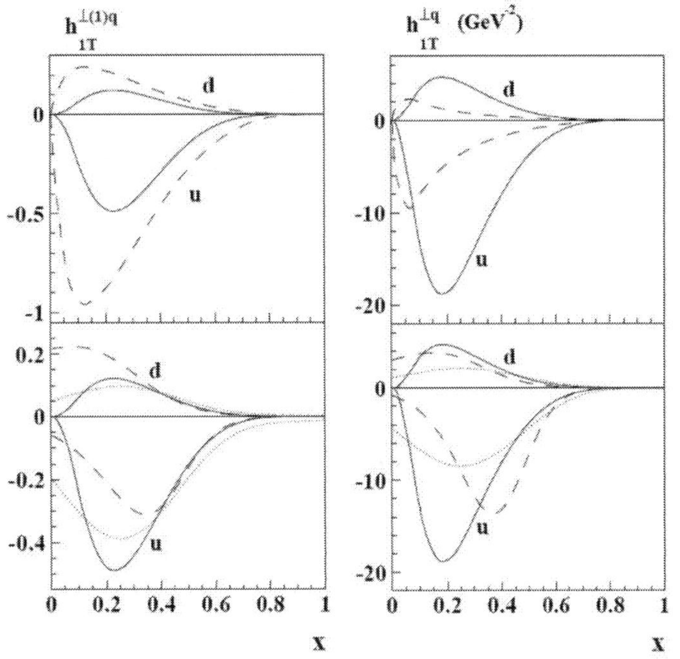

Figure 6. Pretzelosity distribution (left) and the transverse moment of the pretzelosity distribution (right) calculated in different models from Ref. [50] [APS copyright]. Solid curves: light cone calculation with the momentum wave function by Schlumpf [51]. Dashed curves in the upper panels: light cone calculation with the hypercentral model wave function [52, 7]. Dashed curves in the lower panels: results from the spectator model of Ref. [49]. Dotted curves: results from the bag model of Ref. [49].

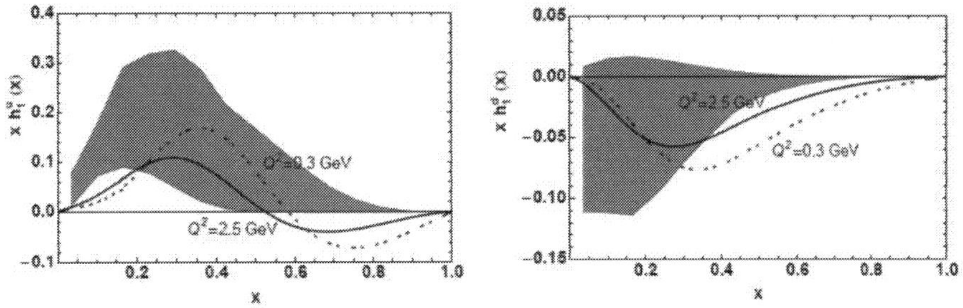

Figure 7. The transversity distribution $xh_1(x)$ for up (left) and down quark (right) from Ref. [61] [APS copyright]. Dashed (Solid) curves for the scalar-vector diquark model before (after) the evolution at LO. Shaded area: uncertainty band in the phenomenological extraction from Ref. [38] .

of sign in the transversity distribution for the up quark, a feature that seems not to be shown by other models. The use of models instead of phenomenological parametrizations, may help to undertand information that are encoded in the TMDs, like their relations with the angular

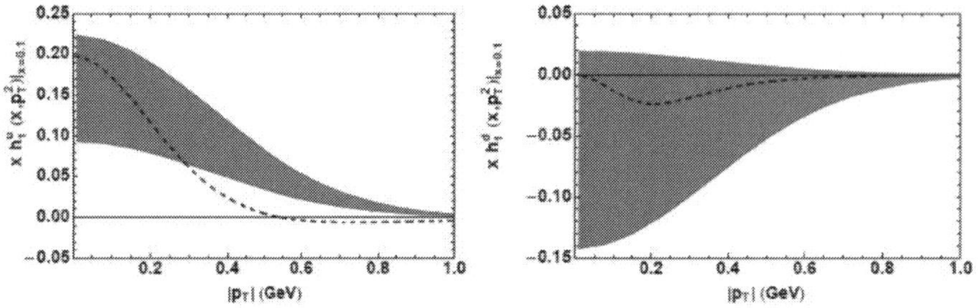

Figure 8. The p_T dependence of the transversity distributions for up (left) and down quark (right) from Ref. [61] [APS copyright]. Dashed curves for the scalar-vector diquark model. Shaded area: uncertainty band in the phenomenological extraction from Ref. [38] .

momenta of the partons and the nucleon deformations. Moreover from an experimental side they could be useful to estimate the size of an observable in different processes and kinematical regimes and to set up Monte Carlo simulations [62].

3.2. Fragmentation functions

A complementary information on the internal structure of hadrons is encoded in the Fragmentation Functions (FFs) involved in the description of semi-inclusive processes with one or more hadrons of the final state detected. In the case of two final hadrons they are called Dihadron Fragmentation Functions (DiFF) and were proposed for the first time in Refs. [63] [64] [65] [66] and then systematically analyzed at leading twist in Ref. [67] and at twist-3 in Ref. [68]. Like the PDFs also the FFs are universal in the sense that they are process independent. They parameterize the process of hadronization, the unknown mechanism by means of which partons (quarks or gluons) turn into confined hadrons in the final state which are observed by the detectors. Clearly they are non perturbative objects, thus they should be modeled or parametrized. Recently a lot of work has been dedicated to DiFF [67][68][69][70][71][72][73]. In principle, although experimentally difficult, the detection of two final state hadrons can give access to information on the spin correlations in the hadronization process in terms of polarized FFs.

As a beautiful example Radici, Jakob and Bianconi have quantitatively demonstrated [70] that in the process $ep \to e'\pi^+\pi^-$ X, with the target proton transversely polarized, it is possible to extract a spin asymmetry that is proportional to the product of the transversity and the DiFF that describes the fragmentation of the polarized parton into two pions. This can be an alternative way to extract the transversity with the advantage in respect to the Collins effect that the process is collinear, the transverse momenta can be integrated out, and the evolution is under control. The first process has been measured at HERMES at $Q^2 = 2.5\ GeV^2$ [75], while the annihilation process will be measured at BELLE at $Q^2 = 100\ GeV^2$. In this context, Bacchetta and Radici have calculated this Dihadron fragmentation function within a diquark spectator model [72], and given prediction for the asymmetry [72] measured at HERMES and finally derived evolution equations for the DiFF in Ref. [73].

In figure 9 (from Ref. [74]), one can compare the predictions by Bacchetta and Radici [72] and the HERMES data [75].

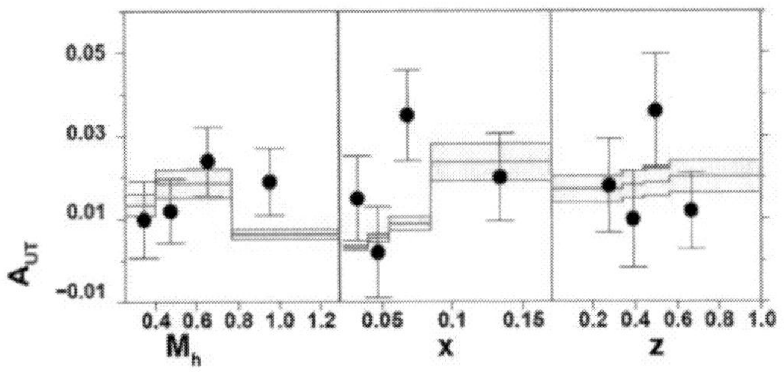

Figure 9. Spin asymmetry for semi-inclusive production of a pion pair in deep-inelastic scattering on a transversely polarized proton as a function of the invariant mass of the pion pair, M_h (GeV), of the light-cone momentum fraction of the initial parton, x, of the energy fraction carried by the pion pair with respect to the fragmenting parton, z, as from Ref. [74] [APS copyright]. Data from Ref.[75]. The uncertainty band is a fit to the data based on the DiFF spectator model of Ref.[72] and on the h_1 parametrization of Ref.[39].

4. Exotic meson states

The exotic meson states are hadrons with zero baryonic number which are not reproducible in the usual quark-antiquark scheme. The exotics which are mainly studied in literature are tetraquarks, hybrids and glueballs.

The tetraquarks are mesons composed of two constituent quarks and two constituent antiquarks. There are three different types of models for tetraquark systems: the uncorrelated, the diquark-antidiquark and the quasi-molecular ones. The group of Genova has developed an algebraic model description for the first two schemes of light tetraquarks for which there are experimental evidences from KLOE, E791 and BES.

The hybrids are mesons with gluonic components. The most commonly studied hybrids are those containing a couple of quark and antiquark and a single (effective, quasi) gluon. There seems to be some evidences of hybrids states in the light quark sector: experiment E852 at BNL have observed $\pi_1(1400)$, $\pi_1(1600)$ and $\pi_1(2000)$ with exotic J^{PC} quantum numbers. The Indiana-Genova collaboration has used the non relativistic reduction of Coulomb gauge QCD to compute the spectrum of the low mass hybrid mesons containing a heavy quark-antiquark pair.The gluon degrees of freedom are treated in the mean field approximation calibrated to the gluelump spectrum. New experiments specifically devoted to the heavy hybrids are to be carried on at Jlab from the Gluex and the Clas collaborations, at PANDA and at BESIII.

The last type of exotic mesons are the glueballs, gluonic colour singlet bound states. There are evidences from the Crystal Barrel collaboration regarding a light scalar glueball $f_0(1500)$ and the PANDA experiment has been designed to discover glueballs up to masses of 4.5 GeV. The group of Trento has studied this resonance in the Interacting Instanton Liquid Model [76].

4.1. Tetraquark states

The quark-antiquark assignment to P-waves [77] has never worked for the lowest lying scalar mesons, $f_0(980)$, $a_0(980)$, $\kappa(800)$ and $f_0(600)$ [78]. Already in the seventies Jaffe [78] suggested the tetraquark structure of the scalar nonet and proposed a four quark bag model. Recently Maiani *et al.* in Ref. [79] have suggested that these mesons could be described as a tetraquark nonet, in particular as a diquark-antidiquark system.

In Ref. [80] a complete classification scheme of the two quark-two antiquark states in terms of $SU(6)_{sf}$ has been published, as well as an evaluation of the tetraquark spectrum for the lowest scalar meson nonet, obtained both in the tetraquark uncorrelated case and in the diquark-antidiquark case using algebraic model tecniques.

In the construction of the classification scheme symmetry principles have been used without introducing any explicit dynamical model. There are two constraints: the tetraquark wave functions should be a colour singlet, as all physical states, and the tetraquarks states must be antisymmetric for the exchange of the two quarks and the two antiquarks.

The allowed $SU(3)_f$ representations for the $qq\bar{q}\bar{q}$ system are $2[1] \oplus 4[8] \oplus [10] \oplus [\overline{10}] \oplus [27]$. The allowed isospin values are $I = 0, \frac{1}{2}, 1, \frac{3}{2}, 2$, while the hypercharge values are $Y = 0, \pm 1, \pm 2$. The values $I = \frac{3}{2}, 2$ and $Y = \pm 2$ are exotic, which means that they are forbidden for the $q\bar{q}$ mesons. The allowed $SU(2)_s$ representations are $2[1] \oplus 3[3] \oplus [5]$. The tetraquarks can have an exotic spin $S = 2$. In Appendices A and B of Ref. [80] all the flavor and spin states in the $qq\bar{q}\bar{q}$ configuration are explicitly written in terms of the single quark and antiquark states. The flavor states are also written in the ideal mixing hypothesys, i.e. as a superposition of the SU(3)-symmetrical states in such a way to have defined strange quark and antiquark numbers.

The spatial, flavor, color and spin parts with given permutational symmetry have been combined together to obtain completely antisymmetric states under the exchange of the two quarks and the two antiquarks. The resulting states are listed in Table III of Ref. [80]. The possible flavor, spin and J^{PC} values can be found in Table V, VI, VII and VIII of Ref. [80] for different orbital angular momenta.

The mass splitting of the candidate tetraquark nonet, i.e. $f_0(980)$, $a_0(980)$, $\kappa(800)$ and $f_0(600)$, see Ref. [80] has been evaluated by means of a mass formula for normal mesons [81] extended to tetraquark systems. The masses of the other mesons belonging to the same tetraquark nonet do not seem in very good agreement with the experimental values, even if, before reaching any conclusion, new experiments, especially on the poorly known $\kappa(800)$ and $f_0(600)$, are mandatory.

The constituent diquark is thought as two correlated constituent quarks with no internal spatial excitations. Thus the tetraquark mesons are described in the diquark-antidiquark limit as composed of a constituent diquark, (qq), and a constituent antidiquark, $(\bar{q}\bar{q})$. The diquark-antidiquark states in this limit are a subset of the tetraquark states previously derived. The spectrum generating algebra is $U(4) \otimes SU(3)_f \otimes SU(2)_s \otimes SU(3)_c$ and the mass formula can be written as:

$$M^2 = (M_{qq} + M_{\bar{q}\bar{q}})^2 + a \cdot n + b \cdot L_{12-34} + c \cdot S_{tot} + d \cdot J, \quad (2)$$

where M_{qq} and $M_{\bar{q}\bar{q}}$ are the diquark and antidiquark masses, n is a vibrational quantum number, L_{12-34} the relative orbital angular momentum, S_{tot} the total spin and J the total angular momentum. The masses for the candidate tetraquark nonet obtained with this model are much closer to the experimental values than the masses obtained in the tetraquark model. However it should be again stressed that, before reaching any conclusion, new experiments are necessary, also to be sure of the existence of all the states of the scalar nonet. If the existence of only some states of the nonet will be confirmed a different kind of clusterization will emerge, and this limit has still to be studied in an algebraic framework.

4.2. The hybrids states

Low energy gluons are responsible for the distinctive features of QCD such as confinement and chiral symmetry breaking. For the purpose of getting insights into the non-perturbative features of QCD, it is possible to design idealized systems that are sensitive to the pure Yang-Mills sector and to bypass many of the complications of the full QCD. A gluelump is an example of such an idealized state. It is a state of the gluon field bound to a static, localized octet source which

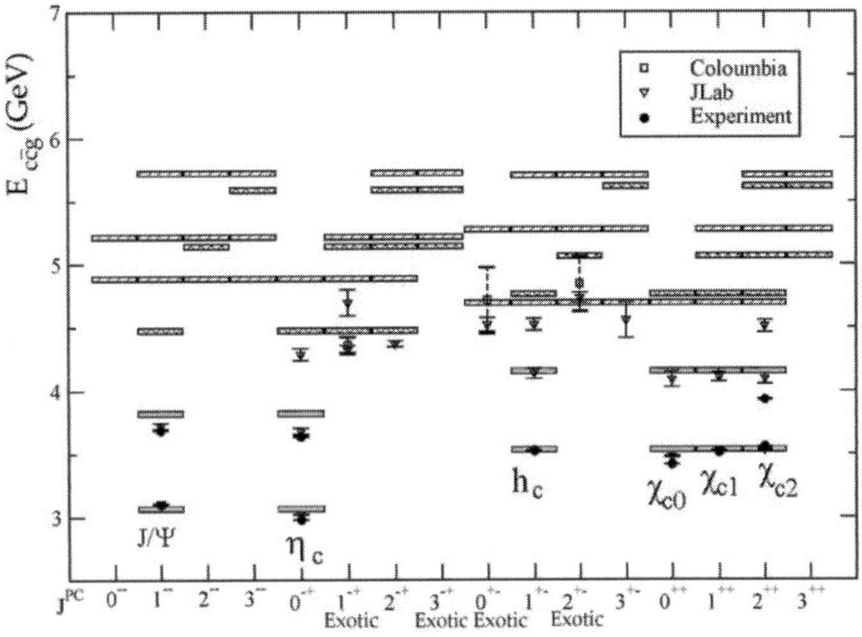

Figure 10. Charmonium (full grey boxes) and charmonium hybrid spectrum (single dashed and double dashed boxes) compared with data (where available) or lattice computations. Single dashed boxes are the $c\bar{c}g$ hybrids dominated by the P-wave quarks, all other have the $Q\bar{Q}$ pair in the relative S-wave orbital. Full grey boxes show results for the spin-averaged masses of ordinary charmonium states. The names of the ordinary charmonium families are also indicated. From Ref. [85][APS copyright].

can be constructed, for example as a quark-antiquark pair placed at an origin. Since the quark and the antiquark are at zero relative separation the system is rotationally symmetric, invariant under parity and charge conjugation. Thus gluelump states can be classified by the same J^{PC} quantum numbers as ordinary mesons.

The gluelump spectrum is studied [82] in the variational approach to Coulomb gauge QCD [83, 84] and massive quasi-gluon excitations are used to construct the Fock space representation. The calculated gluelump spectrum reproduce the spectrum obtained from lattice QCD computations. In particular, the ordering of the levels found in lattice QCD calculations is confirmed and it turns out to be non-trivially related to the non abelian structure of the Coulomb interaction in the Coulomb gauge which leads to three body interaction in the equation for the gluelump energies.

This numerical calculation has been used also to compute the spectrum of low mass hybrid mesons containing a heavy quark-antiquark pair [85], since in this case we can think the states as an external product of the gluelump states and the quark-antiquark states. Using the non relativistic reduction of Coulomb gauge QCD the spectrum of the low mass hybrid mesons containing a heavy quark-antiquark pair is numerically calculated. The gluon degrees of freedom are treated in the mean field approximation calibrated to the gluelump spectrum. Like in the gluelump case the resulting potential from Coulomb gauge QCD has not only a two body interaction, but also a non irreducible three body term linking all three constituents(quasi-particles)– quark, antiquark and gluon. The explicit expression of the two and three body potential can be found in Ref. [85]. It is this three-body interaction, related to the non-Abelian part of the QCD Coulomb interaction, that is responsible for producing the inverted parity ordering of the charmoniom hybrids spectra, as also seen in LQCD results. The numerical results, for charmonium hybrids, are given in figure 10 and compared with LQCD results.

4.3. Glueballs

The group of Trento has used the Interacting Instanton Liquid Model IILM (for a review on the IILM see Ref. [76]) as a tool to study the role played by the chiral interactions in the lowest-lying vector and axial vector meson resonances, i.e a_1 and ρ [86]. Using the same model, then the group of Trento also analized the properties of the glueballs [87], in particular of the lightest resonance.

Instantons have been long argued to be the dominant fluctuations generating the zero-mode zone of the Dirac operator, hence providing the correlations which break spontaneously chiral symmetry. In the IILM, the QCD path integral over all gluon configurations is replaced by an effective theory in which the gauge fields accounted for are those generated by integrating over the positions, color orientations, and sizes of singular-gauge instantons. The principal flaw of the instanton liquid model is that it does not predict a linearly rising potential between static color sources, at large distances. Thus, although this model describes well the non-perturbative QCD correlations at the chiral symmetry breaking scale $\Lambda_\chi \sim 1$ GeV, it fails at the confining scale $\sim \Lambda_{QCD} \sim 0.2$ GeV. Because of the lack of confinement, the IILM can be used as a tool to single-out the contribution of chiral forces in the different hadronic systems by studying where the model works well. In this context, the a_1 and ρ resonances represent good test systems: being in between the pion, strongly related to chiral symmetry breaking, and highly excited states, dominated by color confinement, they are expected to be sensitive to both chiral and confining interactions. In addition the splitting between these two resonance masses can be a measure of the effect of chiral symmetry breaking.

In Ref. [86] momentum projected correlation functions are computed at Euclidean times up to ~ 1.2 fm and the effective mass plot analysis reported. This is done not only for stable states, as in lattice QCD simulations, but also for resonances with a finite width. In the former case the effective mass plot displays a flat plateau in the large Euclidean time limit, while in the latter one displays an approximatively linear and mild fall-off behaviour in the intermediate region. The effective mass plot analysis is applied in particular to determine the mass of the lowest-lying vector and axial-vector mesons, using several values of the quark mass. The results [86] show that the instanton-induced chiral forces are sufficiently strong to generate the ρ and the a_1 mesons even in absence of confinement. On the other hand, the obtained mass values for the vector and axial vector mesons are a 30% greater than the corresponding experimental values,in particular $M_\rho \simeq 1$ GeV and $M_{a_1} \simeq 1.7$ GeV. Thus the masses of the vector and axial mesons receive significant contributions also from confinement, of the order of 30%. It is important to remind that on the contrary the masses of the lowest states like the pion and the nucleon, as completely dominated by chiral forces, were previously exactly reproduced in this scheme [88].

Lattice studies have found that the lightest glueball should be in the mass range 1.500 – 1.750 GeV and that it should be an unusually compact object,with a size of the order ~ 0.2 fm. Thus, it is reasonable to expect that the dominant forces, responsible for its binding, should act at the chiral symmetry breaking scale, rather than at the confinement scale. In this respect, the IILM should be a suitable tool to study the lightest glueball. Introducing the effective energy $E_{eff}(\tau, \mathbf{p})$ as the logarithmic derivative of the momentum-projected correlation function $G_S(\tau, \mathbf{p})$ [87],

$$E_{eff}(\tau, \mathbf{p}) = -\frac{d}{d\tau} \log G_S(\tau, \mathbf{p}), \qquad (3)$$

the spectrum contains a scalar glueball bound-state of mass $M_{0^{++}}$, only if at large Euclidean times the effective energy $E_{eff}(\tau, \mathbf{p})$ becomes independent on τ and converges to the kinetic energy of such a state, propagating with momentum $\mathbf{p}$:

$$\lim_{\tau \to \infty} E_{eff}(\tau, \mathbf{p}) = \sqrt{\mathbf{p}^2 + M_{0^{++}}^2}. \tag{4}$$

Correlation functions in the IILM can be computed numerically using Monte Carlo methods and the results show (see Ref. [87]) that the momentum projected correlator displays a plateau at large Euclidean times, signature of a bound state existence within this scheme[87]. The masses extracted using different quark masses are in the range $M_{0^{++}}^{IILM} \simeq 1.68 - 1.78$ GeV [87], consistent with the predictions of lattice QCD simulations. A complementary analysis was also carried on in the SIA (Single Instanton Approximation) model. This method offers analytic insight, and is based on the extraction of the mass from effective energy evaluated at finite momentum $\mathbf{p}$. Additional evidence [87] for a bound-state was found with an estimated mass $M_{0^{++}}^{SIA} \simeq 1.29 - 1.42$ GeV, i.e. consistent with the value calculated in lattice gauge theory.

5. Lattice QCD calculations of hadron observables

The European Twisted Mass Collaboration (ETMC) has started a systematic program of calculations of three-point correlation functions using the unquenched gauge configurations produced for different values of the lattice spacing and volumes, adopting the tree-level improved Symanzik gauge action and the twisted mass fermionic action with two dynamical flavors tuned at maximal twist (see Ref. [89]). Their aim is the determination of the electromagnetic and weak form factors of light and heavy-light mesons and baryons. Recently [90] they have calculated the electromagnetic pion form factor for pion masses between 260 MeV and 580 MeV. The lattice spacings are lower than 0.1 fm and the lattice volumes greater than 2 fm^3.

The extrapolation to the physical point is carried out using chiral perturbation theory at NNLO. The pion mass, its decay constant and form factor calculated in the [0.05,0.8]GeV^2 range of Q^2, are all simultaneously fitted. The low-energy constants extracted from the chiral fit are in agreement with those obtained in a chiral analysis of pion-pion scattering data [91]. The comparison between the chiral extrapolation of the ETMC results (full dots) and the experimental data (open markers) is shown in figure 11, where the dashed lines correspond to the chiral fit uncertainty (statistical plus systematic). As one can check, there is a very good agreement between the chiral extrapolation of the ETMC lattice results [90] and the experimental data [59]. The use of all-to-all quark propagators computed with a stochastic method, as well as of twisted boundary conditions on the valence quark fields, has allowed to achieve a remarkable statistical precision and very low values of the four momentum transfer, giving the possibility of an accurate chiral extrapolation of the calculated form factor to the physical point.

6. Quark-antiquark and/or meson cloud effects

We have already discussed the open problem of the missing strength in the theoretical prediction for the electromagnetic excitation of the baryon resonances with constituent quark models, that indicates that some fundamental mechanism is lacking in the dynamical description of the nucleon structure. This mechanism can be identified with the production of a quark-antiquark pair [18, 20]. A strong interest on the possible influence of such quark-antiquark and/or meson cloud effects is emerging in the literature. The problem is how to include this feature into the description of hadrons. A possibility is to introduce higher Fock components in the hadron wave function. This means that, at the hadron scale, the baryon acquires a $4q\bar{q}$ configuration

$$\mid \Psi\rangle = \Psi_{qqq} \mid qqq\rangle + \Psi_{4q\bar{q}} \mid qqq\bar{q}q\rangle \tag{5}$$

The consistent inclusion of higher Fock components in CQMs is a hard task, but, as a preliminary approach one can introduce models in which the physical nucleon N is considered

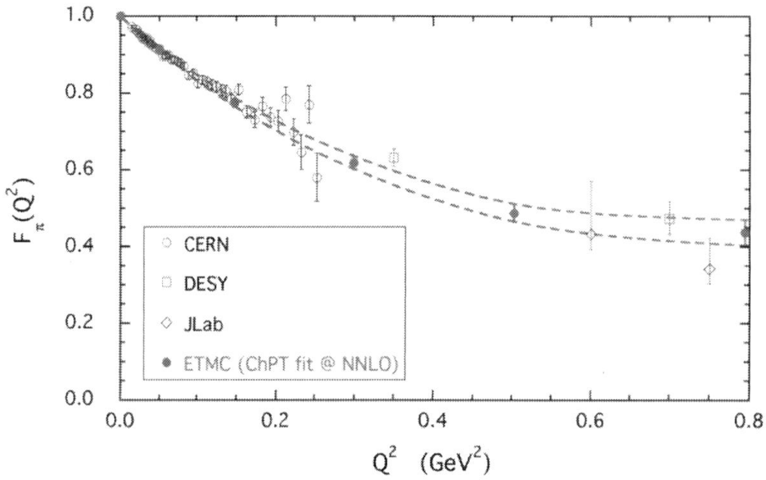

Figure 11. Charge form factor of the pion versus Q^2. The chiral extrapolation of the ETMC lattice results (full dots) Ref. [90] is compared with the experimental data (open markers) Ref. [92]. The dashed lines correspond to the region selected at 1σ level by the NNLO chiral analysis.

to be made of a bare nucleon dressed by a surrounding meson cloud

$$\mid N\rangle = \Psi_{3q} \mid N(qqq)\rangle + \sum_{B,M} \Psi_{(3q)(q\bar{q})} \mid B(qqq)M(q\bar{q})\rangle + ... \quad (6)$$

An approach of this type has been used by the Pavia group to calculate GPDs[93]. The virtual photon can scatter either on a valence quark of the bare nucleon, or on one of the quarks in the baryon-, B, or meson-, M, clusters that form the higher Fock state (BM). The calculations are done considering the possibility of two higher Fock states, pion-nucleon and Delta-nucleon, using a light cone hamiltonian with hypercentral interaction [52, 7] and a direct pion-nucleon(Delta) coupling. The GPDs are calculated by means of the convolution formalism. The results for the isoscalar and isovector unpolarized and helicity GPDs are shown in figure 12 using light cone wave functions from the hypercentral model for the nucleon [52, 7] and a gaussian for the pion.

In another similar approach in a Beijin-Genoa collaboration, using the hCQM [94] wave functions, the pion cloud has been added by means of the pion-nucleon coupling taken from [95]. The preliminary results show that the higher Fock components give relevant contributions both to the elastic form factors of the nucleon and to helicity amplitudes [94].

The contributions of the higher Fock components to the pion form factor have been calculated by the Rome group [96]. The photon vertex is dressed by means of a covariant vector meson dominance and a direct pion absorption by quarks is considered. Among the various terms contributing to the electromagnetic amplitude, the so called instantaneous terms seem to be of relevant importance. An important aspect of the approach is that a unified description of both time- and space-like form factors can be obtained. The theoretical results [96], in comparison to the experimental data, are shown on figure 13. The agreement is impressive. A similar calculation has been done also for the nucleon form factors [97].

All these approaches describe the pairs as mesons that couples to the hadrons. A sistematic inclusion of a quark antiquark pair-creation mechanism at the quark level is not yet available. A possibility is given by a recent work aiming at unquenching the quark model [98]. The $q\bar{q}$ creation mechanism is introduced at the quark level by means of a QCD inspired $3P0$ mechanism. The construction of the formalism and the difficult linked problems have been solved by means of

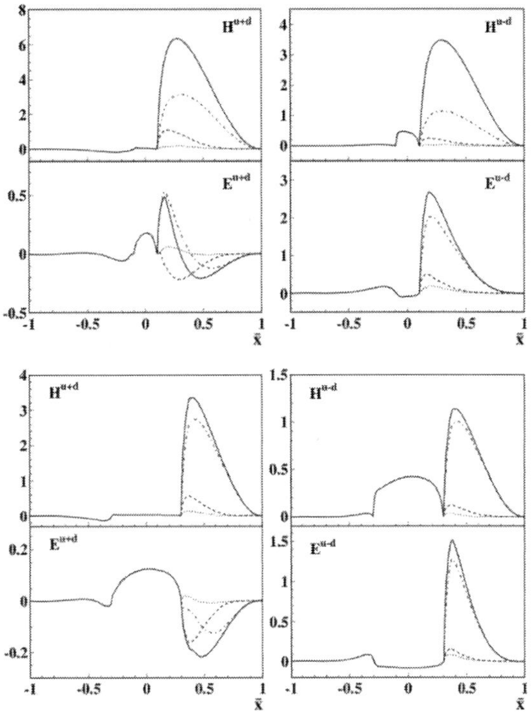

Figure 12. Isoscalar $(u + d)$ and isovector $(u - d)$ of the averaged $H-$ and $E-$ type GPDs calculated in a meson-cloud convolution for $t = -0.5\ GeV^2$ and $\xi = 0.1$ (upper part) or $\xi = 0.3$ (lower part) from Ref. [93] [APS copyright].

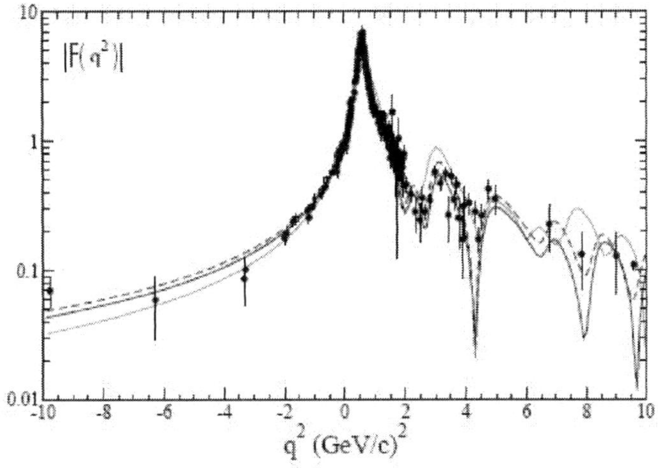

Figure 13. The pion form factor in the time- and space-like regions as from Ref. [96] [APS copyright] compared with data (details and data in the article).

group theory. In fact the sum over all big towers of intermediate states has been performed thanks to group theoretical methods and the permutational symmetry for all identical quarks has been taken into account. All the possible intermediate states are considered, since as Isgur has shown some years ago, no truncated Fock space can preserve the good CQM results for meson spectroscopy and the OZI hierarchy. Preliminary results for the proton spin show that one half is carried by valence quarks and the other one half is due to the orbital angular momenta of the quark-antiquark pairs [98].

Acknowledgements

The author is indebted with the colleagues working in the Universities and/or INFN Sections of Genova, Roma1-2-3, Pavia, Perugia, Trento for supplying their results and sending materials useful for the preparation of her talk and this contribution.

References

[1] 2004 *Hadronic Physics from Lattice QCD*, ed A M Green (Singapore:Word Scientific)
[2] Bijnens J 2007 *Prog. Part. Nucl. Phys.* **58** 521
[3] Morpurgo G 1965 *Physics* **2** 95
[4] Isgur N and Karl G 1979 *Phys. Rev.* D **18** 4187; *Phys. Rev.* D **19** 2653
[5] Capstick S and Isgur N 1986 *Phys. Rev.* D **34** 2809
[6] Bijker R, Iachello F and Leviatan A 1994 *Ann. Phys. (N.Y.)* **236** 69
[7] Ferraris M, Giannini M M, Pizzo M, Santopinto E and Tiator L 1995 *Phys. Lett.* B **364** 231
[8] Glozman L Ya and Riska D O 1996 *Phys. Rep.* C **268** 263; Glozman L Ya, Plessas W, Varga K and Wagenbrunn R F 1998 *Phys. Rev.* D **58** 094030.
[9] Löring U, Kretzschmar K, Metsch B Ch and Petry H R 2001 *Eur. Phys. J.* A **10** 309; Löring U, Metsch B Ch and Petry H R 2001 *Eur. Phys. J.* A **10** 395; 447
[10] Thomas A W and Weise W 2001 *The structure of the nucleon* (Berlin:Wiley-VCH)
[11] Schaefer T and Shuryak E 1996 *Phys. Rev.* D **53** 6522; 1998 *Rev. Mod. Phys.* **70** 323; Diakonov D 2003 *Prog. Part. Nucl. Phys.* **51** 173
[12] Cardarelli F, Pace E, Salmé G and Simula S 1995 *Phys. Lett.* B **357** 267.
[13] De Sanctis M, Santopinto E and Giannini MM 1998 *Eur. Phys. J.* A **1** 187.
[14] Pace E, Salmé G, Cardarelli F and Simula S 2000 *Nucl. Phys.* A **666** 33.
[15] De Sanctis M, Giannini M M , Repetto L and Santopinto E 2000 *Phys. Rev.* C **62** 025208.
[16] Boffi S, Glozman L Y, Klink W, Plessas W, Radici M and Wagenbrunn R F 2002 *Eur. Phys. J.* A **14** 17
[17] De Sanctis, Giannini M M, Santopinto E and Vassallo A 2007 *Phys. Rev.* C **76** 062201(R)
[18] Aiello M, Ferraris M, Giannini M M, Pizzo M and Santopinto E 1996 *Phys. Lett.* B **387** 215
[19] De Sanctis M, Santopinto E and Giannini MM 1998 *Eur. Phys. J.* A **2** 403
[20] Aiello M, Giannini M M and Santopinto E 1998 *J. Phys. G: Nucl. Part. Phys.* **24** 753
[21] Koniuk R and Isgur N 1980 *Phys. Rev.* D **21** 1868
[22] Copley L A, Karl G and Obryk E 1969 *Phys. Lett.* **29** 117
[23] Giannini M M and Santopinto E to be published
[24] Burkert V 2003 *Baryons 2002* (Singapore: World Scientific) 29
[25] Tiator L, Drechsel D, Kamalov S, Giannini M M, Santopinto E and Vassallo A 2004 *Eur. Phys. J.* A **19** 55
[26] Drechsel D, Hanstein O, Kamalov S and Tiator L 1999 *Nucl. Phys.* A **645** 145
[27] Kamalov S, Yang SN, Drechsel D, Hanstein O and Tiator L 2001 *Phys. Rev.* C **64** 032201
[28] Yao W -M *et al* 2006 *Particle Data Group J. Phys.* G **33** 1
[29] Sivers D W 1990 *Phys. Rev.* D **41** 83; 1991 *Phys. Rev.* D **43** 261
[30] Anselmino M, Boglione M and Murgia F 1995 *Phys. Lett.* B **362** 164
[31] Anselmino M and Murgia F 1998 *Phys. Lett.* B **442** 470
[32] Mulders P J and Tangerman R D, 1996 *Nucl. Phys.* B **461** 197; Boer D and Mulders P J 1998 *Phys Rev.* D **57** 5780
[33] Brodsky S J, Hwang D S and Schmidt I 2002 *Phys. Lett.* B **530** 99
[34] Brodsky S J, Hwang D S and Schmidt I 2002 *Nucl. Phys.* B **642** 344
[35] Collins J C 2002 *Phys. Lett.* B **536** 43
[36] Anselmino M, Boglione M, D'Alesio U, Leader E, Melis S and Murgia F 2006 *Phys. Rev.* D **73** 014020
[37] D'Alesio U and Murgia F 2004 *Phys. Rev.* D **70** 074009

[38] Anselmino M, Boglione M, D'Alesio U, Kotzinian A, Murgia F, Prokudin A and Turk C 2007 *Phys. Rev.* D **75** 054032
[39] Anselmino M, Boglione M, D'Alesio U, Kotzinian A, Melis S, Murgia F, Prokudin A and Turk C 2008 *16th International Workshop on Deep Inelastic Scattering (DIS 2008)* (*Preprint* hep-ph/08070173)
[40] Anselmino M *et al* 2009 *Eur. Phys. J.* A **39** 89
[41] Anselmino M *et al* 2005 *Phys. Rev.* D **72** 094007
[42] Boglione M, D'Alesio U, Murgia F 2008 *Phys. Rev.* D **77** 051502
[43] Adam D L *et al* (E704 Collaboration) 1991 *Phys. Lett.* B **264** 462, Adams J *et al* (STAR Collaboration) 2004 *Phys. Rev. Lett.* **92** 171801; Adler SS *et al*(PHENIX Collaboration) 2005 *Phys. Rev. Lett.* **95** 202001
[44] Arapetian A *et al* (HERMES Collaboration) 2004 *Phys. Rev. Lett.* **94** 012002; Alexahakhin V Y *et al* (COMPASS Collaboration) 2005 *Phys. Rev. Lett.* **94** 202002; Ageev E S*et al* (COMPASS Collaboration) 2007 *Nucl. Phys.* B **765** 31
[45] Qiu J-W and Sterman G 1998 *Phys. Rev.* D **59** 014020; Kouvaris C, Qiu J-W, Vogelsang W and Yuan F 2006 *Phys. Rev.* D **74** 114013
[46] Bacchetta A, Bomhof C, D'Alesio U, Mulders P J and Murgia F 2007 *Phys. Rev. Lett.* **99** 212002
[47] Bacchetta A, Bomhof C J, Mulders P J and Pijlman F 2005 *Phys. Rev.* D **72** 034030
[48] Mulders PJ and Tangerman RD 1996 *Nucl. Phys.* B **461** 197; Erratum 1997 *Nucl. Phys.* B **484** 538; Boers D Mulders PJ 1998 *Phys. Rev.* D **57** 5780; Goeke K, Metz A and Schlegel 2005 *Phys. Lett.* B **618** 90; Bacchetta A, Diehel M, Goeke K, Metz A, Mulders P J and M Schlegel 2007 *J. High Energy Phys.* JHEP02(2007)093
[49] Jacob R, Mulders PJ and Rodrigues J 1997 *Nucl. Phys.* A **626** 937
[50] Pasquini B, Cazzaniga S and Boffi S 2008 *Phys. Rev.* D **78** 034025
[51] Schlumpf F 1992 doctoral thesis, University of Zurigo, *Preprint* hep-ph/9211255
[52] Faccioli P, Traini M and Vento V 1999 *Nucl. Phys.* A **656** 400
[53] Pasquini B, Pincetti M and Boffi S 2005 *Phys.Rev.* D **72** 094029; 2007 *Phys. Rev.* D **76** 034020
[54] Brodsky S J, Hwang D S and Schmidt I 2002 *Phys. Lett.* B **530**99
[55] Gamber L P, Godstein G R and Schlegel M 2008 *Phys. Rev.* D **77** 094016
[56] Yuan F 2003 *Phys. Lett.* B **575** 45
[57] Cherednikov O I, D'Alesio U, Kochelev N I and Murgia F 2006 *Phys. Lett.* B **642** 39
[58] Courtoy A, Fratini F, Scopetta S and Vento V 2008 *Phys. Rev.* D **78** 034002
[59] Lu Z and Ma B -Q 2004 *Phys. Rev.* D **70** 094044
[60] Meissner S, Metz A and Goeke K 2007 *Phys. Rev.* D **76** 034002
[61] Bacchetta A, Conti F and Radici M 2008 *Phys. Rev.* D **78** 074010
[62] Bianconi A and Radici M 2005 *Phys. Rev.* D **71** 074014; 2005 *Phys. Rev.* D **72** 074013; 2006 *Phys. Rev.* D **73** 034018
[63] Konishi K Ukawa A and Veneziano G 1978 *Phys. Lett.* B **78** 243
[64] Collins J C and Ladinsky G A 1994 *Preprint* hep-ph/9411444; Collins J C Heppelmann S F and Ladinsky G A 1994 *Nucl. Phys.* B **420** 565
[65] Artru X and Collins J C 1996 *Z. Phys.* C **69** 277
[66] Jaffe R L Jin X and Tang J 1998 *Phys. Rev. Lett.* **80** 1166
[67] Bianconi A, Boffi S, Jakob R and Radici M 2000 *Phys. Rev.* D **62** 034008
[68] Bacchetta A and Radici M 2004 *Phys. Rev.* D **69** 074026
[69] Bianconi A, Boffi S, Jakob R and Radici M 2000 *Phys. Rev.* D **62** 034009
[70] Radici M, Jakob R and Bianconi A 2002 *Phys. Rev.* D **65** 074031
[71] Bacchetta A and Radici M 2003 *Phys. Rev.* D **67** 094002
[72] Bacchetta A and Radici M 2006 *Phys. Rev.* D **74** 114007
[73] Ceccopieri F A, Radici M and Bacchetta A 2007 *Phys. Lett.* B **650** 81
[74] Bacchetta A, Ceccopieri F A, Mukherjee A and Radici M 2009 *Phys. Rev.* D **79** 034029
[75] Airapetian A*et al* (*HERMES Collaboration*) 2008 *J. High Energy Phys.* JHEP06(2008)017
[76] Schaefer T and Shuryak E V 1998 *Rev. Mod. Phys.* **70** 323
[77] Tornqvist N A 1995 *Z. Phys.* C **68** 647
[78] Jaffe R L 1977 *Phys. Rev.* D **15** 267
[79] Maiani L, Piccinini F, Polosa A D and Riquer V 2004 *Phys. Rev. Lett.* **93** 212002
[80] Santopinto E and Galatà G 2007 *Phys. Rev.* C **75** 045206
[81] Iachello F and Mukhopadhyay N C and Zhang L 1991 *Phys. Lett.* B **256** 295
[82] Guo P, Szczepaniak A P, Galata G, Vassallo A and Santopinto E 2008 *Phys. Rev.* D **77** 056005
[83] Szczepaniak A P and Swanson E S 2002 Phys. Rev. D **65** 025012
[84] Feuchter C and Reinhardt H 2004 Phys. Rev. D **70** 105021; Reinhardt H and Feuchter C 2005 Phys. Rev. D **71** 105002

[85] Guo P, Szczepaniak A P, Galata G, Vassallo A and Santopinto E 2008 *Phys. Rev.* D **78** 056003
[86] Cristoforetti M, Faccioli P and Traini M 2007 *Phys. Rev.* D **75** 054024
[87] Tichy M C and Faccioli P 2007 *Preprint* hep-ph/0711.3829
[88] Cristoforetti M, Faccioli P, Negele J and Traini M 2007 *Phys. Rev.* D **75** 034008
[89] Boucaud Ph *et al* (*ETM Coll.*) 2007 *Phys. Lett.* B **650** 304; 2008 *Comput. Phys. Commun.* **179** 695
[90] Frezzotti R, Lubicz V and Simula S 2008 (*ETMC Coll.*) (*Preprint* hep-lat/08124042) ; Simula S (*ETMC Coll.*) 2007 *PoS LATT2007* 371 (*Preprint* hep-lat/07100097)
[91] Colangelo G, Gasser J and Leutwyler H 2001 *Nucl. Phys.* B **603** 125
[92] Amendolia S R *et al* 1986 *Nucl. Phys.* B **277** 168; Ackermann H *et al* 1978 *Nucl. Phys.* B **137** 294; Brauel P *et al* 1979 *Z. Phys.* C **3** 101; Volmer J *et al* (*JLab F(pi) Coll.*) 2001 *Phys. Rev. Lett.* **86** 1713; Tadevosyan V *et al* (*JLab F(pi) Coll.*) 2007 *Phys. Rev.* C **75** 055205; Horn T *et al* (JLab F(pi)-2 Coll.) 2006 *Phys. Rev. Lett.* **97** 192001; Horn T *et al* 2008 *Phys. Rev.* C **78** 058201 ; Huber G M *et al* 2008 *Phys. Rev.* C **78** 045203
[93] Pasquini B and Boffi S 2006 *Phys. Rev.* D **73** 094029
[94] Chen D Y, Dong Y B, Giannini M M and Santopinto E 2007 *Nucl. Phys.* A **782** 62
[95] Theberge S and Thomas A W 1983 *Nucl. Phys.* A **393** 252
[96] De Melo J P B C, Frederico T, Pace E and Salmé G 2006 *Phys. Rev.* D **73** 074013
[97] De Melo J P BC, Frederico T , Pace E, Pisano S and Salmé G 2009 *Phys. Lett.* B **671** 153; 2007 *Nucl. Phys.* A **782** 69
[98] Santopinto E and Bijker R 2008 *Proc. 6th Int. Conf. on Perspective in Hadronic Physics*, ed S Boffi *et al* (Trieste) *AIP Conf. Proc.* vol 1056 (Melville: American Inst. of Phys.) pp 95-97 ; 2008 *Few-Body Syst.* **44** 95

Transverse-momentum-dependent parton distributions in a spectator diquark model

F Conti[1,2], **A Bacchetta**[3] **and M Radici**[2]
[1]Dipartimento di Fisica Nucleare e Teorica, Università di Pavia, I-27100 Pavia, Italy
[2]Istituto Nazionale di Fisica Nucleare, Sezione di Pavia, I-27100 Pavia, Italy
[3]Theory Center, Jefferson Lab, 12000 Jefferson Av., Newport News, VA 23606, USA

E-mail: `francesco.conti@pv.infn.it`

Abstract. Within the framework of a spectator diquark model of the nucleon, involving both scalar and axial-vector diquarks, we calculate all the leading-twist transverse-momentum-dependent parton distribution functions (TMDs). Naive Time-odd densities are generated through a one-gluon-loop rescattering mechanism, simulating the final state interactions required for these functions to exist. Analytic results are obtained for all the TMDs, and a connection with the light-cone wave functions formalism is also established. The model parameters are fixed by reproducing the phenomenological parametrizations of unpolarized and helicity parton distributions at the lowest available scale. Predictions for the other parton densities are given and, whenever possible, compared with available parametrizations.

1. Introduction

Transverse-momentum-dependent parton distributions (TMDs) describe the probability to find, in a hadron, a parton with light-cone longitudinal momentum fraction x and transverse momentum $\boldsymbol{p}_T$ with respect to the direction of the parent hadron momentum, thus encoding low energy non-perturbative information on the QCD partonic structure of the nucleon.

In recent years, great interest has spread around these functions, since intrinsic transverse momentum effects can shed light on various phenomena, such as spin-orbit correlations and single spin asymmetries, and were also shown to be relevant for a correct understanding of multi-jet events at hadron colliders, thus playing a crucial role in LHC physics program [1]. This in turn triggered both theoretical and experimental efforts, aimed at unraveling the TMDs sector. Concerning theory, it was e.g. demonstrated [2] how TMDs which are odd under naive time-reversal transformations (for brevity, T-odd) are not forbidden by QCD time-reversal invariance, and can indeed be non-zero and appear at leading-twist. From the experimental point of view, we still have poor phenomenological information on TMDs, especially concerning their $\boldsymbol{p}_T$ dependence, even though the analysis of azimuthal spin asymmetries, e.g. in semi-inclusive deep inelastic scattering (SIDIS), allowed an extraction of the Sivers function [3], a T-odd TMD describing how the parton distribution of unpolarized quarks is distorted by the transverse polarization of the parent hadron, and a first attempt to extract the T-odd Boer-Mulders function, i.e. the distribution of transversely polarized partons in an unpolarized hadron, was also presented [4]. All of the above studies assume a flavor-independent Gaussian $|\boldsymbol{p}_T|$-distribution, though there is no compelling physical reason for this choice.

On account of the difficulty to study parton distributions from first principles, it is of great importance to develop simple models able to reproduce the essential qualitative features of

TMDs and to gain insight into their physical content, e.g. concerning partons orbital angular momentum, as well as to perform numerical estimates and simulations on observables of interest, thus guiding further experimental investigations. Among other few TMDs models [5] known in literature, the spectator diquark model [2] here employed owes its appeal in combining relative simplicity and useful versatility, allowing to obtain analytic results for all TMDs. We include both scalar and axial-vector diquarks and we further distinguish between isoscalar (*ud*-like) and isovector (*uu*-like) spectators; axial-vector diquarks are essential for a realistic flavor analysis, since the isospin triplet diquark *uu* flavour combination requires a $S = 1$ spin state. The relative phases necessary to produce T-odd structures are generated by approximating the gauge link operator, ensuring color gauge invariance for the correlators of the theory, with a one gluon-exchange mechanism. We also focus on a possible interpretation of the model results in terms of light-cone wave functions, which in turn gives a more general and physical dress to the results and opens the way to a wider range of applications. The model free parameters are fixed by reproducing the phenomenological parametrizations of unpolarized and longitudinally polarized parton distribution functions (PDFs) at the lowest available scale, and their values employed to obtain numerical results and comparison with available parametrizations for some parton distribution functions of primary interest. As we shall see, an overall good agreement with phenomenology is achieved.

2. Model results for Time-even TMDs

In the framework of parton model, the fundamental theoretical object to study quarks distributions into the proton is the quark-quark correlator, which is the diagonal matrix element, over a bound hadronic state with momentum P and spin S, of a bilocal operator built up with light-front quark field operators:

$$\Phi(x, \boldsymbol{p}_T; S) = \int \frac{d\xi^- d\boldsymbol{\xi}_T}{(2\pi)^3}\, e^{ip\cdot\xi}\, \langle P, S|\bar{\psi}(0)\,\mathcal{U}_{[0,\xi]}\,\psi(\xi)|P, S\rangle\Big|_{\xi^+=0}\,, \tag{1}$$

having used light-cone coordinates (so that the quark momentum is given by $p \equiv [p^+, p^-, \boldsymbol{p}_T] = [xP^+, (p^2 + \boldsymbol{p}_T^2)/2xP^+, \boldsymbol{p}_T]$); the gauge link operator, $\mathcal{U}_{[0,\xi]} = \mathcal{P}\,\exp(-ig\int_0^\xi dw \cdot A(w))$ connects the two different space-time points 0, ξ by all possible ordered paths followed by the gluon field A, coupling to the quark field with strength g, and its presence is mandatory in order to ensure color-gauge invariance, while its physical origin lies in the residual interactions between the active quark and the spectator system.

The spectator approximation consists in inserting a completeness relation and truncating, at tree-level, the sum over final states to a single on-shell spectator state with mass M_X and diquark quantum numbers; Eq. (1) can then be explicitly written as

$$\Phi(x, \boldsymbol{p}_T; S) \sim \frac{1}{(2\pi)^3}\,\frac{1}{2(1-x)P^+}\,\bar{\mathcal{M}}^{(0)}(S)\,\mathcal{M}^{(0)}(S)\,, \tag{2}$$

where m is the active quark mass, and the on-shell condition $(P-p)^2 = M_X^2$ for the spectator diquark implies an off-shell condition for the quark (M being the hadron mass):

$$p^2 \equiv \tau(x, \boldsymbol{p}_T) = -\frac{\boldsymbol{p}_T^2 + L_X^2(m^2)}{1-x} + m^2\,, \quad \text{with} \quad L_X^2(m^2) = xM_X^2 + (1-x)\,m^2 - x(1-x)M^2\,. \tag{3}$$

In order to match proton spin, the diquark can have either spin 0 (scalar, $X = s$) or spin 1 (axial-vector, $X = a$), as well as isospin 0 (isoscalar *ud*-like system) or isospin 1 (isovector *uu*-like system). The tree-level "scattering amplitude" $\mathcal{M}^{(0)} = \langle P - p|\,\psi(0)|P, S\rangle$, which is indeed a Dirac spinor, is constructed from the Feynman diagram in figure 1 (left), supposing a nucleon-quark-scalar diquark vertex $\mathcal{Y}_s = i g_s(p^2)\mathbf{1}$ and a nucleon-quark-axial vector diquark vertex $\mathcal{Y}_a^\mu = i[g_a(p^2)/\sqrt{2}]\gamma^\mu\gamma_5$, where $g_X(p^2)$ is a suitable form factor. Moreover, dealing with axial-vector diquarks, there is an intrinsic model arbitrariness in the expression of the sum over polarization states, $d^{\mu\nu} \equiv \sum_{\lambda_a} \varepsilon^{*\mu}_{(\lambda_a)} \varepsilon^{\nu}_{(\lambda_a)}$, and consequently of the diquark propagator (intervening in T-odd TMDs calculation) as well. In Ref. [6], a complete systematics has been presented: analytic results are derived for all leading-twist TMDs with pointlike, dipolar and exponential form factors, and three different choices of the axial-vector diquark polarization sum, with different physical contents, are probed. Here we will limit ourselves to the expressions obtained using a dipolar form factor, $g_X^{dip}(p^2) = (p^2 - m^2)/|p^2 - \Lambda_X^2|^2$ (with Λ_X a cut-off parameter), which smoothly suppress the influence of high $|\boldsymbol{p}_T|$, and choosing the so-called light-cone polarization sum (see Ref. [6] for details), physically corresponding to the propagation of only diquark transverse polarization degrees of freedom (the other choices also involve longitudinal as well as scalar polarizations). The theoretical motivation behind this last preference is that it guarantees, unlike the other two, that in DIS processes the axial-vector diquark (which, being charged, can interact with the probing virtual photon) only contributes to the longitudinal structure function F_L (the same situation which naturally occurs for the scalar case): this surely implies a violation of the Callan-Gross relation, but leaves unchanged the leading-order interpretation of the transverse structure function F_T as a charge-weighted sum of quark distributions, thus limiting the phenomenological impact represented by the presence of the diquarks.

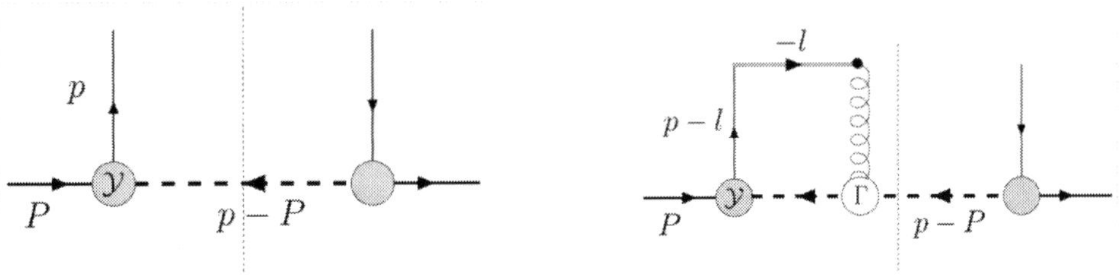

Figure 1. Left: tree-level spectator model diagram for the calculation of T-even leading-twist parton densities for an active quark (solid line), with a scalar or axial-vector diquark (dashed line). Right: interference between the one-gluon exchange diagram in eikonal approximation and the tree level diagram.

By properly projecting the tree level model correlator of Eq. (2) on various Dirac structures, all Time-even leading twist TMDs can be calculated explicitly. Focusing on the unpolarized distribution, for instance, we can write (with $\gamma^+ = (\gamma^0 + \gamma^3)/\sqrt{2}$)

$$\begin{aligned} f_1(x, \boldsymbol{p}_T^2) &= \frac{1}{4}\,\mathrm{Tr}\Big[\left(\Phi(x, \boldsymbol{p}_T; S) + \Phi(x, \boldsymbol{p}_T; -S)\right)\gamma^+\Big] + \text{h.c.} \\ &= \frac{1}{(2\pi)^3}\,\frac{1}{8(1-x)P^+}\,\mathrm{Tr}\left[\left(\bar{\mathcal{M}}^{(0)}(S)\,\mathcal{M}^{(0)}(S) + \bar{\mathcal{M}}^{(0)}(-S)\,\mathcal{M}^{(0)}(-S)\right)\gamma^+\right] + \text{h.c.} \end{aligned} \quad (4)$$

thus obtaining, with the choices just discussed, the following results:

$$f_1^{q(s)}(x, \boldsymbol{p}_T^2) = \frac{g_s^2}{(2\pi)^3}\,\frac{[(m + xM)^2 + \boldsymbol{p}_T^2]\,(1-x)^3}{2\,[\boldsymbol{p}_T^2 + L_s^2(\Lambda_s^2)]^4}\,, \quad (5)$$

$$f_1^{q(a)}(x, \boldsymbol{p}_T^2) = \frac{g_a^2}{(2\pi)^3}\,\frac{[\boldsymbol{p}_T^2\,(1+x^2) + (m + xM)^2\,(1-x)^2]\,(1-x)}{2\,[\boldsymbol{p}_T^2 + L_a^2(\Lambda_a^2)]^4}\,, \quad (6)$$

for the scalar and axial-vector contributions, respectively.

Besides this diagrammatic approach, T-even TMDs can also be derived by exploiting the so-called overlap representation in terms of light-cone wave functions (LCWFs) [7]; these are the frame independent interpolating functions between hadrons and quark/gluon degrees of freedom, and within our model can be defined as follows:

$$\psi^{\lambda_N}_{\lambda_q}(x, \boldsymbol{p}_T) = \sqrt{\frac{p^+}{(P-p)^+}} \frac{\bar{u}(p, \lambda_q)}{p^2 - m^2} \mathcal{Y}_s U(P, \lambda_N) , \tag{7}$$

$$\psi^{\lambda_N}_{\lambda_q \lambda_a}(x, \boldsymbol{p}_T) = \sqrt{\frac{p^+}{(P-p)^+}} \frac{\bar{u}(p, \lambda_q)}{p^2 - m^2} \varepsilon^*_\mu(P-p, \lambda_a) \mathcal{Y}^\mu_a U(P, \lambda_N) , \tag{8}$$

for scalar and axial-vector diquarks respectively, in terms of the quark and nucleon free spinors, u and U, and of the axial-vector diquark polarization vector ε^μ (see Ref. [7, 8]). The indices λ_N, λ_q and λ_a refer to the helicity of the nucleon, the quark and the axial-vector diquark respectively, and are constrained by angular momentum conservation to the spin sum rule $\lambda_N = \lambda_q(+\lambda_a)+L_z$, where L_z is the projection of the relative orbital angular momentum between the quark and the diquark. The explicit spectator model expressions for LCWFs can then be used to recover the results in eqs. (5) and (6) in terms of LCWFs overlaps:

$$f_1^{q(s)}(x, \boldsymbol{p}_T^2) = \frac{1}{16\pi^3} \frac{1}{2} \sum_{\lambda_N=\pm} \sum_{\lambda_q=\pm} |\psi^{\lambda_N}_{\lambda_q}|^2 , \tag{9}$$

$$f_1^{q(a)}(x, \boldsymbol{p}_T^2) = \frac{1}{16\pi^3} \frac{1}{2} \sum_{\lambda_N=\pm} \sum_{\lambda_q=\pm} \sum_{\lambda_a=\pm} |\psi^{\lambda_N}_{\lambda_q \lambda_a}|^2 , \tag{10}$$

and similar formulae hold for every T-even TMD: the approach is indeed completely general. We refer to [6] for details; for the sake of completeness and in view of its importance, we report here also the model results for the chiral-odd TMD transversity distribution, i.e. the transverse spin distribution in a tranversely polarized hadron:

$$h_1^{q(s)}(x, \boldsymbol{p}_T^2) = \frac{g_s^2}{(2\pi)^3} \frac{(m+xM)^2 (1-x)^3}{2\,[\boldsymbol{p}_T^2 + L_s^2(\Lambda_s^2)]^4} , \tag{11}$$

$$h_1^{q(a)}(x, \boldsymbol{p}_T^2) = -\frac{g_a^2}{(2\pi)^3} \frac{\boldsymbol{p}_T^2\, x(1-x)}{[\boldsymbol{p}_T^2 + L_a^2(\Lambda_a^2)]^4} . \tag{12}$$

By integration over $\boldsymbol{p}_T$, finally, analytic model expressions can be derived also for the usual integrated PDFs, $f_1(x)$, $g_1(x)$ and the transversity $h_1(x)$.

3. Model results for Time-odd TMDs

It is a well known fact that, by means of tree level amplitudes only, no Time-odd effect can be simulated, since there is no residual interaction between the active quark and the spectators, i.e. we lack the interference between two competing channels, coupled to the same final state, required to produce the complex amplitude whose imaginary part would give the T-odd contribution. We could anyway go further and consider the single-gluon-exchange scattering amplitude (in eikonal approximation) $\mathcal{M}^{(1)}$, as the leading-twist $\mathcal{O}(\alpha_s)$ approximation of the gauge link operator [9]; its interference with the tree-level term $\mathcal{M}^{(0)}$ thus allows to compute the Sivers and Boer-Mulders T-odd TMDs in the model. For the Sivers case, e.g., we have:

$$\frac{\varepsilon_T^{ij} p_{Ti} S_{Tj}}{M} f_{1T}^{\perp}(x, \boldsymbol{p}_T^2) = -\frac{1}{(2\pi)^3} \frac{1}{8(1-x)P^+} \mathrm{Tr}\left[\left(\mathcal{M}^{(1)}(S)\bar{\mathcal{M}}^{(0)}(S) - \mathcal{M}^{(1)}(-S)\bar{\mathcal{M}}^{(0)}(-S)\right)\gamma^+\right] + \mathrm{h.c.} , \tag{13}$$

where the transverse tensor $\varepsilon_T^{ij} = \varepsilon^{\mu\nu ij}\, n_{+\mu} n_{-\nu}$ only non-zero components are $\varepsilon_T^{12} = -\varepsilon_T^{21} = 1$. After introducing proper Feynman rules for the loop integral involved in $\mathcal{M}^{(1)}$, its imaginary part can be computed applying Cutkosky rules for the corresponding box diagram (see figure 1, right), by cutting, i.e. putting on shell, the eikonal and the axial-vector diquark propagators; the final model results read (considering, as before, dipolar form factor and light-cone polarization sum):

$$f_{1T}^{\perp\, q(s)}(x, \boldsymbol{p}_T^2) = -\frac{g_s^2}{4} \frac{M\, e_c^2}{(2\pi)^4} \frac{(1-x)^3\, (m + xM)}{L_s^2(\Lambda_s^2)\, [\boldsymbol{p}_T^2 + L_s^2(\Lambda_s^2)]^3}\,, \tag{14}$$

$$f_{1T}^{\perp\, q(a)}(x, \boldsymbol{p}_T^2) = \frac{g_a^2}{4} \frac{M\, e_c^2}{(2\pi)^4} \frac{(1-x)^2\, x\, (m + xM)}{L_a^2(\Lambda_a^2)\, [\boldsymbol{p}_T^2 + L_a^2(\Lambda_a^2)]^3}\,, \tag{15}$$

for the Sivers function (with $e_c^2 = 4\pi C_F \alpha_s$, where $1_3\, C_F = \sum_a t^a t^a = 1_3\, (4/3)$, t^a being the SU(3) Gell-Mann color matrices), while for the Boer-Mulders function we find: $h_1^{\perp\, q(s)}(x, \boldsymbol{p}_T^2) = f_{1T}^{\perp\, q(s)}(x, \boldsymbol{p}_T^2)$ and $h_1^{\perp\, q(a)}(x, \boldsymbol{p}_T^2) = -\, f_{1T}^{\perp\, q(a)}(x, \boldsymbol{p}_T^2)/x$.

As already seen for T-even TMDs, T-odd densities as well do admit an overlap representation in terms of tree level LCWFs, but while for T-even functions this simply appeared as a sum of products of LCWFs with the same arguments, we are now forced to consider convolution integrals of LCWFs with different arguments, including a suitable operator in order to describe the needed final-state-interactions (FSI) produced by the gluon rescattering. By comparing the overlap expressions with those obtained within the Feynman diagram approach, we found that this FSI kernel has a universal form (i.e., the same for scalar and axial-vector diquarks, and for Sivers and Boer-Mulders functions), reminiscent of the virtual gluon propagator:

$$G(x, x', \boldsymbol{p}_T, \boldsymbol{p}'_T) = -\, i\, \frac{C_F\, \alpha_s}{2\pi\, (\boldsymbol{p}_T - \boldsymbol{p}'_T)^2}\, \delta(x - x')\,. \tag{16}$$

4. Numerical results

In order to give predictive power to the model, we must fix its free parameters, namely the quark mass m, the nucleon-quark-diquark coupling g_X, the diquark mass M_X, and the cutoff Λ_X in the nucleon-quark-diquark dipolar form factor, for $X = s, a$ scalar and axial-vector diquarks. A good fit to phenomenological parametrizations also requires to distinguish between the two isospin states of the axial-vector diquark, i.e. the axial-vector isoscalar state with $I_3 = 0$ (corresponding to the ud system, label a) and axial-vector isovector diquark with $I_3 = 1$ (uu system, label a'). The model scale is another free parameter, but we choose to identify it with the lowest possible Q^2 at which parametrizations are available: hence, we fitted model results to the f_1^u and f_1^d ZEUS parametrizations [10] (ZEUS2002) at $Q_0^2 = 0.3$ GeV^2 and to the g_1^u and g_1^d leading-order (LO) data from Ref. [11] (GRSV2000) at $Q^2 = 0.26$ GeV^2. The global normalizations are fixed by imposing that $\int dx \int d^2\boldsymbol{p}_T\, f_1^{q(s/a)}(x, \boldsymbol{p}_T^2) = 1$.

The relation between quark flavors, i.e. the functions entering the fit, and diquark types, i.e. our model results, is written as

$$f_1^u = c_s^2\, f_1^{u(s)} + c_a^2\, f_1^{u(a)}\,, \qquad f_1^d = c_a'^2\, f_1^{d(a')}\,. \tag{17}$$

where the weight factors c_s, c_a and c'_a are also considered free parameters, and their values fixed by the global fit. The reason for this approach is twofold: first, this model involves non-zero (and relativistically enhanced) LCWFs with $L_z = 1$, hence the quark-diquark system can have a non-vanishing relative orbital angular momentum, and thus the proton wave function no longer displays a simple $SU(4)$ symmetry; secondly, a fundamental limitation of the diquark model consists in the intrinsic impossibility to simultaneously fulfil the quark number and momentum sum rules: as a matter of fact, on the one hand the total number of quarks "seen" in this model is only one, since the other two are always confined into the diquark state, and on the other hand the charged diquarks can be probed in DIS, thus contributing to the overall momentum. On the basis of such considerations, we decided to avoid imposing the momentum sum rule and to let the fit choose the values of the parameters c_X (see figure 4 and table I of [6] for the results of the fit and the corresponding values of the model parameters).

We now present a brief selection of plots, in order to highlight some crucial features of our results. In figure 2 we show the $\boldsymbol{p}_T^2$ dependence of the unpolarized TMD: we observe that f_1^u displays a non-monotonic behavior for $x \leq 0.02$, which is due to the presence of LCWFs with non-zero orbital angular momentum; indeed, the contribution of LCWFs with $L_z = 1$, dominant at high x, falls linearly with $\boldsymbol{p}_T^2$ as $\boldsymbol{p}_T^2 \to 0$, unlike the one from $L_z = 0$ LCWFs. This simple example shows how the study of the transverse-momentum-dependence of TMDs can already expose some effects due to orbital angular momentum. We also note that in our model the average quark transverse momentum decreases as x increases, and that d quarks on average carry less transverse momentum than u ones. Though dealing with a relatively simple model, we thus already acknowledge how the widely accepted assumption of a factorized, flavor-independent quark transverse-momentum distribution is inadequate. A significant violation of the x-$\boldsymbol{p}_T$ factorization has also been recently obtained in a chiral quark soliton model [17].

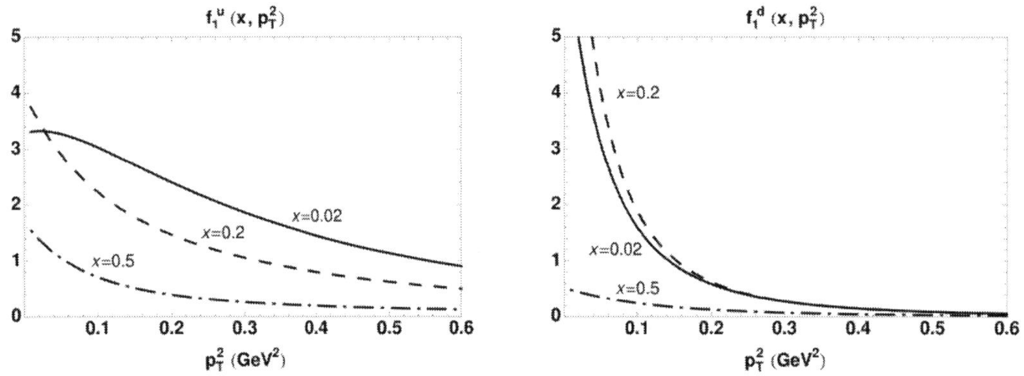

Figure 2. The $\boldsymbol{p}_T^2$ dependence of $f_1(x, \boldsymbol{p}_T^2)$ for u (left) and d quark (right), for different values of x.

In figure 3 we show the results for up and down quarks transversity distribution $h_1(x) \equiv \int d^2\boldsymbol{p}_T\, h_1(x, \boldsymbol{p}_T^2)$. Model functions are represented by the dashed lines and assumed to refer to the model scale $Q_0^2 = 0.3$ GeV^2, while the solid lines are the curves obtained after applying the DGLAP evolution at LO up to the scale $Q^2 = 2.5$ GeV^2 [12]; the latter scale pertains to the parametrization of Ref. [13], whose uncertainty band is depicted as a shaded area. We find reasonable agreement with the phenomenological data, with the maxima in the correct positions and a somewhat too small result for the up quark at small x. An interesting feature, typical of this model, is the change of sign at $x \sim 0.5$ in h_1^u (it cannot be reproduced by parametrizations, whose functional form does not allow for a sign change): indeed, the axial-vector diquark contribution is negative (see Eq. (12)) and becomes dominant over the scalar one at sufficiently high x. In figure 4 the $|\boldsymbol{p}_T|$-shape (at $x = 0.1$) of $h_1(x, \boldsymbol{p}_T^2)$ is shown, together with the parametrization in Ref. [13]: we have again a satisfying agreement between model results and phenomenology, at least below $|\boldsymbol{p}_T| \sim 0.3$ GeV/c (moreover, we stress that in this case no TMD-evolution was applied to model functions). Also $h_1^u(x = 0.1, \boldsymbol{p}_T^2)$ changes sign, at $|\boldsymbol{p}_T| \sim 0.5$ GeV/c, due to the dominance of the negative S=1 diquark contribution at high $|\boldsymbol{p}_T|$.

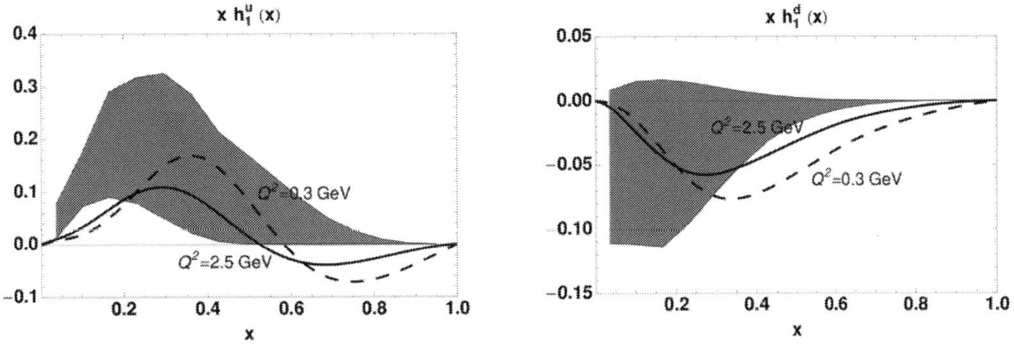

Figure 3. The transversity distribution $x\,h_1(x)$ for u (left) and d quark (right). Dashed (solid) line for the model results before (after) DGLAP evolution at LO up to the scale of the parametrization from Ref. [13], whose uncertainty band is represented by the shaded area.

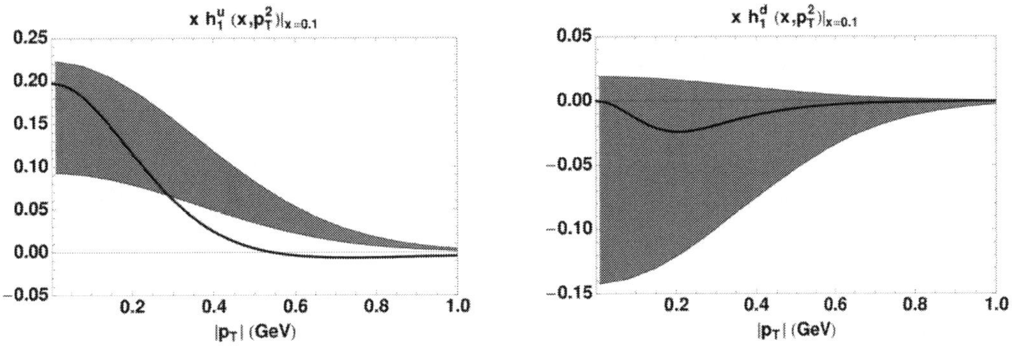

Figure 4. Same as in the previous figure, but for the $|\boldsymbol{p}_T|$-dependence of transversity at $x = 0.1$.

Turning to T-odd functions, they are mainly studied in connection with single spin asymmetries (SSA). In figure 5 we report the model results for the first $\boldsymbol{p}_T$-moment of the Sivers function, defined as $f_{1T}^{\perp(1)}(x) \equiv \int d^2\boldsymbol{p}_T\,(\boldsymbol{p}_T^2/2M^2)\,f_{1T}^{\perp}(x,\boldsymbol{p}_T^2)$, compared with two different available parametrizations [14, 15] extracted from SIDIS data on transverse SSA. The spectator model predicts the correct signs for the u and d quark functions (also in agreement with lattice simulations [16]), and the maxima are reached at about the same x (~ 0.3) as the parametrizations, but the theoretical curve underestimates the experimental one for down quark, while the agreement is more satisfactory for the up quark (anyway, also in this case no scale-evolution was applied). Finally, figure 6 shows the model results for the spin density of unpolarized quarks with flavor q in transversely polarized protons, defined as

$$f_{q/p^\uparrow}(x,\boldsymbol{p}_T) = f_1^q(x,\boldsymbol{p}_T^2) - f_{1T}^{\perp q}(x,\boldsymbol{p}_T^2)\,\frac{(\hat{\boldsymbol{P}}\times\boldsymbol{p}_T)\cdot\boldsymbol{S}}{M}\,, \tag{18}$$

revealing the expected asymmetry in transverse-momentum space (along $\pm\hat{\boldsymbol{p}}_{\boldsymbol{y}}$, for a proton transverse polarization along $+\hat{\boldsymbol{x}}$, and choosing $\hat{\boldsymbol{P}} = -\hat{\boldsymbol{z}}$), connected with the Sivers function. The distortion, towards positive (negative) values of p_y for u (d) quarks, is in agreement with similar lattice results [16] and with the signs of the anomalous magnetic moments κ_q.

Although no reliable experimental parametrization is yet available for the Boer-Mulders function, also in this case we checked the agreement between our model functions signs (negative for both u and d) and the lattice calculations of Ref. [16].

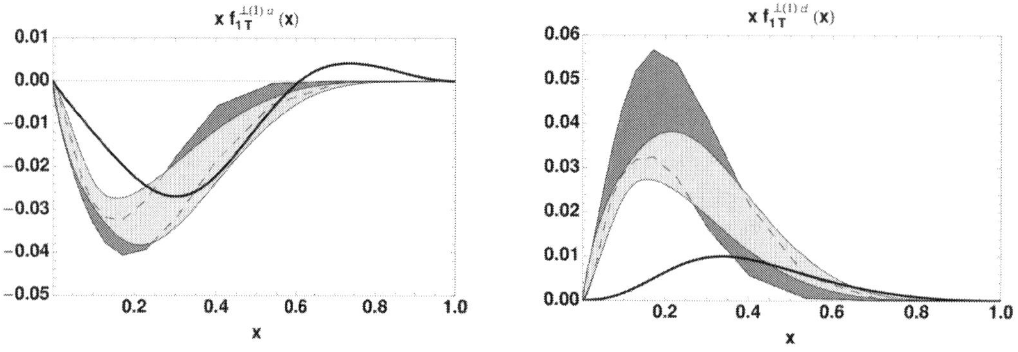

Figure 5. The first $\boldsymbol{p}_T$-moment $x\, f_{1T}^{\perp(1)}(x)$ of the Sivers function for u (left) and d (right) quark. Solid line for the spectator diquark model results. Darker shaded area for the uncertainty band of parametrizations from Ref. [14], lighter one from Ref. [15].

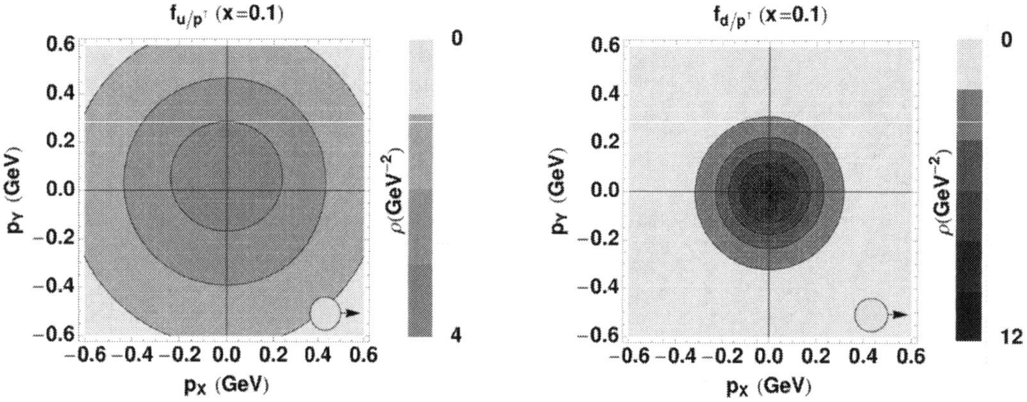

Figure 6. Model results for the spin density of unpolarized quarks in transversely polarized protons in $\boldsymbol{p}_T$ space at $x = 0.1$, for u (left) and d (right) quarks. Proton transverse polarization along $+\hat{\boldsymbol{x}}$.

References

[1] Hautmann F, Jung H 2008 *JHEP* 0810:113
[2] Brodsky S, Hwang D S and Schmidt I 2002 *Phys. Lett.* B **530** 99
[3] Anselmino M *et al* 2005 *Preprint* `hep-ph/0511017`
[4] Zhang B, Lu Z, Ma B-Q and Schmidt I 2008 *Phys. Rev.* D **77** 054011
[5] Pasquini B, Cazzaniga S and Boffi S 2008 *Phys.Rev.* D **78** 034025
[6] Bacchetta A, Conti F, Radici M 2008 *Phys.Rev.* D**78** 074010
[7] Brodsky S J, Hwang D S, Ma B-Q and Schmidt I 2001 *Nucl. Phys.* B **593** 311
[8] Lepage G P, Brodsky S J, 1980 *Phys. Rev.* D **22** 2157
[9] Boer D, Mulders P J and Pijlman F 2003 *Nucl. Phys.* B **667** 201
[10] ZEUS, Chekanov S *et al* 2003 *Phys. Rev.* D **67** 012007
[11] Gluck M, Reya E, Stratmann M and Vogelsang W 2001 *Phys. Rev.* D **63** 094005
[12] Hirai M, Kumano S and Miyama M 1998 *Comput. Phys. Commun.* **111** 150
[13] Anselmino M *et al* 2007 *Phys. Rev.* D **75** 054032
[14] Anselmino M *et al* 2009 *Eur. Phys. J.* A **39** 89
[15] Collins J C *et al* 2005 *Preprint* `hep-ph/0510342`
[16] QCDSF, Gockeler M *et al* 2007 *Phys. Rev. Lett.* **98** 222001
[17] Wakamatsu M 2009 *Preprint* `arXiv:0903.1886[hep-ph]`

Monte Carlo simulations of QCD thermodynamics in the PNJL model

M Cristoforetti, T Hell and W Weise

Physik-Department T39, Technische Universität München, James-Franck-Str. 1, D-85747 Garching

Work supported in part by BMBF, GSI and the DFG Excellence Cluster "Origin and Structure of the Universe"

E-mail: mcristof@ph.tum.de

Abstract. We apply a Monte Carlo method to the two flavor Polyakov loop extended Nambu and Jona-Lasinio (PNJL) model. In such a way we can go beyond mean field calculations introducing fluctuations of the fields. We study the impact of fluctuations on the thermodynamics of the model. We calculate the second derivatives of the thermodynamic grand canonical partition function with respect to the chemical potential and present a comparison with lattice data also for flavor non-diagonal susceptibilities.

1. Introduction

Results of QCD thermodynamics from lattice computations are reproduced surprisingly well with a quasiparticle model, an extension of the Nambu – Jona-Lasinio model with inclusion of Polyakov loop dynamics (the PNJL model) at the mean field level [1, 2, 3, 4, 5, 6, 7, 8]. A better understanding of these results requests the investigation of fluctuations in the PNJL model. This can be done by numerical simulations of the thermodynamics using standard Monte Carlo techniques (MC-PNJL). The advantage over other methods is that Monte Carlo calculations automatically incorporate fluctuations to all orders.

We present new results of such computations and discuss the role and importance of fluctuations beyond mean field. It turns out that the temperature dependence of the second non-diagonal derivative of the thermodynamic grand canonical partition function with respect to quark chemical potentials is particularly sensitive to such fluctuations. While the mean field PNJL predicts a vanishing coefficient, our calculation agrees well with available lattice data.

2. The PNJL partition function

The Euclidean action of the two-flavor PNJL model, at finite baryon and isospin chemical potential, is given by [9, 7]

$$\begin{aligned} \mathcal{S}_E(\psi,\bar\psi,\phi) &= \int_0^{\beta=1/T} \mathrm{d}\tau \int \mathrm{d}^3x\, [\bar\psi(i\not{D} + \gamma_0\tilde\mu - m_0)\psi \\ &\quad + G((\bar\psi\psi)^2 + (\bar\psi i\gamma_5\vec\tau\psi)^2] - \beta\int \mathrm{d}^3x\, \mathcal{U}(\phi,\beta), \qquad (1) \end{aligned}$$

Journal of Physics: Conference Series **168** (2009) 012021 doi:10.1088/1742-6596/168/1/012021

where ψ is the $N_f = 2$ doublet quark field and $m_0 = \text{diag}(m_u, m_d)$ is the quark mass matrix. The quark chemical potential matrix is defined as $\tilde{\mu} = \text{diag}(\mu + \mu_I, \mu - \mu_I)$, where μ_I is the isospin isovector chemical potential. In the PNJL model quarks interact with a background color gauge field $A_4 = iA_0$, where $A_0 = \delta_{\mu 0} g \mathcal{A}_a^\mu t^a$ with $\mathcal{A}_a^\mu \in \text{SU}(3)_c$ and $t^a = \lambda^a/2$. The field A_4 is related to the traced Polyakov loop by:

$$\Phi = \frac{1}{N_c}\text{tr}_c L \quad \text{with } L = \exp\left(i \int_0^\beta d\tau A_4\right) \quad \text{and } \beta = \frac{1}{T}. \tag{2}$$

In the Polyakov gauge, the matrix L is given in a diagonal representation

$$L = \exp(i(\phi_3\lambda_3 + \phi_8\lambda_8)). \tag{3}$$

The dimensionless effective fields ϕ_3 and ϕ_8 introduced here are identified with the Euclidean gauge fields in temporal direction divided by the temperature, $A_4^{(3)}/T$ and $A_4^{(8)}/T$. These two fields are a parametrisation of the diagonal elements of $\text{SU}(3)_c$.

The effective potential $\mathcal{U}$ models the confinement-deconfinement transition in pure gauge QCD on mean-field Ginzburg-Landau level. In this paper we consider an ansatz for the effective potential given in [5, 10] motivated by the SU(3) Haar measure:

$$\frac{\mathcal{U}(\Phi, \Phi^*, T)}{T^4} = -\frac{1}{2}a(T)\Phi^*\Phi + b(T)\ln[1 - 6\Phi^*\Phi + 4(\Phi^{*^3} + \Phi^3) - 3(\Phi^*\Phi)^2], \tag{4}$$

where the temperature dependent prefactors are given by

$$a(T) = a_0 + a_1\left(\frac{T_0}{T}\right) + a_2\left(\frac{T_0}{T}\right)^2 \quad \text{and} \quad b(T) = b_3\left(\frac{T_0}{T}\right)^3. \tag{5}$$

The parameters are chosen such that the critical temperature of the first order transition is equal to $T_0 = 270$ MeV, as given by lattice calculations, and that $\Phi^*, \Phi \to 1$ as $T \to \infty$.

After performing a bosonization of the PNJL Lagrangian introducing the auxiliary fields σ and π the partition function can be rewritten as

$$\begin{aligned} \mathcal{Z} &= \mathcal{N}\int \mathcal{D}\phi\mathcal{D}\sigma\mathcal{D}\pi \exp[-\mathcal{S}[\sigma(\vec{x}), \pi(\vec{x}), \phi(\vec{x})]] \\ &= \mathcal{N}\int \mathcal{D}\phi\mathcal{D}\sigma\mathcal{D}\pi \exp\left[\frac{1}{2}\text{Tr}\ln[S^{-1}] - \beta\int d^3x\Big(\mathcal{U}(\phi, \beta) + \Big(\frac{\sigma^2 + \vec{\pi}^2}{2G}\Big)\Big)\right]. \end{aligned} \tag{6}$$

with the inverse quark propagator

$$S^{-1} = \begin{pmatrix} -\partial\!\!\!/ + (\mu + \mu_I - iA_4)\gamma_0 - M & i\gamma_5\pi \\ i\gamma_5\pi & -\partial\!\!\!/ + (\mu - \mu_I - iA_4)\gamma_0 - M \end{pmatrix},$$

and $M = m_0 - \sigma$.

In order to introduce fluctuations in the PNJL model we consider the total volume of our system as composed of a certain number of subregions, $V = \sum_i V_i$, and require that the fields are completely correlated if they lie in the same subvolume, and decorrelated otherwise [11]. In this way the fields can fluctuate in the sense that they can have different values from one subvolume to another. Physically this means that our fields are only weakly dependent on the position and can be considered constant inside subvolumes of a given dimension (figure 1).

From this assumption it follows that the action $\mathcal{S}[\sigma(\vec{x}), \pi(\vec{x}), \phi(\vec{x})]$ can be decomposed into a sum of contributions related to the i-th volume

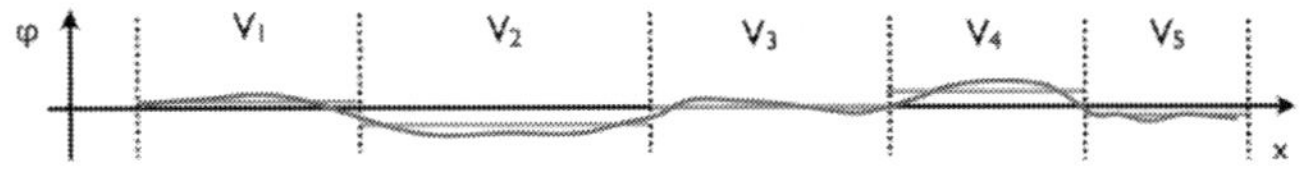

Figure 1. (Color online) Approximation of the space dependence of the fields: the behavior of the field (blue line) is approximated in our model using a mean value (red lines) on subvolumes. In this way we permit small fluctuations of the field.

$$\mathcal{S}[\sigma(\vec{x}),\pi(\vec{x}),\phi(\vec{x})] = \sum_i \mathcal{S}[\sigma(\vec{x}),\pi(\vec{x}),\phi(\vec{x})]|_{\vec{x}\in V_i} = \sum_i \mathcal{S}_i[\sigma,\pi,\phi]. \tag{7}$$

The last equality emphasizes the strong correlation of the fields inside each subvolume. We assume therefore that the fields are constant for all $\vec{x} \in V_i$, hence, the partition function becomes

$$\mathcal{Z} \;=\; \prod_i \Big(\int \mathcal{D}\phi_i \mathcal{D}\sigma_i \mathcal{D}\pi_i \exp[-\mathcal{S}_i(\sigma,\pi,\phi)]\Big). \tag{8}$$

Computing the partition function (8) we introduce fluctuations in the PNJL model.

3. Monte Carlo method

In order to evaluate (8), we adopt standard Monte Carlo techniques. Using the Metropolis algorithm we start from a random configuration C and suggest a new configuration C' that will be accepted with probability

$$p = \min\{1, \exp(-S[C'])/\exp(-S[C])\}. \tag{9}$$

Writing Eq. (8), we point out that the fields involved in each subvolume are independent from the fields in all the other volumes. As a consequence, if we update a given configuration moving only the fields in the k-th subregion the action changes according to

$$\begin{aligned} S(\phi_1,...,\phi'_k,...\phi_n) &= \sum_{i\neq k}^{n} S(\phi_i) + S(\phi'_k) \\ S(\phi_1,...,\phi_k,...,\phi_n) &= \sum_{i=1}^{n} S(\phi_i) \\ S(\phi_1,...,\phi'_k,...,\phi_n) - S(\phi_1,...,\phi_k,...,\phi_n) &= S(\phi'_k) - S(\phi_k) \end{aligned} \tag{10}$$

this means that the Metropolis algorithm acts independently on each single volume. If we thus assume that every V_i will be of a fixed size, $V_i = \tilde{V}$, Eq. (8) can be rewritten in momentum space as

$$\begin{aligned} \mathcal{Z} \;=\; & \Big(\int \mathcal{D}\phi \mathcal{D}\sigma \mathcal{D}\pi \exp\Big[\beta\tilde{V}\frac{1}{2}\sum_n \int \frac{\mathrm{d}^3p}{(2\pi)^3} \mathrm{Tr}\ln[S^{-1}(i\omega_n,\vec{p})] \\ & -\beta\tilde{V}\Big(\mathcal{U}(\phi,\beta) + \Big(\frac{\sigma^2+\pi^2}{2G}\Big)\Big)\Big]\Big)^n \qquad n\to\infty. \end{aligned} \tag{11}$$

The calculation of the partition function using Monte Carlo can be performed by generating a set of configurations for a given subvolume. We only need to fix the size of this volume. Here we adopt the standard temperature dependence of the volume used in lattice calculations,

$$a = \frac{1}{N_t T} \;\rightarrow\; V = N_s^3 a^3 = \frac{N_s^3}{N_t^3 T^3}, \tag{12}$$

where a is the lattice spacing, N_t is the number of lattice sites in the Euclidean time direction and N_s the number of lattice sites in the space dimensions. It follows that

$$V \propto k/T^3, \tag{13}$$

where k will be conveniently chosen. In this paper we perform calculations at different values of k in order to understand the impact of k on thermodynamic quantities. In the next sections we will see how this Monte Carlo version of the PNJL model (MC-PNJL) works both in pure gauge and in the $N_f = 2$ case.

4. Pure Gauge MC-PNJL

In this section we apply the Monte Carlo approach to pure gluodynamics, i.e. to the Polyakov loop sector of the model. Contributions due to fluctuations in this case are very small, although we prefer to start with this simple case, because it is instructive for the understanding of the method.

We start with a generic configuration that in our system corresponds to A_3 and A_8 at a given temperature. Using Monte Carlo we generate points in the configuration space close to the minimum of the action. In particular applying cooling we will reach exactly the minimum. In figure 2 we can see the difference between cooling and Metropolis in the case of $T = T_c = 0.27$ GeV. For cooling (left picture) the action must decrease, and in few steps we fall into one of the three minima of the potential. On the other hand, using the full Metropolis the action can also slightly increase. As a consequence, the space of the available configurations is much larger (right picture).

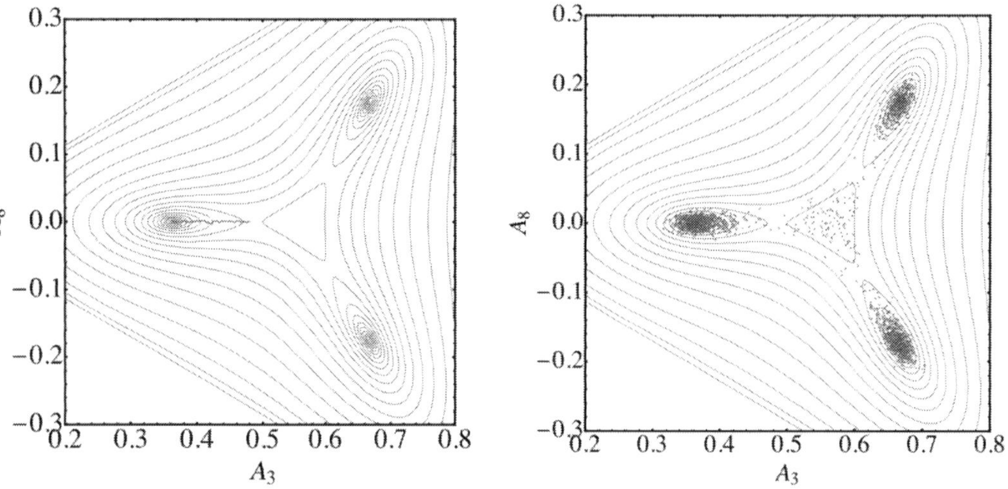

Figure 2. (Color online) Comparison between the cooling (left) and Metropolis (right) procedure applied to the Polyakov loop effective potential at $T = T_c = 0.27$ GeV.

To fix the parameters of the effective potential we have computed the pressure, energy and entropy density at different temperatures including fluctuations, trying to fit the available lattice data (figure 3 left). In the right panel of figure 3 we compare the Polyakov loop obtained using the mean-field method (dashed line) with our results (red points) and the lattice data (black points) we can see that there is practically no difference between the two methods: both agree well with lattice data.

Once established how this method works in the case of the pure gluodynamics we can move to the more interesting case of the two flavors MC-PNJL model.

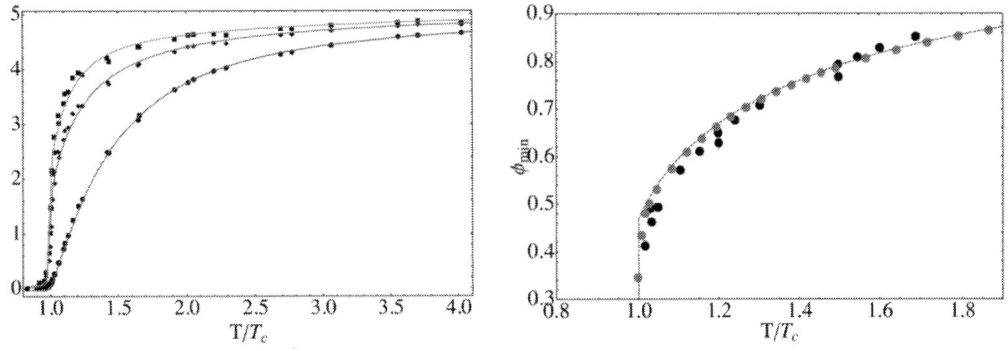

Figure 3. (Color online) Left: fit of Lattice data [12] for pressure, energy and entropy density. Right: Comparison of the Polyakov loop dynamics with the data in [13]

5. Two flavors MC-PNJL model

The starting point studying the thermodynamics for $N_f = 2$ is the partition function (11). The degrees of freedom in this case are the A_3 and A_8 components of the gauge field, and the two bosonic fields σ and π. As explained before, evaluating the path integral, we need to fix the size of the volume $\tilde{V}$ as a function of the temperature. According to Eq. (13) we only have to fix the value of k. Since the value of this parameter is not known a priori, in this paper we consider four different values, $k = 95, 135, 500, 5000$. In this way we can study the dependence of the observables as a function of the volume size.

5.1. Chiral and deconfinement transitions

Since our aim is the investigation of the QCD phase diagram, the first thing we need to study is how the chiral and deconfinement transitions are affected by the introduction of fluctuations in the model. This can be achieved evaluating the chiral condensate and the Polyakov loop for the different volumes and comparing that with the mean-field result. This comparison is presented in figure 4: the presence of fluctuations does not modify the behavior of the Polyakov loop, the five different sets of data overlap perfectly. For the chiral condensate we can notice only small dependence on the temperature below the critical temperature. In analogy to the pure gauge case we can conclude for $N_f = 2$ that the order parameters of the transitions are not affected by the presence of fluctuation.

This last observation is indeed important when we move to the case of finite chemical potential, because this means that we can use mean-field approximation to determine the critical temperature: corrections due to the presence of fluctuations are negligible, and this permits to overcome the sign problem which is not present in mean field calculations.

6. Taylor expansion at finite chemical potential

The Taylor expansion approach to QCD thermodynamics at non-zero quark chemical potential represents one of the possibilities to overcome the sign problem on the lattice: instead of performing explicit calculations at $\mu_q \neq 0$ one expands the thermodynamic potential in a Taylor series around zero chemical potential,

$$\frac{\Omega(T,\mu)}{T^4} = \frac{1}{VT^3} \ln \mathcal{Z} = \sum_{n=0}^{\infty} c_n(T) \left(\frac{\mu}{T}\right)^n, \tag{14}$$

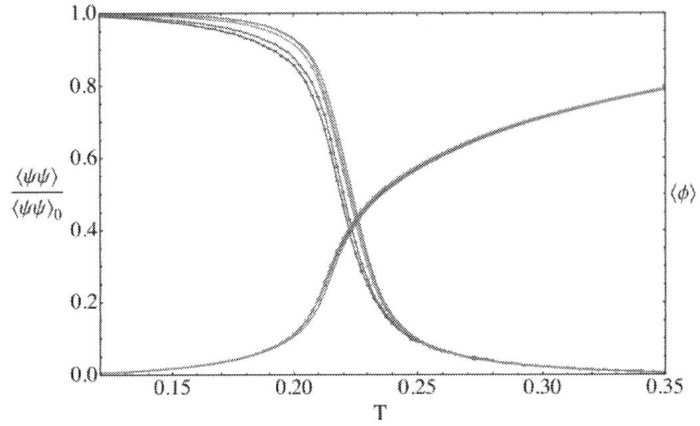

Figure 4. (Color online) Chiral condensate and Polyakov loop: dependence on the volume size. The red lines are the mean field calculation, while moving from the blue to the green line we go from $k = 95$ to $k = 5000$.

with

$$c_n(T) = \frac{1}{n!}\frac{\partial^n \Omega}{\partial(\mu/T)^n}\Big|_{\mu=0} = \frac{1}{n!VT^3}\frac{\partial^n \ln \mathcal{Z}}{\partial(\mu/T)^n}\Big|_{\mu=0} \tag{15}$$

and n even due to the CP symmetry of the system.

The comparison between lattice data and MC-PNJL calculations for these coefficients represents a crucial test of the model. In particular the second non-diagonal coefficient is the most important because, as explained in the introduction, to obtain a non-vanishing result for c_2^{ud} we must take into account fluctuations of the fields. We know that the size of fluctuations in our model depends on the size of our correlation volume, therefore, as before, we evaluate the Taylor coefficients for different values of the parameter k.

To perform the calculations using Monte Carlo, we need to represent the derivative of the logarithm of the partition function in an explicit form. To do that, we start from the definition of our partition function,

$$\mathcal{Z}[T, \mu, \mu_I] = \int \mathcal{D}f\mathcal{D}A \exp\Big[\frac{V}{T}\ln\det M[T, \mu, \mu_I, f, A] - S_g[A]\Big]. \tag{16}$$

From this expression we have

$$\begin{aligned}\frac{\partial \ln \mathcal{Z}[T, \mu, \mu_I]}{\partial \mu} &= \frac{\partial}{\partial \mu}\ln \int \mathcal{D}f\mathcal{D}A \exp\Big[\frac{V}{T}\ln\det M[T, \mu, \mu_I, f, A] - S_g[A]\Big] \\ &= \frac{1}{\mathcal{Z}[T, \mu, \mu_I]}\frac{V}{T}\int \mathcal{D}f\mathcal{D}A \frac{\partial \ln\det M[T, \mu, \mu_I, f, A]}{\partial \mu} e^{-S[T,\mu,\mu_I,f,A]} \\ &= \frac{V}{T}\Big\langle \frac{\partial \ln\det M[T, \mu, \mu_I, f, A]}{\partial \mu}\Big\rangle. \end{aligned} \tag{17}$$

Proceeding in the same way for the second derivative we obtain for the c_2 and isovector c_2^I coefficients

$$\begin{aligned} c_2^{(I)}(T) &= \frac{1}{2}\frac{T^2}{VT^3}\frac{\partial^2 \ln \mathcal{Z}[T, \mu, \mu_I, f, A]}{\partial \mu_{(I)}^2} \\ &= \frac{1}{2T^2}\Big\langle \frac{\partial^2 \ln\det M[T, \mu, \mu_I, f, A]}{\partial \mu_{(I)}^2}\Big\rangle, \end{aligned} \tag{18}$$

where we have neglected the expectation values which involves odd derivatives, due to the symmetry property of the determinant.

The definition of the flavor diagonal and non-diagonal coefficients we are interested in is given by

$$c_n^{uu} = \frac{c_n + c_n^I}{4}, \qquad c_n^{ud} = \frac{c_n - c_n^I}{4}. \tag{19}$$

In figure 5 we show the results for c_2^{uu} and c_2^{ud}. For the diagonal coefficient we can see that the different choices of k affect the matching with the lattice data (black points) only slightly. The picture changes totally when looking at the non-diagonal coefficient: in this case, at increasing volume the signal reduces. This is exactly what we expect because we know that with growing V the fluctuations responsible for the non vanishing expectation value disappear. Thus we recover the vanishing mean-field result (dashed light-blue line).

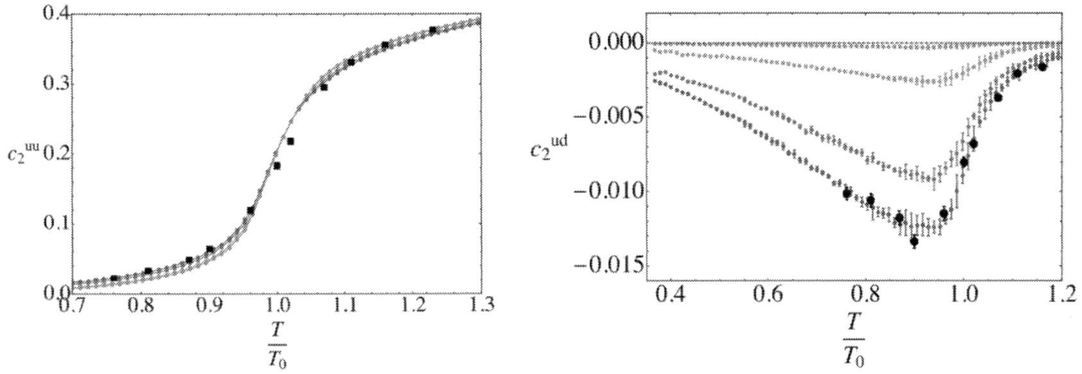

Figure 5. (Color online) Second order diagonal and non-diagonal coefficients c_2^{uu} and c_2^{ud}. Black points are lattice data from [14]; the colored lines show the model predictions (k increase from blue to red)

It is quite surprising, though, that for the smallest values of k we can reproduce the lattice data with reasonable accuracy. This seems to indicate that the PNJL model can capture indeed the dynamics of the fluctuations at the origin of the non-vanishing c_2^{ud}.

7. Conclusions

In this work we applied standard Monte Carlo techniques in order to go beyond mean-field in the Polyakov loop extended Nambu – Jona-Lasinio model calculations. Fluctuations introduced in this way strongly depend on a new parameter that fixes the minimal volume $\tilde{V}$ as defined in section 2. Inside this volume the fields are totally correlated while they are totally decorrelated on distances larger than $\tilde{V}^{1/3}$. Studying the QCD thermodynamics we found that the introduction of fluctuations does not alter the behavior of the traced Polyakov loop in the pure gauge sector nor the behavior of the Polyakov loop and chiral condensate in $N_f = 2$. On the other hand these fluctuations are crucial in the evaluation of the Taylor expansion coefficients of the pressure. In particular, using the MC-PNJL we can reproduce very well the lattice data for the second non-diagonal coefficient which vanishes in mean field approximation.

Collecting all this results we have that, the MC-PNJL model on the one hand includes correctly the dynamics associated with confinement and chiral symmetry breaking, on the other hand we can reproduce quite well lattice data. So it runs for as a good model in the investigation of the full phase diagram.

References

[1] Fukushima K 2003 *Phys. Rev.* D **68** 045004
[2] Fukushima K 2004 *Phys. Lett.* B **591** 277
[3] Hatta Y and Fukushima K 2004 *Phys. Rev.* D **69** 097502
[4] Ratti C, Thaler M A and Weise W 2006 *Phys. Rev.* D **73** 014019
[5] Roessner S, Ratti C and Weise W 2007 *Phys. Rev.* D **75** 034007
[6] Ratti C, Roessner S and Weise W 2007 *Phys. Lett.* B **649** 57
[7] Roessner S, Hell T, Ratti C and Weise W 2008 *Nucl. Phys.* A **814** 118
[8] Hell T, Rossner S, Cristoforetti M and Weise W 2009 *Phys. Rev.* D **73** 014004
[9] Zhang Z and Liu Y X 2007 *Phys. Rev.* C **75** 064910
[10] Ratti C, Roessner S, Thaler M A and Weise W 2007 *Eur. Phys. J.* C **49** 213
[11] Megias E, Ruiz Arriola E and Salcedo L L 2006 *Phys. Rev.* D **74** 114014
[12] Boyd G *et al* 1996 *Nucl. Phys.* B **469** 419
[13] Kaczmarek O, Karsch F, Petreczky P and Zantow F 2002 *Phys. Lett.* B **543** 41
[14] Allton C R *et al* 2005 *Phys. Rev.* D **71** 054508

The SPES project: an ISOL facility for exotic beams

G Prete[1], A Andrighetto[1], L Biasetto[1], M Manzolaro[1], F Gramegna[1], A Lombardi[1], A Pisent[1], J Esposito[1], E Fagotti[1], M Cinausero[1], P Mastinu[1], L Calabretta[2], and SPES collaboration[3]

[1] INFN Laboratori Nazionali di Legnaro, Viale dell'Università 2, 35020 Legnaro, Italy
[2] INFN Laboratori Nazionali del Sud, via Santa Sofia 62, 95125 Catania, Italy
[3] www.lnl.infn.it/~spes

E-mail: prete@lnl.infn.it

Abstract. SPES (Selective Production of Exotic Species) is an INFN project to develop a Radioactive Ion Beam (RIB) facility as an intermediate step toward EURISOL. The SPES project is part of the INFN Road Map for the Nuclear Physics development in Italy and is supported by the whole Italian Nuclear Physics community and mainly by LNL and LNS the INFN National Laboratories of Nuclear Physics in Legnaro and Catania. The Laboratori Nazionali di Legnaro (LNL) was chosen as the site for the facility construction due to the presence of the PIAVE-ALPI accelerator complex, which will be used as re-accelerator for the RIBs. The SPES project is based on the ISOL method with an UCx Direct Target and makes use of a proton driver of at least 40 MeV energy and 200 microA current. Neutron-rich radioactive beams will be produced by Uranium fission at an expected fission rate in the target in the order of 1013 fissions per second. The key feature of SPES is to provide high intensity and high-quality beams of neutron rich nuclei to perform forefront research in nuclear structure, reaction dynamics and interdisciplinary fields like medical, biological and material sciences. The exotic isotopes will be re-accelerated by the ALPI superconducting linac at energies up to 10AMeV for masses in the region of A=130 amu with an expected rate on target of 109 pps.

1. Physics case

Starting from a nucleus on the stability line and adding successively neutrons one observes that the binding energy of the last neutron decreases steadily until it vanishes and the nucleus decays by neutron emission. The position in the nuclear chart where this happens defines the neutron drip line. It lies much farther away from the valley of stability than the corresponding drip line associated with protons, owing the absence of electrical repulsion between neutrons. The location of the neutron drip line is known only for nuclei with mass up to around 30.

The interest in the study of nuclei with large neutron excess is not only focused on the location of the drip line but also on the investigation of the density dependence of the effective interaction between the nucleons for exotic N/Z ratios. In fact, changes of the nuclear density and size in nuclei with increasing N/Z ratios are expected to lead to different nuclear symmetries and new excitation modes. While in the case of some very light nuclei a halo structure has been identified, for heavier nuclei the formation of a neutron skin has been predicted.

The evolution of nuclear properties towards the neutron drip line depends on how the shell structure changes as a function of neutron excess. This evolution has consequences on the ground state properties (spin, parity, and electromagnetic moments) and on the single-particle and collective excitations. In particular, studies of neutron-rich nuclei beyond doubly magic ^{132}Sn are of key importance to investigate the single-particle structure above the N=82 shell closure and find out how the effective interaction between valence nucleons behaves far from stability.

New modes of collective motion are also expected in connection with the formation of a neutron skin, namely oscillations of the skin against the core, similar to the soft dipole mode already identified in the case of very light halo nuclei. Presently, neither the thickness nor the detailed properties of the neutron skin of exotic nuclei are known. This information is needed to enable a quantitative description of compact systems like neutron stars, where exotic nuclei forming a Coulomb lattice are immersed in a sea of free neutrons, a system which is expected to display the properties of both finite and infinite (nuclear matter) objects.

Despite the large number of experimental studies, so far it is still not possible to predict reliably the limits of nuclear stability or the behaviour of the Nuclear Equation of State (NEOS) at low and high baryon densities. In particular, the asymmetry term in the NEOS is largely unknown but in the region close to saturation. However, it is just this energy which plays an important role in setting the stability limits. For this reason, it is quite challenging to investigate the behaviour of nuclear matter far from stability. Although the SPES energy range is somewhat limited for studies of this kind, the neutron-rich ion beams of SPES will allow one to further extend the investigation of the NEOS along the isospin coordinate, in a region where it is largely unknown at low as well as high excitation energy.

2. Facility description

The basic elements of the ISOL facilities are the primary accelerator, the production target coupled to the ion source (TIS), the charge breeder (or the charge exchange system), the beam transport system and the re-accelerator. According to the requirements of the experimental needs a High Resolution Mass Spectrometer (HRMS) can be part of the transport system.

SPES is designed to have a Cyclotron as primary accelerator able to supply at least 40 MeV 0.2 mA proton beam onto a UCx direct target to produce a fission rate of about 1013 fission/s. Thus, a total beam power of 8 kW has to be managed. A surface ionization source will be used with the possibility to add a laser device to improve the purity of the ionized exotic species. For this purpose a HRMS with a mass resolution 1/20000 is also planned. To reach the charge state and ion velocity that fit the requirement for injection into the PIAVE-ALPI acceleration system a Charge Breeder and 2 High Voltage platforms (HV 250kV) will be used. The first platform will host the TIS and first stage mass separator, the second the Charge Breeder. The description of the facility is reported in the Technical Design Report at the LNL web site [1].

As the facility will handle radioactive species, special care is devoted to the radiation protection safety and several systems are added to prevent radiation hazards. A Control System will integrate in a homogeneous architecture the many subsystems necessary for the operation of the facility: from the accelerator control to the radiation and safety survey.

Several factors have to be considered to determine the intensities and the ion species available for experiment in an ISOL facility. The production of isotopes inside the primary target is the first ingredient but a crucial point, as we are dealing with radioactive species, is the target release time, i.e. the time needed by the reaction products to reach the ionization source from inside the target grains, where they are produced. The in-target beam intensity at SPES has been determined starting from the fission fragment production yield calculated with the MCNPX [2] transportation Monte Carlo code in which the target geometry is included. The following diffusion and effusion of the exotic species inside the target was evaluated with both GEANT4 [3] and RIBO [4] Monte Carlo codes. The calculations have been tuned using the available experimental data from ISOLDE, ORNL and PNPI and the complete geometry of our target has been included. Finally, source ionization and extraction, charge breeding, beam transport and re-acceleration efficiencies have to be considered. Following the literature, we assumed 1+ and N+ (charge breeder) ionization efficiencies equal to 90% (1+) and 12% (N+) for Kr and Xe, 30% (1+) and 4% (N+) for Zn, Sr, Sn, I and Cd. The typical Linac ALPI transmission efficiency is 50%.

The final estimated beam currents for the SPES facility are shown in figure 1 for some interesting species.

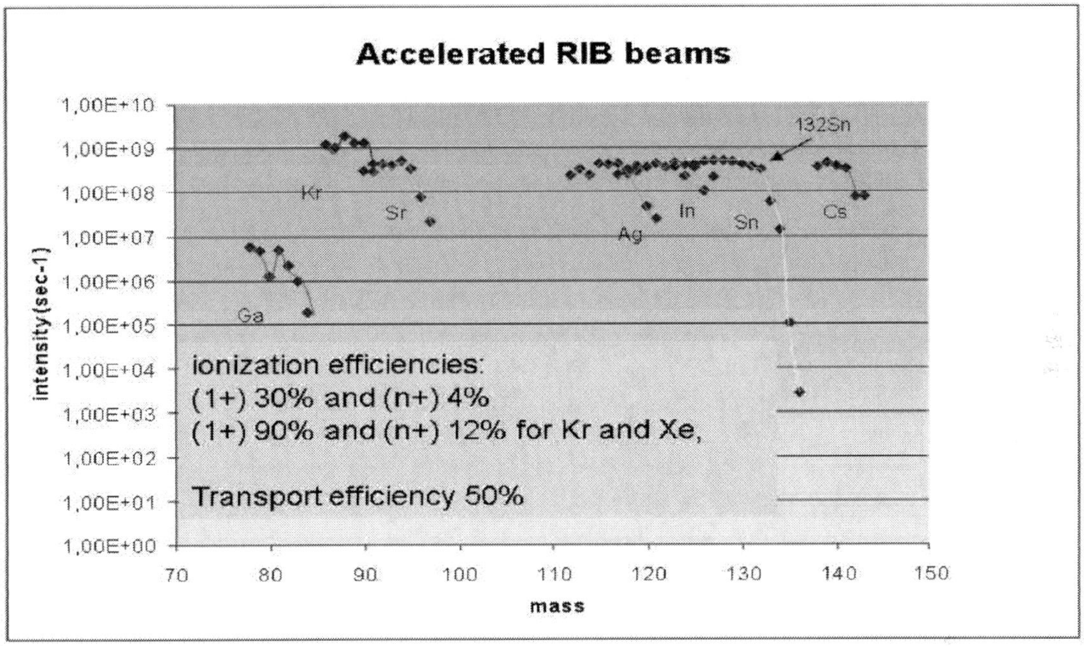

Figure 1. Expected on-target intensities calculated considering emission, ionization and acceleration efficiencies (see text) for different isotopes.

3. Proton driver

A proton driver based on a cyclotron with energy 40-50 MeV and current 0.2 mA fulfils the requirements for the SPES project as the direct target is actually designed for 8 kW power. A driver with a capability of 50 kW (70 MeV, 0.75 mA) with the possibility of a current upgrade reaching 1.5 mA and a beam power of 100kW is indeed very interesting for the development of the SPES project, as further developments of SPES will be in the direction to increase the maximum sustained power in the target, with the aim to increase the exotic beam intensity and to follow the EURISOL trend for a 100 kW direct target.

A commercial cyclotron, with characteristics which fulfill the needs for the SPES project, was recently developed by IBA: the Cyclone? 70 (C70) and it is actually in operation at the ARRONAX (Acclrateur Recherche Radiochimie Oncologie Nantes) project [5]. The cyclotron is equipped with two exit ports allowing for dual proton beam extraction; this allows at the same time RIB production and the development of applied physics based on 70 MeV proton beam of current as high as 0.5 mA.

Journal of Physics: Conference Series **168** (2009) 012022 doi:10.1088/1742-6596/168/1/012022

4. Target system

The UCx target is made by 7 disks (each 1 mm thick and 4 cm diameter) to optimize power dissipation and release time of the fission products. The gap between the disks allows an efficient cooling of the system by thermal radiation. The total amount of the U fissile material is only 28 g.

A detailed study has been performed to evaluate the thermo-mechanical behaviour with two codes: the generally used ANSYS [6] code and one provided by ENEA [7]. Experimental tests of the target principle were performed at the HRIBF facility (ORNL-USA). The main result is that, in the adopted configuration, the target does not melt and to reach the operating temperature it is necessary to supply external power.

A strong R&D program is under development on the Direct Target subjects for material, characterization techniques and prototyping. The possibility to produce disks of carbides with the right dimensions has been proved developing and characterizing LaC and UCx pellets.

Collaborations with ISOLDE (CERN) and HRIBF (ORNL) have been established as well as participation to the EURISOL-DS Task3. A detailed discussion of the target status and development can be found in Ref. [8].

The TIS system is developed following the EXCYT and ISOLDE design. The choice of the ion source kind to be used has primarily been dictated by efficiency and secondarily by its capability of selective ionization. We consider three kinds of ion sources for SPES: the Surface Ion Source, the Forced Electron Beam Induced Arc Discharge (FEBIAD) and the Resonant Ionization Laser Ion Source (RILIS). All of these three sources are used at ISOLDE and they constitute a good reference point for further SPES goals in the ion-source development. The first version of the SPES TIS will be equipped with a Surface Ionization Source with the option to couple a Resonant Ionization Laser Ion Source as a second step in source development.

5. Beam transport and re-acceleration

The secondary beam line transport system will handle the radioactive beam from the output of the ionization source to the low-energy experimental area and to the re-accelerator complex. One of the main problems to operate an ISOL facility is the beam purification since the extracted species are transported according to their M/q value. Due to the low rigidity of the beam, electrostatic quadrupoles can be used to focus and transport the beam. This guarantees a reliable beam handling and a very simple procedure to set the beam transport line.

The beam extracted from the source with 50 kV extraction potential, will cross through a first stage of M/Z purification, which allows trapping the largest amount of radioactive contaminant. According to other facilities, and to satisfy the previous constraint, we plan to use a small Wien filter, placed on the first HV platform just beyond the source. Furthermore a small magnetic dipole, like in the EXCYT design, can be also used. A mass resolving power (M/?M) of 300 for this analytical magnet is acceptable. It will be followed by a 1/20000 High Resolution Mass Spectrometer (HRMS) which allows the isobar selection. To improve selection capability the HRMS shall operate at an input energy in the order of 200 keV. To fulfil this requirement the HV platform, where both target and first mass separator are mounted, is operated at 200 kV supplying 1+ beam at total energy of 250 keV.

To optimize the reacceleration, a Charge Breeder will be developed to increase the charge state to N+ before the injection of the exotic beam in the PIAVE Superconductive RFQ, which represents the first re-acceleration stage before the final injection in ALPI.

The Charge Breeder acts as a trap where the 1+ ions are stopped and re-extracted with increased charge state. To fulfil these requirements the Charge Breeder is mounted on a second HV platform operated at 250 kV; this allows to stop the incoming ions and to give the right energy to the out-coming ones. The scheme of the transport line is shown in figure 2.

The re-acceleration of the exotic species will be performed by the acceleration complex

Journal of Physics: Conference Series **168** (2009) 012022 doi:10.1088/1742-6596/168/1/012022

PIAVE-ALPI. The PIAVE injector is in regular operation at LNL since fall 2006. It is based on an ECR Ion Source (placed on a 350 kV platform), and on super-conducting RFQ able to accelerate ions with $A/q \leq 8.5$ up to 1.2 AMeV. For the SPES beams a transfer line from the Charge Breeder will be added. No main difficulties are expected as the ions coming from the Charge Breeder have similar characteristics as that ones produced in the present ECR. The ALPI acceleration capability allows to push the RIBs energy up to 10AMeV for masses in the region of A=130 amu.

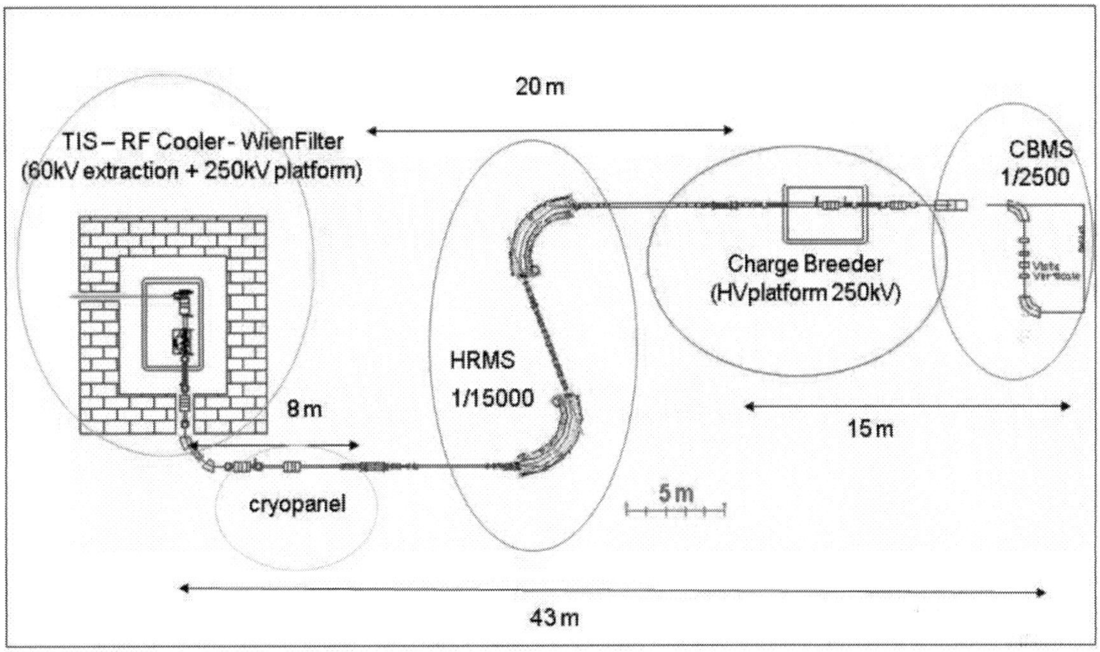

Figure 2. Scheme of the transport line for the SPES exotic beams. For details see the text.

References

[1] *SPES Technical Design Report* 2008 ed G Prete and A Covello INFN-LNL 223 http://www.lnl.infn.it/ spes/TDR2008/tech_design08_index.htm
[2] *LA-CP-05-0369 2005 MCNPXTM Users manual, Version 2.5.0* ed Denise B Pelowitz
[3] *GEANT4: CERN Program Library W5013 Geant_detector description and simulation tool*
[4] Santana Leitner M UPC-ETSEIB / CERN 2005, Ph.D. thesis http://ribo.web.cern.ch/rib
[5] *Proc. of the 18th Int. Conf. on Cyclotrons and their Applications* 2007, http://accelconf.web.cern.ch/accelconf/c07/
[6] http://www.ANSYS.com
[7] Cevolani S FIS-P815-022 (ENEA report)
[8] Andrighetto A, Antonucci C M *et al* 2007 *Eur. Phys. J.* ST **150** 273; Andrighetto A *et al* 2008 *Proc. Zakopane Conference on Nuclear Physics* to be published in *Acta Physica Polonica*

Journal of Physics: Conference Series **168** (2009) 012023 doi:10.1088/1742-6596/168/1/012023

EURISOL: an European Isotope Separation On-Line radioactive ion beam facility

A Bonaccorso[1] **and G Prete**[2]

[1]INFN, Sezione di Pisa, Largo Pontecorvo 3, 56127 Pisa, Italy.
[2]INFN, Laboratori Nazionali di Legnaro, viale dell'Università 2, 35020 Legnaro, Italy

E-mail: prete@lnl.infn.it, bonac@df.unipi.it

Abstract. The EURISOL program aims at the construction, beyond the year 2013, of the "next-generation" European Isotope Separation On-Line (ISOL) Radioactive Ion Beam (RIB) facility. It will extend and amplify the exciting work presently being carried out at the first generation RIB facilities in Europe and other parts of the world. The EURISOL facility will play a complementary role to the recently approved upgrade of the GSI facility at Darmstadt, Germany, the European "next-generation" In-Flight RIB facility FAIR.

1. Introduction

In the report [1] published by the NuPECC Working Group entitled "Radioactive Nuclear Beam Facilities", it was stated that the next generation of radioactive ion beam facilities should aim at having an intensity of accelerated radioactive beams being a factor 1000 higher than any facility presently running or in the commission stage. Such facilities should clearly be complementary if more than one is needed to cover the foreseen physics issues, and together they should be second to none worldwide. The scenario emerging from the NuPECC report is that the European option will at least include an In-Flight Facility. The facility should be based on in-flight fragmentation or fission of high-intensity heavy ions. This facility is the future GSI accelerator complex, which could consist of a high intensity, high-energy synchrotron with an accumulator ring. The energy range of the RIBs would here cover the region from a few hundred MeV/u to about 1 GeV/u.

EURISOL - an Isotope Separation On-Line (ISOL) post-accelerator facility which would be a very powerful complement to the proposed GSI In-Flight Facility. The aim is to produce exotic beams in the neutron rich side with the good beam characteristics due to a linac re-accelerator with energies up to 150 MeV/u (A=130) and intensities in the order of 1012 pps for ^{132}Sn.

The construction of the EURISOL facility is expected beyond 2015 after the development of "second generation" facilities as SPIRAL2 and SPES addressing some of the main technological challenges. The main structure of EURISOL will consists of: (i) a driver accelerator; (ii) a neutron converter to generate high-energy neutrons for exotic isotope production; (iii) a high power (Multi Mega Watt - MMW) target-ion-source unit for two-step isotope production (neutron induced), and several direct target stations (100 kW) in which exotic isotopes are directly produced by the driver beam impinging on the target material.

Targeting for ISOL facility is traditionally a very active R&D area as shown by the continuous effort devoted to this item by the ISOLDE facility at CERN. Neutron rich exotic nuclei are essentially produced by U or Th fission and the target material is mainly natural U or Th

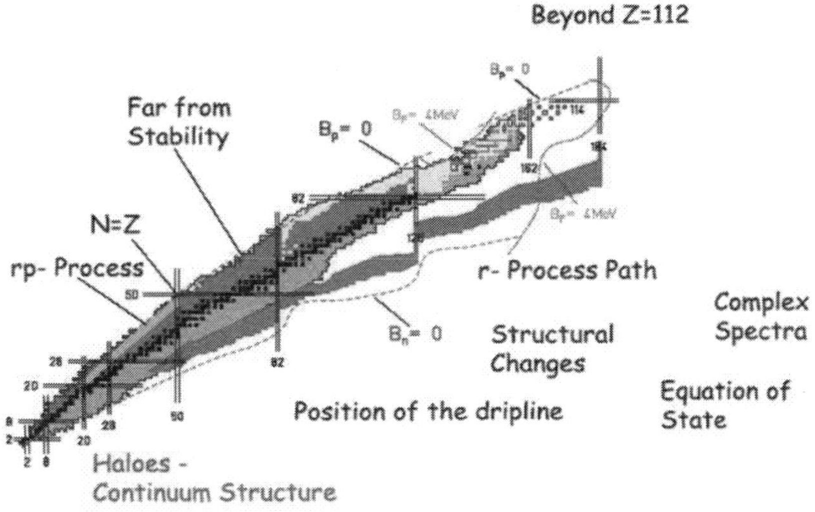

Figure 1. Some of the burning issues at present-day radioactive beam facilities.

in oxide or carbide form. The Charge Breeder to increase ion charge state from 1+ to n+ for an effective acceleration process : At the REX-ISOLDE facility, an Electron Beam Ion Source (EBIS) is used, while the SPIRAL facility at GANIL will employ an Electron Cyclotron Resonance (ECR) ion source. A Charge Breeder R&D project is concerned with determining the optimal solution. For the post-accelerated beams, it would be interesting to cover the entire energy range up to a maximum energy of about 100 MeV/u, i.e. in the fragmentation regime. Note that to run a complete physics program, one should also have a "low energy" output. It is part of the facility the development of dedicated up-to-date experimental instrumentation.

2. The EURISOL concept

2.1. Furthering target and accelerator technology

A 4-year Design Study began in 2005 in order to work on the technologically challenging aspects of the project, the instrumentation and the radiation safety issue. Synergy with other projects is being examined, including a feasibility study for the new "beta-beam" neutrino proposal, forming an integral part of the Design Study. After this, possible sites will be ready for a full Engineering Design, to be followed by construction of the facility. Researches and engineers of several European laboratories are collaborating in twelve tasks to further the EURISOL design. The EURISOL layout consists of superconducting linear accelerator providing protons of energy 1 GeV and an impressive power of 5 MW, but also capable of accelerating deuterons, ^{3}He and ions up to mass 40. The beams will impinge simultaneously on two types of targets, either

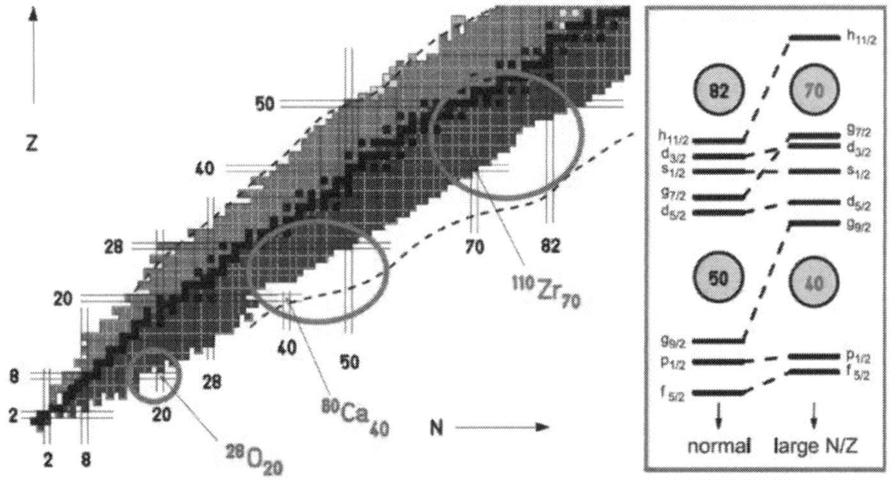

Figure 2. Nuclear chart in the light- and medium-mass region. The circles indicate areas with possibly new magic numbers. The right-hand side shows the single-particle energies for nuclei close to stability and for nuclei with a large N/Z ratio.

directly or after conversion of the protons into neutrons through a loop containing 1 ton of mercury surrounded by kilograms of fissile material. The unstable nuclei produced diffuse out of the target, are ionized and selected, and can be used directly at low energy or reaccelerated by another linear accelerator to energies up to 150 MeV per nucleon in order to induce nuclear reactions. Prototypes of some of the most critical parts of the facility are being built within the Design Study.

2.2. Multi Mega Watt target station

Technical preparatory work and demonstration of principle for a high-power station for production of beams of fission fragments using the mercury proton-to-neutron converter-target and cooling technology is carried out in collaboration with the communities working on spallation neutron sources, accelerator-driven systems and neutrino factories. The converter will be surrounded by large amounts (4 Kg) of fissile material.

2.3. EURISOL driver

The EURISOL Driver is a superconducting (SC) linac that includes 5 different sections: injector, low-β section, medium-β section, high-β section and CW beam splitter. The lengths of the injector and of the SC part are 12 m and 247 m, respectively, while the 4-way beam splitting section is 55 m long. An additional extraction sector is located at the beginning of the high-β section, giving a beam line for 264 MeV deuterons. All cryostats are planned to work at 4.5K.

2.4. EURISOL post-accelerator complex

Three independent post-accelerators, fed by separate target ion-source systems, should be incorporated into the facility. This will allow to feed three different experimental areas in parallel: a very-low-energy area (< 1 MeV/u), a low-energy area (between 1 and 5 MeV/u), and a high-energy area (from 5 MeV/u up to the maximal energy). A maximum energy of 150 MeV/u for the benchmark $^{132}Sn^{25+}$ beam is envisaged. Moreover, maximum energies in the vicinity of 80-100 MeV/u would be very welcome for "heavier" beams. RFQ and Super Conductive Linacs are the key items to develop the EURISOL post acceleration system.

2.5. Safety issues

The increase of the RIB intensities by a factor of 100-1000 compared to presently operating facilities will give rise to challenging issues in terms of nuclear safety and radioprotection. In the framework of the radiation characterization and safety studies detailed special attention is devoted to estimate the prompt and residual radiation fields in the MMW target and its surroundings. An important safety and radioprotection objective is to demonstrate the feasibility of the EURISOL project over the entire life cycle, including decommissioning. The large quantities of radioactive wastes produced in the bulk shielding of the MMW target at the end of the EURISOL facility operational lifetime require radiological characterization. In this context it is important to assess the issues related to radioactive wastes arising during the facility operation.

3. EURISOL physics

Over the past two decades, RIB physics has been driven by remarkable discoveries which have added new elements to the nuclear paradigm. As yet we have only explored a relatively small part of the nuclear landscape, especially on the neutron-rich side where the limit of stable nuclei is only known for the lightest elements. We may expect that more discoveries will be made in the light of the new developments with the guide of a theory framework of growing predictive power. There are, at present, already a number of identified challenges and exciting issues that will be addressed in the coming years (see figure 1). One of the main motivations for a new high-performance nuclear physics facility will be the physics at the drip-lines. The largest unknown nuclear territory is on the neutron-rich side of stability but there are also many interesting physics issues on the neutron-deficient side namely close to the proton dripline. The heavy-element region represents a special subject of interest for the characterization of the valley of stability in the super-heavy region. At present, we are approaching a delineation of the one-proton drip-line, i.e. the limit of stability for proton-rich, odd-Z nuclei. The observation of proton radioactivity gives some definite points where the drip-line is situated. The heaviest known case is ^{185}Bi (N=102, Z=83) [2]. However, the two-proton drip-line, the limit for proton-rich, even-Z nuclei, has been reached only up to nickel (Z=28). The situation is even more dramatic on the neutron-rich side of the valley of stability, where, except for the lightest nuclei, we are still far from reaching the neutron drip-line. Indeed, the neutron drip-line is at present known only for elements up to fluorine (Z=9). Studying nuclei far from the stability line allows to verify the shell structure in these unknown regions. The concept of the mean field and the consequent shell structure encountered in nuclei lies at the heart of our current understanding of the nuclear many-body system. Yet direct empirical verification of this structure has only been possible up to now for nuclei on, or a few nucleons away from, the line of stability, since it is only with nucleon transfer reactions that we can directly probe the single-particle structure. For the remainder of the nuclear chart we rely on extrapolation or interpolation in regions between the well-known magic numbers. In fact, even this traditional textbook property of nuclei is predicted to break down far from stability, as nuclei become weakly bound and increasingly diffuse at the surface. Experimental evidence of vanishing shell effects at N=20 and N=28 for nuclei with large neutron excess has already been found by many experimenters. In these cases, the neutron-proton interaction is thought to be responsible for changes in the single-particle orbital spacings and consequently for shell effect weakening [3]. See figure 2. In addition to the effect described above, it is expected that the surface of medium-mass, very neutron-rich nuclei would be essentially composed of diffuse neutron matter [4]. As a consequence, the derivative of the mean-field potential is thought to be weaker, reducing the spin-orbit surface term. The magicity at N=40 and N=70 would be enhanced as a result, whereas the N=50 and N=82 gaps would be reduced, turning the mean-field description of nuclei back towards one which corresponds more closely to a pure harmonic-oscillator (HO) potential. A specific feature of HO magic numbers is that a change of parity occurs across the gap. It is therefore expected that excitations which preserve the parity symmetry across a gap, for example electric quadrupole excitations, would be strongly hindered. These modifications of the shell ordering would provide new magic numbers and new ways to generate collectivity. It is well known from studies of nuclei close to the line of β-stability that nucleons preferentially form pairs (proton pairs, neutron pairs) in the nucleus under the influence of the short-range nucleon-nucleon attractive force. However, surprisingly little is known about the basic properties of the particle-particle force as it relates to pairing. In most models, schematic average pairing interactions [5] are usually treated by means of the BCS approximation. They perform remarkably well when applied to nuclei in the neighbourhood of the valley of stability. Here, pairing can be treated as small perturbation acting in the neighbourhood of the Fermi surface. The development of EURISOL will allow many very neutron-rich nuclei to be produced for the first time. Such drip-line nuclei are unique laboratories for studies of neutron pairing properties [6, 7], and thus of effective interactions, in a neutron-rich environment. The exploration of a series of isotopes and isotones, progressively moving away from the region of stability, has brought many surprises (e.g. new shapes at low energy, phase and shape coexistence, intruder states). Learning about the evolution in moving even further out may give access to a better understanding of these phenomena. Moreover, experiments may reveal the missing elements needed to get to a unified understanding of these phenomena, or indicate hints of new symmetries and interactions governing the nuclear many-body system. One interesting possibility is to continue a systematic investigation of their β-decay and, in particular, of their β-delayed particle emission processes. Such decay modes are typical for exotic nuclei; they are to a large extent understood experimentally and will therefore provide very powerful probes for investigations of the nuclear structure at the drip-lines. Experimentally one has up to now observed many different types of β-delayed particles [8] i.e. p, 2p, n, xn, d, t and α. It is obvious that such decay modes become more frequent as one goes away from the β-stability line, since the Q-values become larger and the particle separation energies lower.

3.1. Superheavy elements

Nuclear chemists and physicists strive to complete the Mendeleiev table of the elements by discovering ever heavier elements and studying their chemical and physical properties. The high intensity radioactive beams from EURISOL will lead to the production and study of new isotopes of these elements and possibly to the discovery of the elusive long lived "magic" super-heavies that are expected to be located in a region of mass $\sim$300 amu and Z$\sim$114-118. It is not possible to reach this region without the use of reactions with neutron rich beams. Up to now isotopes with Z in the interesting region have been produced but the mass is not high enough, more neutrons are necessary.

3.2. Exotic radioactivity
The targets at the EURISOL facility will allow a number of hitherto unknown exotic nuclei to be produced. One interesting possibility is to continue a systematic investigation of their radioactive decay. Recently a new type of radioactivity has been discovered in very neutron-deficient nuclei, where two protons are emitted simultaneously by the nucleus.

4. Nuclear astrophysics
4.1. The r-process
Approximately half of the nuclear species in nature beyond iron are produced via neutron capture processes on very short time scales in neutron-rich astrophysical environments, i.e. the so-called r-process. Only under such conditions is it possible that highly unstable nuclei near the neutron drip-line are produced, also leading - after decay back to stability - to the formation of the heaviest elements in nature like Th, U and Pu. Despite its importance, the exact stellar site where the r-process occurs is still a mystery. The key to its understanding will probably only be obtained from a close interaction between astronomy, cosmos-chemistry, nuclear physics and astrophysical modeling of explosive scenarios. The questions of neutron capture, β-decay and mass measurement in the regions of closed neutron shells will be carefully investigated with the high intensity RIBs of EURISOL.

4.2. Neutron stars
Neutron stars are the remnants of core collapse supernovae. They are the most compact stellar object after black holes. Indeed, pulsars (space lighthouses), and magnetars, generating the most intense magnetic field of the Universe, are neutron stars. The modeling of their inner crust is essential in order to understand the cooling process of the star, and also observational irregularities (glitches).

The crust is composed of neutron-rich nuclei immersed in a neutron gas. Vortices (quantum tornadoes) at work in the nuclear matter of the neutron star could be the explanation for glitches. The investigation of exotic nuclei is essential to understand the crucial role of superfluidity in these systems. EURISOL will open a new field of experiments on neutron-rich nuclei aimed at a better understanding of neutron stars.

4.3. X-ray bursts
The energy generated by nuclear processes on the surface of accreting neutron stars is observed as short X-ray bursts if the nuclear burning is unstable. In such scenarios, matter is accreted for hours or days until a thermonuclear explosion is triggered by the ignition of the triple-alpha reaction and the break-out reactions from the hot CNO cycles into the rapid proton-capture process (rp-process) - a sequence of (α,p), (p,γ) and β^+ reactions. There are many open questions concerning X-ray bursts. In many cases, the motivation is to obtain information on neutron stars and on the properties of matter under extreme conditions. To answer these questions, experimental masses and electron-capture rates on neutron-rich nuclei are now needed and can be provided by a facility like EURISOL.

5. Fundamental interactions
5.1. Beyond the standard model
The study of nuclear decay modes has played a crucial and undeniable role in determining the basic structure of fundamental interactions. In particular, β-decay has contributed to establishing such aspects of particle physics as parity violation, the nature of the neutrino or lepton number conservation, and has thus provided the experimental foundations for a large part of the Standard Model (SM) of electroweak interactions. Precision measurements

in nuclear β-decay and in atomic transitions constitute simple means to search for signatures of new interactions or small violations of the fundamental symmetries. At low energies the search for new physics beyond the SM is a very exciting activity presently carried out at several ISOL facilities, as well as at cold and ultra-cold neutron sources around the world. The importance of this field, where the atomic nucleus is used as a laboratory for the tests of fundamental conservation laws, has been recognized in the description of the EURISOL proposal and is considered to be one of the four key areas of modern science on which RIB facilities will have a major impact.

5.2. Beta beam

Neutrinos come in three varieties. One of the major discoveries of the last decade is spontaneous oscillations between these neutrino families, which imply that neutrinos are not massless as previously assumed. In order to learn more about these elusive particles and perform stringent tests of quantum symmetries, neutrino physicists need a new type of neutrino beam called beta-beam. The neutrinos would be produced by the radioactive beta decay of massive amounts of unstable nuclei, for example ^{6}He and ^{18}Ne, accelerated close to the speed of light. The seed nuclei would be produced by EURISOL and the beta-beam facility, elaborated within the Design Study, would be a natural extension of EURISOL.

6. Key experiment

The Physics and Instrumentation Task of the EURISOL Design Study, lead by R. Page of the University of Liverpool, has compiled a list of around 25 key experiments which could be performed at EURISOL. These have emerged from the presentations at a workshop held at ECT*, Trento in January 2006 and discussions at the subsequent Task meetings. Since the science case for EURISOL is continuously evolving the list should not be regarded as being definitive, but rather as a snapshot of the diverse and exciting physics opportunities that will be available with the EURISOL facility.

- Super-allowed beta decay and the weak-interaction standard model
- Correlation measurements in nuclear beta decay to search for physics beyond the standard model
- In-beam spectroscopy of heavy elements
- Synthesis and decay of the heaviest elements
- Optical spectroscopy of the heaviest elements
- Neutron capture cross sections of radioactive nuclei
- The r-process path between the N=50 and N=82 shells
- Ground-state two-proton radioactivity
- Beta-delayed two-neutron emission
- Structure beyond the neutron drip line: $^{26-28}$O
- Mass of ^{78}Ni ground state
- Magnetic moments of isomers in the ^{78}Ni region
- Charge radius of ^{78}Ni
- ^{44}Ti Abundance as a Probe of Nucleosynthesis in Core Collapse Supernovae
- One or two neutron or well-defined cluster (like alpha-particle) break-up
- Isospin Dependence of Correlations
- Systematics of Isoscalar Giant Resonances in Exotic Nuclei
- Mapping of Single Particle Energies using Transfer Reactions

- The density dependence of the symmetry energy
- Neutron-proton effective mass splitting
- Isospin dependent phase transition
- Isospin fractionation and isoscaling
- Fundamental tests with low energy beta-beams
- Nuclear structure studies with low energy beta-beams

We present in the following some of these experiments, particularly relevant for our Italian community interests.

6.1. The r-process path between the N=50 and N=82 shell

A systematic study of the basic nuclear structure properties of neutron-rich nuclei on the process path between the N=50 and N=82 major neutron shells is proposed for measurements at the Nuclear Astrophysics Sector of EURISOL. Measurements of these basic properties will provide the fundamental information for extension to neutron-rich nuclei of the nuclear structure and reaction models needed for full-scale r-process nucleosynthesis studies.

Nuclear structure properties of neutron-rich nuclei are of paramount importance for the understanding and modeling of the r-process, responsible for the synthesis of approximately half of the elements heavier than iron. Presently available experimental information on basic properties such as masses and decay schemes of neutron-rich nuclei do not include nuclei in the r-process path in the regions between the N=50, 82 and 126 shells. The observables we are interested in are: nuclear mass (S_n, and Q_β), half-life (β-decay $T_{1/2}$), β-delayed neutron emission probability (P_n). Additional important observables are: energies and J^π of excited states, ground-state deformation, neutron-capture rates. The proposed experiment will consist of a systematic study of basic nuclear structure properties of neutron rich isotopes of (possibly) all elements from Fe to Sn, covering all waiting-point nuclei in the r-process path between the N=50 and N=82 neutron shells. Depending on beam intensities obtainable, the following isotopes will be studied (mass number range given for each element): $^{68(*)-74}$Fe, $^{73-75}$Co, ^{78}Ni(*), $^{79(*)-81}$Cu, $^{80(*)-84}$Zn, $^{81(*)-87}$Ga, $^{84(*)-90}$Ge, $^{87(*)-95}$As, $^{90(*)-98}$Se, $^{91(*)-101}$Br, $^{96(*)-106}$Kr, $^{101(*)-107}$Rb, $^{102(*)-110}$Sr, $^{105-115}$Y, $^{108-118}$Zr, $^{111-123}$Nb, $^{114-124}$Mo, $^{121-125}$Tc, $^{124-126}$Ru, ^{127}Rh, ^{128}Pd, ^{129}Ag(*), ^{130}Cd(*),^{131}In(*), ^{132}Sn(*). Nuclides marked with (*) have been already measured and will be used as benchmark cases.

Detection system will consist of a penning trap mass spectrometer for mass measurements (see e.g. ISOLTRAP), laser ion-source for half-life measurements (e.g. RILIS at ISOLDE). Multi-coincidence set-ups will be used with various detection systems.

From the theoretical point of view, modern nuclear structure models for the nuclear mean field, will be the main ingredient of the study. Also, capabilities for large-scale shell-model calculations (needs the development of appropriate residual interactions to be employed in studies of nuclei far from stability). Full dynamical calculation of r-process nucleosynthesis with inclusion of the nuclear physics input obtainable from the experimental data.

6.2. Neutron capture cross sections of radioactive nuclei

Radioactive nuclei produced in the various nucleosynthesis processes play a special role in nuclear astrophysics. Alive radioactivity in the Universe can nowadays be detected with satellite based gamma-ray observatories (RHESSI, INTEGRAL) via the characteristic decay lines or by mass spectroscopy of stellar dust grains. This information can be used for testing stellar models like supernova explosions or for obtaining information about the age and the chemical evolution of the Universe. For this purpose it is however required to measure stellar reaction rates, which lead to the production and destruction of these radioactive isotopes. In some cases (e.g. for (p,γ)

and ((α,γ) reactions) the measurements can be performed in inverse kinematics by producing low energy radioactive beams. In cases where this possibility does not exist (e.g. (n,γ) reactions) isotopically enriched radioactive targets have to be used for determining the reaction rates.

There is a long list of radioactive isotopes, which can be observed alive in the solar system. The explanation of the detected amount is a challenging test for stellar models and nucleosynthesis calculations (for details see the overview paper by Wasserburg *et al* [9]). The respective (n,γ)) rates are the crucial nuclear physics input for such calculations. So far, experimental data for these rates are almost completely missing. As examples for the many required data, this proposal focuses on the two radioactive nuclei ^{60}Fe and ^{107}Pd. ^{60}Fe: The production of ^{60}Fe in massive stars depends strongly on the very uncertain ^{59}Fe(n,γ))^{60}Fe and ^{60}Fe(n,γ))^{61}Fe cross sections. The long-lived ^{60}Fe, with $T_{1/2}$=1.5×10^6 yr has been detected in deep sea sediments [10] and is likely to originate from a nearby supernova, which triggered the formation of the solar system. This isotope can also be observed by the γ-ray satellite INTEGRAL [11] via the decay of the daughter nucleus ^{60}Co. This will be important to constrain the ^{60}Fe output of supernovae. So far, no experimental data are available for the ^{59}Fe(n,γ)^{60}Fe cross section, whereas the available theoretical predictions between 1.8 mb and 3.4 mb deviate by almost 100%. ^{107}Pd: Spectroscopic observations of elements on the surface of a dozen very old, metal-poor halo stars (see e.g. [12]) found very good agreement with the solar r-process pattern for nuclei above barium (A>138). This can be interpreted as product of an early, very robust r-process, which always operates in the same way, independent of the initial conditions. Nuclei below barium, however, are under-produced compared to the solar r-abundances. For the discussion whether a second r-process is needed to explain the observations, Silver is most interesting since its observed abundance is particularly low. This could point to a problem with the neutron capture cross sections of the related isotopes. In case of Ag, the cross sections of the two stable isotopes ^{107}Ag and ^{109}Ag are known with uncertainties of 3%. However, a large fraction of elemental Ag is produced by the radiogenic contribution of ^{107}Pd, which has a half-life of 6.5×10^6 years. ^{107}Pd was also found in several iron meteorites but the amount cannot be consistently explained with current models taking into account the correlation to other radioactive nuclei (^{60}Fe, ^{26}Al, ^{41}Ca) [1]. So far, there is only one measurement of the neutron capture cross section of ^{107}Pd [13], which was performed with a fission product Pd sample with a very low (16%) enrichment in ^{107}Pd. At present, many neutron capture experiments on radioactive samples cannot be performed because the required masses of a few hundred milligrams of isotopically enriched samples are not available. In addition, backgrounds from the activities of such samples are prohibitive as well. The advent of new, high flux neutron sources (e.g. n_TOF at CERN or the Frankfurt neutron generator at the Stern-Gerlach Zentrum, FRANZ) will allow for precision measurements on minute samples of 1016 atoms only. Therefore, we propose to produce isotopically enriched samples by ion implantation of radioactive beams. The proposed experiment will consist in the implantation of radioactive nuclei in a thin carbon backing to produce isotopically enriched samples.

6.3. β-delayed two-neutron emission

Very little is known on nucleon-nucleon correlations in the atomic nucleus. One way to study these correlations is search for correlated emission of two nucleons. The best probes from a theoretical point of view is most likely the emission of two neutrons, as the neutrons are not perturbed by the Coulomb barrier. When moving further and further away from the valley of stability, the Q values for β-decay increase more and more. Close to the drip line, β-delayed particle emission is observed. Even further away, β-delayed two-nucleon emission can be observed and studied. β-delayed two-proton emission has been observed experimentally for 9 different nuclei, but only the decay of ^{31}Ar has been studied to some extent. On the neutron-rich side, β-delayed two-neutron emission has been observed for 7 nuclei with branching ratios ranging from

1% to 10%. However, correlations between the two neutrons have not been searched for in any of these nuclei. These correlations may yield valuable information about the pairing of nucleons inside the atomic nucleus, which is not accessible otherwise. In particular, two-neutron emission has a decisive advantage over two-proton emission, which is that the Coulomb barrier does not affect the two neutrons and a possible correlation should be observable outside the nucleus. Beyond their interest for correlation studies, the decay characteristics of these nuclei are also of interest for the modeling of the astrophysical rapid-neutron caption process. To identify new two-neutron emitters and to study their decays, these isotopes have to be implanted into a catcher, which is surrounded by a high-efficiency, high-granularity neutron detection system. The observables to be measured are the half-life of the nucleus, the energy of the neutrons and in particular the angle between the two neutrons. In addition, to complete the β-decay scheme, γ-radiation should also be observed.

The known β-2n emitters are ^{11}Li, ^{17}B, ^{17}C, ^{30,31}Na, ^{32,33}Na, but many others are expected close to the neutron drip line. These isotopes can be produced e.g. by fragmentation reactions or by deep inelastic reactions. After implantation in the center of the detection set-up, the neutrons will be detected and their energy and angular correlation will be determined. High-intensity neutron-rich stable or radioactive beams at about 100-150 MeV/nucleon for fragmentation or deep inelastic reactions

To produce the aforementioned nuclides by projectile fragmentation, a high-resolution, high-acceptance fragment separator is needed to separate the exotic species from the bulk part of less exotic nuclei. At the end of this separator, a high-efficiency, high-granularity neutron set-up is needed to give access to the individual energies of the two neutrons and their emission angle.

The interpretation of the experimental data requires sophisticated theoretical models which describe coherently the nuclear structure part and the nuclear dynamics of the emission process. Most of the models available today treat only one of the two parts reasonably well. These models have to be refined and new concepts like time-dependent approaches have to be implemented. In particular, the experimental observables to study e.g. the pairing force are not very well established.

6.4. Structure beyond the neutron drip line: $^{26-28}O$

Intense beams of very neutron-rich Ne isotopes will be employed to explore the unbound isotopes of oxygen, including doubly-magic ^{28}O. In addition to their low-lying structure, n-n and 4n correlations will also be explored. The character and structure of the very neutron-rich isotopes of oxygen, $^{26-28}$O have been long standing issues in nuclear structure. Indeed, ^{28}O remains arguably the only doubly-magic system that remains to be observed. Numerous experiments performed over the last 15 years have demonstrated that ^{24}O is almost certainly the last particle stable oxygen isotope. Present research is limited by production rates to improving the ground state mass excess of ^{24}O, searching for the (unbound) first excited states of ^{23,24}O as well as the (unbound) ground state of ^{25}O - to date no bound excited states have been seen or are predicted in ^{23,24}O. Interestingly, based on the current limits set on the energy of the first 2^+ state in ^{24}O, Brown suggests that this nucleus is "doubly magic". Moreover, the same shell model calculations indicate that $^{26-28}$O are only weakly unbound. Given that the neutron-rich F isotopes are particle bound up to at least ^{31}F (^{28,30}F are unbound) - some 6 neutrons beyond ^{24}O - and that ^{28}O (N=20) lies just below the N$\sim$20 island of inversion, $^{26-28}$O represent very sensitive tests for modern nuclear structure models.

The ground and low-lying excited states of $^{26-28}$O, as reconstructed from the ^{24}O fragments and coincident neutrons are the interesting observables in this experiment. It will be performed in the following way: a high-energy beam of $^{28-30}$Ne will be reacted on a thick ($\sim$500 mg/cm^2) Be or C secondary reaction target. The unbound $^{26-28}$O will be populated via two-proton removal/"knockout" reactions ($\sigma \approx 100\mu$b). The beam velocity fragments - ^{24}O and neutrons

Journal of Physics: Conference Series **168** (2009) 012023 doi:10.1088/1742-6596/168/1/012023

- will be detected at very forward angles. The relative-energy spectra for $^{26-28}$O will be reconstructed from the measured momenta of the fragments.

$^{28-30}$Ne beams of intensities of 104 pps or more, with energies of ~100-150 MeV/nucleon will be required. Charged fragments will be detected using a large-gap, sweeper magnetic spectrometer, coupled with a position sensitive Si-CsI array - the measurement of A and Z for masses up to 30 and Z up to 10 is required. The neutrons will be detected in a high-acceptance detector array which will provide a measurement of the positions and times-of-flight. To achieve the desired statistical accuracy in a running time of 1 week, an overall (geometric and intrinsic) one-neutron detection efficiency of ~50% is required. In addition, the identification and measurement of multi-neutron events (M_n = 2-4) and the rejection of cross-talk events is essential. Improved shell-model calculations, both classical and in the continuum, of the low-lying structure of the $^{26-28}$O are needed. Ideally these calculations will take into account the results of ongoing experimental studies of the less neutron-rich O and F isotopes.

6.5. Mass of ^{78}Ni ground state

Neutron-rich, doubly magic ^{78}Ni is a key nuclide on the neutron-rich landscape. Its mass is a fundamental observable that will test model predictions.

Shell gaps in exotic nuclei are the subject of intense investigation due to their sensitivity to little understood components of the nuclear force, and their importance, for the neutron-rich nuclides, in astrophysical r-process calculations. The nuclide ^{78}Ni will provide key tests of the most up-to-date model predictions. Here we address the measurement of its mass.

The mass will be measured, as part of a systematic mass measurement program addressing the n-rich landscape. The directly produced ISOL beam of ^{78}Ni will be used, and about 20 ion/s of 50keV ^{78}Ni are needed. An advanced Penning trap system similar to the ISOLTRAP/ISOLDE set-up will be used (cf. MATS design at FAIR, or TITAN at TRIUMF).

On the other hand, from the theoretical point of view, mass model predictions are also needed.

6.6. Magnetic moments of isomers in the ^{78}Ni region

The neutron-rich, doubly magic nuclide ^{78}Ni, and neighbouring nuclides, are expected to have isomers arising from cross-shell excitations. Their magnetic moments will provide benchmark tests of shell-model predictions. Shell gaps in exotic nuclei are the subject of intense investigation due to their sensitivity to little-understood components of the nuclear force, and their importance, for the neutron-rich nuclides, in astrophysical r-process calculations. The nuclide ^{78}Ni, and its neighbours, will provide benchmark tests of the most up-to-date shell-model predictions. Here we address measurements of excited-state magnetic dipole moments.

The magnetic dipole moments will be measured by time-differential perturbed angular distributions (TDPAD) as part of a wider program to characterize isomers in the ^{78}Ni region of the neutron-rich landscape. Secondary fragmentation of a radioactive ^{81}Ga beam will be used. 100 A MeV ^{81}Ga is needed to produce, by fragmentation, 103 ion/s of ^{78}Ni, with about 10% in isomeric states. Beam pulsing, on on a μs time scale, is needed. A dipole magnet and Ge gamma-ray detectors are needed as well as theoretical shell model prediction

6.7. Charge radius of ^{78}Ni

Following up the previous section, here we address the measurement of its charge radius, as part of a wider program to determine, by laser spectroscopy, nuclear radii and moments in the ^{78}Ni region. The charge radius will be measured, as part of a systematic program addressing the n-rich landscape. The directly produced ISOL beam of ^{78}Ni will be used. and an intensity of 20 ion/s of 50 keV ^{78}Ni will be needed.

6.8. ^{44}Ti abundance as a probe of nucleosynthesis in core collapse supernovae

Core collapse supernovae are remarkable astrophysical sites, representing one of the most extreme physics laboratories in Nature. There is immense interest in attempting to elucidate the physics that drives them, and perhaps the single most important diagnostic tool at our disposal is the isotope ^{44}Ti. Unlike any other observable, it combines the specificity of isotopic (not elemental) abundance, can be observed promptly and directly (by γ-ray observing satellites), and its production can be associated with specific aspects of the core collapse mechanism. The quantitative interpretation of these observations urgently requires that several key nuclear reaction rates, such as ^{44}Ti((α,p), and ^{45}V(p, γ), must be measured. Despite there being general agreement on the mechanism by which a massive evolves and explodes, some fundamental uncertainty remains; even the best models fail to generate a robust explosion. Gamma-ray emission from the decay of newly synthesized ^{44}Ti, and ^{44}Ca excesses in pre-solar grains, provide a powerful tool for testing the dynamics used in a particular model. However, for this method to be useful, several key nuclear reactions that determine the abundance of ^{44}Ti urgently need to be determined.

Reaction rate near and within Gamow window are the important observables in this case as well as spins, parities, energies and widths of low energy resonance. Direct measurement of the ^{44}Ti(α,p)^{47}V reaction: a previously attempted measurement [14] was unable to explore the astrophysical interesting energy regime. Extrapolation of the result to lower energies suggests the ^{44}Ti(α,p)^{47}V reaction rate is significantly higher than previously expected (reducing the amount of ^{44}Ti synthesized), but too much uncertainty remains to be useful. An experiment with sensitivity at energies corresponding to those in the astrophysical environment is needed. Indirect measurements of the ^{45}V(p, γ)) reaction such as ^{45}V(p,p) and (for example) ^{45}V(d,p): extremely limited data exist up to now [15] and do not cover the relevant energy region. Shell model and isobaric analogue considerations [16] suggest several relevant states should be suitable for this technique. Beam properties necessary to perform such a test are: low energy ($\sim$0.5-2 MeV/u) beams of ^{45}V and ^{44}Ti and beam currents in excess of 107 pps . On the other hand the detection system will consist of silicon charged particle arrays, similar to the Louvain-Edinburgh Detector Array. Several targets have to be used: a CH_2 target for the ^{45}V(p,p) measurement, a deuterated polyethylene target for the ^{45}V(d,p) measurement, a thin window/windowless gas target for the ^{44}Ti(α,p) measurement, with a thickness of $\sim$1018 atoms cm^2. Hydrodynamical and network calculations are needed to allow interpretation of results as well as shell model calculations to help identify states in ^{46}Cr.

6.9. One or two neutron or well defined cluster (like α-particle) break-up

The goal of this experiment would be to study the interior part of wave functions of nucleons and clusters and to check the limits of validity of mean field and single particle concepts. The long-range and short-range correlations can vary as a consequence of the isospin dependence of the N-N interaction. An example of study of new reaction mechanisms: full kinematics reconstruction deep inelastic experiments.

Spectroscopy of light exotic nuclei has been very successful by using surface reactions such as break-up or transfer. Do spectroscopic factors just represent asymptotic properties of wave functions? Slightly deep inelastic reactions with heavy exotic projectiles can answer this question as well as show the limits of mean field concept validity.

Neutron and core energy or parallel momentum distributions and absolute break-up cross sections plus angular correlations (n-core) would be the observables to be measured. Forward peaked neutron- (or proton-) core coincidence, corresponding to impact parameters for which projectile and target, both heavy nuclei, interact at distances shorter than the strong absorption radius. (This is also a check of the transparency of the optical potential for n-rich nuclei.) Gamma rays from the projectile and possibly the multiplicity of neutrons from target de-excitation.

We need to perform an experiment that would simultaneously explore the reaction mechanism and give information on the projectile structure. This requires a target which is very well known: the best is ^{208}Pb because the neutron-target optical potential is parameterised better than any other and its excited states are well known. The ideal incident energy is around 60 A MeV. A relatively high intensity incident beam is needed. If all detectors have a good efficiency and large solid angle coverage, then even an intensity of 10^3 pps would be acceptable. With an improved set-up of the type employed in [17], for example, we should be able to undertake in triple-coincidence measurements for the following observables: i) the valence neutron in-plane and out-of-plane momentum (θ and ϕ) distributions and the neutron energy or parallel momentum distribution. In the case of a two-neutron halo nucleus, also two-neutron-core angular correlations. ii) The same three observables for the core (or its fragments); γ-rays from the core to distinguish ground state from excited states. iii) Finally the excitation energy of the target from the low-energy neutron multiplicity measured over 4π or the target γ-rays. The resolution for i) and ii) should be good enough to infer the invariant mass (excitation energy E*) of the projectile. The γ-rays from the excited core should also be measured in order to determine the E* of the projectile. The important point would be to define the reaction plane such that the position of the neutron with respect to it would be determined (azimuthal angle). Absolute cross sections should be measured as well. Data should be handled by performing an event-by-event impact parameter analysis. From the angular distribution and energy distribution of the neutron at very large impact parameters (selecting the Coulomb break-up mechanism) the valence neutron separation energy of the projectile may be estimated. For each excited state of the core (all events in coincidence with a certain γ-ray from the core), the peak positions of the core and neutron parallel momentum distributions and the excitation energy of the target via the neutron multiplicity could be used to map the total final energy distribution. For each single particle state thus identified we go down in core impact parameter mapping the wave function up to the distance at which the core does not survive the interaction with the target. This is the point at which strong correlations and clustering effects enter the game. If a single neutron breakup can still be identified in coincidence with each individual cluster, the same game reconstructing the neutron momentum distribution should be played and "spectroscopic factors" (overlaps) could be deduced for the neutron with respect to each cluster. A fairly accurate map of the correlations should thus be obtained. Well-focused, high intensity (10^5 pps), 50-80 A.MeV,^{38}Ne to ^{46}Ar or heavy Ni (depending on availability) will be needed. Experimentally such studies rely heavily on large area, high efficiency, and high-granularity neutron detector arrays, the further development of which should be envisaged for EURISOL. For example a neutron 4π (liquid scintillator) detector (like ORION) but with much lower background for the neutrons from the target and a detector like DEMON for the forward neutrons would be needed, plus γ-ray detectors. Coupling with a large gap spectrometer sweeper magnet would be highly desirable. Theoretical support needed for interpreting such experimental results consists of advanced shell model calculations, developments of break-up models including full kinematics, microscopic optical potentials for neutron rich nuclei.

6.10. Isospin dependent phase transition

The fragmentation phase transition will be quantitatively studied as a function of isospin and charge and located in the phase diagram of asymmetric nuclear matter. Scaling observables from sources differing in charge/isospin will be developed to control data selection criteria. Beam energies overcoming the fragmentation threshold will be needed.

Nuclear matter is known to present at least two major phase transitions: a transition to the quark-gluon plasma at high energy density, and a transition to a nucleonic vapour phase at a temperature of a few MeV. We want to determine quantitatively the low temperature phase diagram of nuclear matter and the characteristics of the expected liquid-gas phase transition. If multi-fragmentation experiments in the past 10 years have established approximate values for the temperature, energy and density of this phase change, its nature and order are still largely unknown, as well as its isospin dependence. Studying the transition with finite nuclei has the extra advantage of also revealing the thermodynamic anomalies, which should be associated with first order phase transitions of any finite system (negative heat capacity, negative susceptibility, negative compressibility, bimodal distributions). Such signals have already been observed with stable beams and need now to be confirmed and studied as a function of the isospin asymmetry. This physics case has strong interdisciplinary connections with atomic, molecular, and cluster physics. In the framework of EURISOL, we expect important synergies with the Limits of Stability subtask. On the theoretical level, the study of multi-fragmentation with exotic beams has also important astrophysical consequences. Indeed multi-fragmentation is a unique laboratory for the formation of inhomogeneous structures due to Coulomb frustration. Such structures have to be correctly modeled for the supernovae explosion process and the cooling dynamics of proto-neutron stars. The main observables needed to characterize the transition and locate it on the phase diagram (ρ_n,ρ_p) are the energy threshold for fragmentation, temperature measurements, inclusive and exclusive charge scaling, IMF multi-charge correlations, fluctuations and bimodality observables. All these observables individually bear intrinsic ambiguities and need to be measured at the same time on the same data set. The onset of fragmentation can be established using well-developed techniques for stable beams: identification of a fragmentation source, calorimetric measurement, fragment multiplicities and velocity correlations. The change of the fragmentation threshold with the source charge and asymmetry allows the charge and asymmetry dependence of level densities, limiting temperature and instability properties to be accessed. New physics will be searched for looking for scaling violations of fragment observables. The different phase transition signals used for stable beams will be crossed and compared with dedicated simulations to confirm and locate the transition line on the phase diagram.

Different isotopes of medium and heavy beams from neutron poor to neutron rich (e.g. ^{56}Ni $\rightarrow$ ^{74}Ni, ^{106}Sn $\rightarrow$ ^{132}Sn, ^{200}Rn $\rightarrow$^{228}Rn) in an energy range 30-100 A.MeV will be used as beams. The detection system will consist of (i) 4π and low threshold complete A and Z identification for IMF (FAZIA); (ii) 4π neutron detector; (iii) high angular resolution $\Delta\theta$<0.5 LCP and neutron arrays for correlation measurements. Extensive realistic simulations of the collision dynamics (HIPSE,MD) will be needed to support data selection criteria. Both microscopic and macroscopic statistical models (LGM, AMD, SMM, MMM) have to be employed to give quantitative predictions on phase transition observables. Dedicated statistical simulations including the experimental constraints have to be developed. A theoretical improvement of evaporation codes for side feeding corrections will be needed.

7. Conclusions

In this paper we have presented some of the results of the first four years of Design Study for the proposed next generation European facility for unstable beam, EURISOL. It is the hope of our community that such a project will be approved and finally realized for the progress of Nuclear Science.

Acknowledgments

Most of the material presented in this paper is contained in documents that can be found at www.eurisol.org. We thank all colleagues involved and therein mentioned for their precious work.

Journal of Physics: Conference Series **168** (2009) 012023 doi:10.1088/1742-6596/168/1/012023

References

[1] Bennet R *et al* April 2000 *Radioactive Nuclear Beam Facilities* NuPECC Report
[2] Woods P J and Davids C N 1997 *Ann. Rev. Nucl. Part. Sci.* **47** 541
[3] Otsuka T *et al* 2001 *Phys. Rev. Lett.* **87** 082502
[4] Dobaczewski J *et al* 1994 *Phys. Rev. Lett.* **72** 981
[5] Lane A M 1964 *Nuclear Theory* (Benjamin, New York)
[6] Dobaczewski J *et al* 1996 *Phys. Rev.* C **53** 2809
[7] Dobaczewski J *et al* 1984 *Nucl. Phys.* A **422** 103
[8] Jonson B and Nyman G 1996 in *Nuclear Decay Modes* ed. Poenaru D N (IOP publishing)
[9] Wasserburg G J, Busso M, Gallino R and Nollett K M 2006 *Nucl. Phys.* A **777** 5
[10] Knie K , Korschinek G , Faestermann T, Dorfi E A, Rugel G and Wallner A 2004 *Phys. Rev. Lett.* **93** 171103
[11] Harris M J *et al* 2005 *Astron. Astrophys.* **433** L49
[12] Sneden C *et al* 2003 *Ap. J.* **591** 936
[13] Macklin R L 1985 *Nuc. Sci. Eng.* **89** 79
[14] Sonzogni A A *et al* 2000 *Phys. Rev. Lett.* **84** 1651
[15] Ahn J K *et al* 2001 *Phys. Rev. Lett.* **87** 132504
[16] Horoi M , Jora R, Zelevinsky V, Murphy A St J, Boyd R N and Rauscher T 2002 *Phys. Rev.* C **66** 015801
[17] Périer Y, Lott B, Galin J, Liénard E, Morjean M, Orr N A , Péghaire A, Quednau B M and Villari A C C 1999 *Phys. Lett.* B **459** 55

Journal of Physics: Conference Series **168** (2009) 012024 doi:10.1088/1742-6596/168/1/012024

Generation IV nuclear energy systems and the need of accurate nuclear data

N Colonna

Istituto Nazionale di Fisica Nucleare, Sezione di Bari, Via Orabona 4, 70126 Bari, Italy

E-mail: nicola.colonna@ba.infn.it

Abstract. To satisfy the world's demand of energy, constantly increasing over the years, a suitable mix of different energy sources has to be envisaged. In this scenario, an important role may be played by nuclear energy, provided that major safety, waste and proliferation issues affecting current nuclear reactors are satisfactorily addressed. In this respect, a large effort is under way since a few years towards the development of advanced nuclear systems that would use more efficiently the uranium resources, and produce a minimal amount of long-lived nuclear waste. The main activity concerns Generation IV reactors, with full or partial waste recycling capability. Their design requires R&D in numerous fields. Among the different needs, it is of fundamental importance to improve the knowledge of basic nuclear data, such as cross-sections for neutron-induced reactions on actinides. The main characteristics and principle of operation of the new generation nuclear systems are here described, together with the related needs of new and accurate nuclear data. Finally, an example of activity currently undergoing in the field is shown, with the recent experimental results obtained at the neutron facility n_TOF at CERN.

1. Introduction

The constant increase in the world population, as well as a global improvement in the life standards, is resulting in a continuously growing demand of energy. Figure 1 shows the world population (left panel) and the primary energy consumption (right panel) over the last 25 years [1]. A comparison with the expectations based on the per-capita consumption of 1980 (green curve) shows that since the beginning of the new millennium, the energy needs are increasing at a faster rate than the world population, mostly due to the economic growth in highly populated areas like Brasil, Russia, India and China. Based on these considerations, the demand of energy is going to continue to raise in the future, and by 2030 it is foreseen to increase by 50 % relative to the current value.

Figure 2 shows the current share of different fuels in the total primary energy supply (TPES), which includes transportation (left panel), and in the electricity production (right panel). At present, a large fraction of the energy needs is supplied by fossil fuels, In fact, oil, coal and natural gas contribute with more than 85 % to the TPES and with 75 % to the electricity production. Among other sources, hydroelectric and nuclear energy significantly contribute to electricity production, while the share of other renewable sources (solar, wind, tide) is around 2 % in electricity production, and almost negligible in the total energy supply. It is clear that the current situation cannot be sustained much longer. The use of fossil fuels, in fact, poses severe environmental problems, due to the large production of CO_2, the main cause of the greenhouse effect and related climatic changes, and of other pollutants. Furthermore, the extraction of fossil

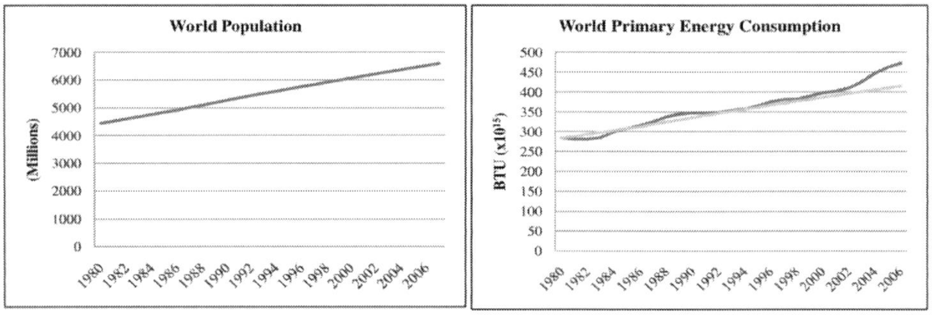

Figure 1. (Color online) Increase of the world's population (left panel), and of the global energy consumption (right panel), over the last 25 years. The actual energy needs (in British Thermal Units), shown by the blue curve, are compared with the trend estimated on the basis of the world's population, assuming a constant per-capita value of 1980 (green curve). The effect of the growth of the BRIC countries in the last few years is evident.

fuels is expected to reach a peak in a few years, with a consequent increase in the costs and uncertainties in the availability of this source.

It is therefore becoming urgent to find alternative energy supplies, that are safe, clean and possibly economic. In the short and medium term, a more efficient use of the energy resources may partially counterbalance the continuous growth of the energy demand. Furthermore, the R&D on renewable energy sources, in particular from solar and wind, may lead to an increased contribution from these sources. However, it is generally believed that an important role in the mix of energy sources of the future could be played by nuclear energy. The contribution of this energy supply may in principle become important even in the field of transportation, if large-scale production of hydrogen by nuclear energy is implemented.

As of today, 439 nuclear power plants are operating, mostly in countries members of the Organization for Economic Co-operation and Development (OECD), contributing by approximately 6 % to the total energy consumption and by 15 % to the electricity production alone. Most of the currently operating reactors are Generation II reactors. Generation III reactors, with enhanced safety and efficiency, are already available while a more advanced class (the so-called Generation III+) are expected to become available in a few years. Most of the world's currently operating nuclear power plants, which started operation in the 70's and were originally certified for a 40-year operation, are expected to apply for a 20-year license extension. As a consequence, by 2030 many of them will be at or near the end of their life and will need to be replaced. Research is therefore now being carried out with the objective of developing future-generation nuclear energy systems and making them available for international deployment by 2030. An international cooperation, the Generation IV Forum [2], was set-up a few years ago aiming at defining the technology roadmap towards the development of advanced nuclear

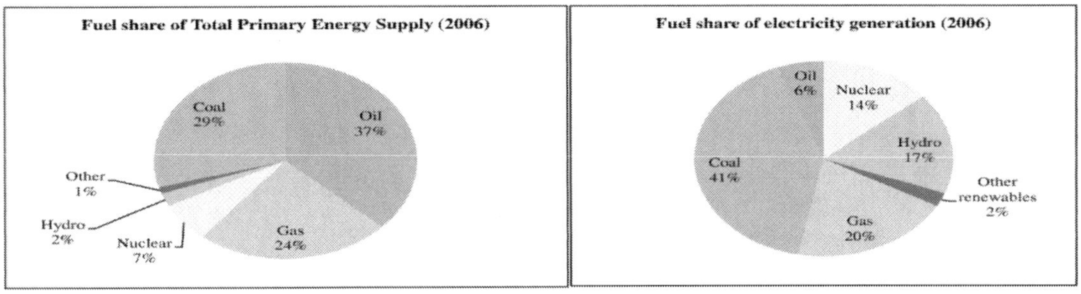

Figure 2. The share of different energy sources to the overall energy consumption (left panel) and to the electricity production. The predominant role of fossil fuels in transportation could be reduced by using nuclear energy to produce hydrogen.

systems that would address safety, waste, economic and proliferation concerns. R&D is required in many areas, such as fuel cycle, fuel preparation and reprocessing, new materials for structural elements and for monitoring systems, risk assessment, etc.. Among the request, an important part concerns improvements in the knowledge of basic nuclear data involved in the reactor's physics. In this talk, the main features of the new generation reactors and the need of accurate nuclear data, mostly on cross-sections for neutron-induced reactions, will be discussed.

2. Current issues in nuclear energy production

The present nuclear reactors are based on the so-called "once-through" process, in which the U fuel, after preparation, goes through the reactor core only once. After the irradiation period, ranging from 8 to 13 months in most reactors, the spent fuel is removed and stored (either in temporary or permanent repositories). The efficiency of the "once-through" cycle, in terms of uranium resources utilization, is quite low. Most importantly, it generates large quantities of nuclear waste for which a suitable repository has to be found.

The main concerns that currently limit a larger use of nuclear energy are related to safety issues in current power plants, to the production of large volumes of radioactive waste to be handled and eventually stored in geological repositories, and to the so-called proliferation issue, that is the possibility to divert part of the fresh or spent fuel for military or terroristic purposes. On a longer time-scale, the issue of the availability of uranium resources may also become a limiting factor in the use of nuclear energy. Estimates of the available resources indicate that, at the current rate, the availability of Uranium may become a problem in 50 to 100 years, but a possible increase in the share of nuclear energy in the near future may lead to availability problems on a shorter time-scale.

Table 1. Main components of the nuclear waste produced in current nuclear reactors, their half-life, and the total amount produced in France.

Component	Isotope	Half-life(years)	Quantity(kg/year)
Fission Fragments	^{135}Cs	2.3× 10^6	400
(39 ton/year)	^{99}Tc	2.1× 10^5	1000
	^{93}Zr	1.5× 10^6	900
	^{129}I	1.0× 10^7	200
	^{107}Pd	6.5× 10^6	250
Plutonioum	^{238}Pu	88	190
(11.4 ton/year)	^{239}Pu	2.4× 10^4	6500
	^{240}Pu	6.5× 10^3	2500
Minor Actinides	^{237}Np	2.1× 10^6	480
(1.1 ton/year)	^{241}Am	430	250
	^{243}Am	7.4× 10^3	140
	^{245}Cm	8.5× $1-^3$	1

2.1. The problem of nuclear waste

The management of the high-level radioactive waste arising from nuclear power production and from the dismantling of nuclear weapons is one of the major public environmental concern in the use of nuclear energy. Together with fission fragments, several transuranic isotopes are built up in a nuclear reactor, as a result of a chain of neutron captures and successive β-decays, starting at ^{238}U and, to a lesser extent, at ^{235}U. Many of these isotopes are α-emitters, characterized by a very long lifetime. They include Pu, Np, Am and Cm isotopes. A list of the yearly production of high-level radioactive nuclear waste in France is reported in table 1. The most important contribution, in terms of long-term radiation hazard, comes from ^{237}Np, ^{241,243}Am and ^{245}Cm, often referred to as "minor actinides".

Contrary to fission fragments, which loose most of their radiotoxicity within a few hundred years, actinides remain potentially dangerous for thousands of years, a period too long for the resistance of man-made barriers. Therefore, their disposal requires deep underground geologically stable repositories. The volume of nuclear waste that needs to be stored, however, is so large that currently considered repositories are quickly becoming insufficient, and at the current rate of nuclear power production, new geological sites with large storage capacity would need to be located and prepared every few years.

Typically, the spent fuel of Pressurized Water Reactors (the most common type of current nuclear power plants) is constituted for 96 % by Uranium, 1 % Plutonium and 3 % Minor Actinides and fission products. In some countries, like France, Great Britain and Japain, the spent fuel is reprocessed to extract U and Pu that can be reutilized. The process reduces the final volume of nuclear waste to be stored, but at present is still costly. For this reason, in other countries (in particular the US) the spent fuel unloaded from the reactor core is stored directly.

A possible solution to the problem of nuclear waste could come from transmutation processes in which long-lived radioactive isotopes are transformed into short-lived or stable ones. The most effective nuclear process is neutron-induced reactions. In particular, long-lived fission fragments (LLFF) can be incinerated by means of neutron capture reactions. A typical example is the case of ^{99}Tc, a fission fragment with half-life of two hundred thousand years. Following neutron capture, ^{100}Tc is formed, which has the very short half-life of 16 s, decaying in the stable ^{100}Ru isotope. For transuranic elements Np, Pu, Am and Cm, neutron-induced fission is in principle

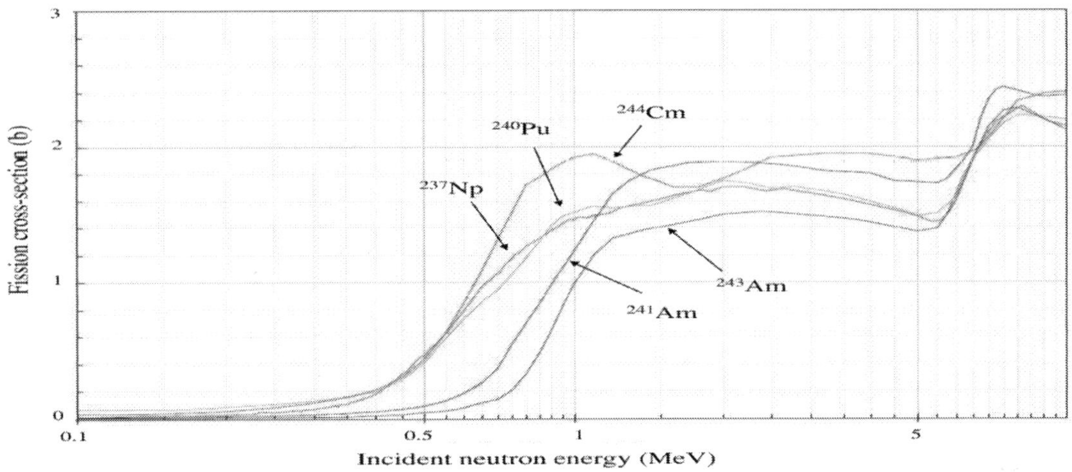

Figure 3. Fission cross-section as a function of the neutron energy for long-lived minor actinides present in nuclear waste. A recycling of these isotopes in the fuel cycle can only be achieved in advanced nuclear systems with a fast neutron spectrum.

a much more efficient process, since it leads also to energy gain and a surplus of neutrons. However, as shown in the figure 3, the fission cross-section of many actinides is characterized by a threshold, around few hundreds keV, which hinders the possibility to use conventional thermal reactors for their transmutation. The possibility to incinerate the high-radiotoxicity component of the nuclear waste is therefore linked to the availability of a fast neutron spectrum.

Several possibilities are being considered for nuclear waste incineration. One of them is the use of subcritical systems driven by accelerators (the so-called Accelerator Driven Systems). In this scheme, the excess neutrons needed to sustain the chain reaction are produced the spallation of high energy proton beams on heavy material (such as Pb). Another possibility is the use of the so-called fusion-fission transmuters, in which the intense neutron fluxes needed for transmutation are produced in fusion systems. Most of the efforts in the field of advanced nuclear systems are however now being devoted towards the development of Generation IV nuclear reactors.

3. Advanced nuclear reactors

The main challenge in the design of advanced nuclear systems is the possibility to recycle a large fraction of the long-lived actinides produced during operation of the reactor. The use of fuel mixtures that include part of the waste would lead to the double advantage of burning (or transmuting) highly radiotoxic isotopes while producing energy. Contrary to once-through process, therefore, recycling would allow on the one hand to reduce the volume of nuclear waste to be permanently stored in geological sites, and on the other hand to better exploit the uranium resources, thus ensuring a long-term sustainability of the nuclear energy option. In the following, the basic principles of Generation IV nuclear reactors are described.

Table 2. A list of the six types of Generation IV nuclear reactors currently being considered for development. The main features of the reactors are also indicated.

Reactor type	Acronym	Neutron spectrum	Main use	Size
Gas-cooled Fast Reactor	GFR	Fast	Electricity & Hydrogen	Medium
Lead-cooled Fast Reactor	LFR	Fast	Electricity & Hydrogen	Large or small
Sodium-cooled Fast Reactor	SFR	Fast	Electricity	Large or medium
Molten Salt reactor	MSR	Thermal	Electricity	Large
			& Hydrogen	Large
Supercritical-Water-Cooled Reactor	SWCR	Thermal	Electricity	Large
Very-High-Temperature Reactor	VHTR	Thermal	Hydrogen	Medium

3.1. Generation IV reactors

The perspectives of a larger use of nuclear energy to fulfil the energy needs of the futures rely on the design of new systems that would solve the issues of safety, economics, waste management and proliferation resistance affecting current generation reactors. Schematically, Generation IV reactors should present the following characteristics: (i) *Sustainability*: the new nuclear energy systems should be characterized by efficient fuel utilization, so to ensure long-term availability of the U resources; (ii) *Safety*: the new systems should have a very low probability of incident and core damage, by relying as much as possible on passive safety systems; furthermore, solution maximizing the physical protection against acts of terrorism should be adopted in their design; (iii) *Environmental impact*: the new systems should minimize the production of nuclear waste and related long-term impact on the environment; (iv) *Economics*: Generation IV nuclear systems should have costs and financial risks comparable to other energy sources; (v) *Non-proliferation*: with the new systems it should become very difficult the diversion or theft of material for military use or for terroristic purposes.

It should be emphasized that the use of Generation IV nuclear systems for energy production would make easier to meet clean-air objectives since, as in current reactors, the emission of CO_2, either in the direct or indirect processes related to fuel extraction and preparation, is much smaller than for other energy sources.

At present, six types of reactors are being considered. They are listed in table 2. Three of them are fast reactors, which are able to burn minor actinides, and that could be eventually loaded with the nuclear waste produced by thermal reactors (including the present ones). The other three types are thermal reactors, characterized by a much improved burn-up and thermal efficiency. An important aspect regards the possibility to build the new reactors in various sizes. While large, monolithic reactors in principle are more convenient due to the scale economy, recent studies have shown that small and medium size reactors could represent a more desirable solution in several cases, for example for countries with small electric grids (< 10 GW), or with a limited investment capacity. Small reactors may also be a solution for producing electricity in remote locations, such as the Poles or islands, or in places far from the electric grid. Finally, small or medium size reactors may allow to minimize the investment risks typically associated with the large capitals needed for the construction of large nuclear power plants.

The main challenge in the development of the Generation IV reactors is to ensure a minimal production of long-lived nuclear wastes and a very efficient utilization of the uranium resources.

For this reason, the basic requirement in the design of some reactors is the possibility of reprocessing and recycling the fuel, which could include together with uranium and plutonium, all other actinides as well. It is important to stress that the new systems would allow the transmutation of the long-lived minor actinides Np, Am and Cm, with a consequent reduction of their inventory. In fact, the main objective of the three fast reactor is electricity production and waste management. Another objective of some Generation IV reactors is the possibility to produce hydrogen by exploiting high temperature processes. In particular, one of the Generation IV reactors chosen for R&D, the Very High Temperature Reactor, is expected to operate with a coolant outlet temperature of 1000^0C, which would enable high efficiency thermochemical water-splitting without carbon emission.

Considering the different goals of future generation reactors, it is unlikely that they can be met by one or two of the designs now being considered for development. More realistic, the objectives of the Generation IV systems will be achieved by employing a combination of the different reactors types. In particular, while all six types of Generation IV reactors are characterized by a higher thermal and burnup efficiency, relative to current systems, efficient transmutation of long-lived waste can only be performed with the three fast reactor (LFR, GFR and SFR), whose operation should therefore allow to reduce inventory of minor-actinides produced by the thermal reactors based on the once-through process. An appropriate mixing of the different Generation IV reactors may therefore be needed to meet in a sustainable way the future nuclear energy demand.

The time-scale towards the development of Generation IV reactors, as envisaged by the Generation IV Forum, will proceed in three phases. The next few years will be devoted to the so-called "viability phase", i.e. to the feasibility study and to the proof-of-principle of the new reactor concepts. This phase is expected to be completed for all six types of new reactors by 2015, although for some reactors (like the Sodium-Cooled Fast Reactor), this first phase may be completed in a shorter time-scale, due to sufficient experience already gained in the past. The second phase regards the study of the performance of the new systems. In this phase, the various parts of the new systems, including fuel recycling, will be developed and optimized. This phase will end up when it can be shown that the performance of the various systems is good enough, so to allow implementation on an industrial scale. Presumably, this phase may be completed by 2020. At that point the third and final phase, of the demonstration, will take place, with the aim to proof the capability of the new and innovative technology to be exploited on the large scale for energy production and/or for other predicted uses. If the demonstration phase is successful, the new systems may be ready for commercialization by industries. As already mentioned, it is desirable that this would occur by 2030, when a large fraction of the current reactors will be about to be shut down.

3.2. Accelerator Driven Systems

Another innovative systems currently being considered, mainly for nuclear waste incineration (via transmutation reactions) are the so-called Accelerator Driven Systems. They refer to a device obtained by combining a sub-critical nuclear reactor with a proton or deuteron accelerator that provides, through the spallation process on a heavy material, the neutron excess necessary to sustain the chain reaction. The main advantage of such a device resides in the fact that it is intrinsically safe, since the chain reaction cannot by itself run out of control (being the neutrons that sustain it produced with an external source that can be easily switched off). Furthermore, the use of high energy primary beams could be advantageous in tailoring the neutron energy spectrum so to match the fission cross-section of minor actinides, thus maximizing their transmutation rate.

Most of the proposed designs of Accelerator Driven Systems assume continuous-wave proton beams with energy $\sim$1 GeV, and rely on Pb as spallation target and, in some cases, as reactor

Journal of Physics: Conference Series **168** (2009) 012024 doi:10.1088/1742-6596/168/1/012024

coolant at the same time. The main objectives of these systems could be to burn transuranium and minor actinides produced by conventional thermal reactors, although part of the energy released in the waste burnup could be used to produce electricity for powering the accelerator.

Several technological issues have to be solved towards the development of Accelerator Driven Systems. In particular, the accelerator should be able to deliver high beam currents, steadily and with only minor interruptions. At present, the accelerator technology is still far from reaching the required performances, although progresses are being made in this respect.

3.3. The Th/U fuel cycle

A final remark regards the possibility to develop new reactors based on innovative fuel cycles. In particular, a topic of great interest is the use of the Th/U fuel cycle in either critical or subcritical systems. This cycle is based on the fertile ^{232}Th isotope, which breeds the fissile ^{233}U by neutron capture and subsequent β-decay of ^{233}Th and ^{233}Pa, according to the following scheme:

$$^{232}\mathrm{Th}(\mathrm{n},\gamma)^{233}\mathrm{Th} \rightarrow ^{233}\mathrm{Pa} \rightarrow ^{233}\mathrm{U}\ , \tag{1}$$

where the first steps has a half-life of 22.3 minutes, and the second one of 27 days. An interesting advantage in using this cycle, as compared to the conventional uranium/plutonium cycle currently used in all operating power plants, is related to the low production of high-mass actinides, due to the absence of ^{238}U which is the seed for the production of most actinides in conventional reactors. As a consequence of the much lower buildup of heavy transuranium isotopes, the radiotoxicity of the spent fuel in reactors based on the Th/U fuel cycle is significantly reduced. Another advantage of this cycle is the large availability of the fuel. In fact, thorium is 3 to 4 times more abundant than uranium in the earth crust, and is widely distributed in nature as an easy mining resource in many geographical areas. Finally, natural thorium is entirely constituted by the fertile ^{232}Th isotope, so that there is no need for isotopic separation in the preparation of the fertile component of the fuel.

4. The need of nuclear data for advanced nuclear systems

In order to reduce uncertainties in the design and operation of new generation reactors, high precision data on the cross-section for neutron-induced reactions on a variety of isotopes are required, from thermal energy to several tens of MeV. In particular, a pressing need exists for new measurements on capture and fission reactions for the main isotopes involved in the Th/U fuel cycle, as well as for long-lived Pu, Np, Am and Cm isotopes. Finally, data are required also for long-lived fission fragments involved in transmutation projects, and for structural material. A list of the requests is contained in the "Nuclear Data High Priority Request List" [3], from the Nuclear Energy Agency, constantly updated with new requests coming from different fields, such as current and advanced reactor systems, various industrial applications, nuclear medicine, radioprotection, space applications, etc..

Neutron cross-sections are available by means of compilations of experimental data [4] or through evaluated nuclear data libraries [5], such as ENDF (mostly compiled in the United States), JEFF (the European library maintained by the NEA), JENDL (the Japanese database) and BROND (the Russian one). Together with the evaluated cross-section, these libraries include information on the reliability of the data, which is mostly linked to the accuracy of the available experimental results. Evaluated data and related uncertainties are the result of a large effort in which experimental data are combined with nuclear model calculations, often based on sophisticated reaction theories.

The present knowledge of the neutron cross-section results largely inadequate for the new applications in the field of emerging nuclear technologies, and they need to be updated with new experimental or theoretical information. Sensitivity studies performed by means of Monte

Table 3. Current and needed accuracy (in %) on neutron cross-sections required for the design of Generation IV fast reactors [6].

Isotopo	Reaction	Energy range	Current accuracy	Needed accuracy
^{238}Pu	fiss	0.2–1.4 MeV	17	3–5
^{239}Pu	capt	2–500 keV	7–15	4–7
^{241}Pu	fiss	450 eV–1.4 MeV	8–20	2–5
^{242}Pu	fiss	0.5–2.2 MeV	19–21	3–5
^{241}Am	fiss	2.2–6 MeV	9	2
^{243}Am	fiss	0.5–6 MeV	12	3
^{244}Cm	fiss	0.5–1.4 MeV	50	5
^{245}Cm	fiss	67–183 keV	47	7

Carlo simulations indicate that for most long-lived fission fragments and minor actinides, the present uncertainties are much larger than needed for the reliable design and safe operation of new generation reactors. Table 3 lists the current and needed accuracy on the cross-section of various reactions involved in the design of advanced nuclear systems. The table applies to different types of Generation IV fast reactors, as well as for ADS, and has been obtained for different core design. The need of improving the current knowledge of nuclear data is evident. A striking example is the case of the neutron-induced fission cross-section of ^{243}Am.

In figure 4 the results of the most important measurements performed up to date on this reaction are shown (coloured symbols). The available data cluster in two groups, with a discrepancy between them of around 15 %. The current evaluations follow the lower cluster. However, a large uncertainty still persists, with the most recent experimental results (collected in 2004) indicating a cross-section higher than all previous results. New and accurate measurements are therefore needed to settle this question.

Following the sensitivity analysis performed for Generation IV fast reactors, as well as for ADS and other projects of nuclear waste incineration [7], a long list of needed data can be made. Capture and fission cross-sections are needed for fertile and fissile isotopes involved in the Th-cycle, in particular ^{232}Th, ^{231}Pa, ^{233}U, ^{234}U and ^{236}U. Similarly, reliable experimental data on capture, fission and inelastic cross-sections, in a wide energy range are required for transuranic isotopes, in particular ^{237}Np, ^{238,240,241}Pu, ^{241,243}Am and ^{244,245}Cm. Furthermore, the incineration scheme of long lived fission products requires accurate capture cross-sections for ^{79}Se, ^{99}Tc, ^{129}I, ^{135}Cs, ^{151}Sm, etc.. Finally, data on structural material being considered as neutron-production target or as coolant, are still far from being accurately known, especially at high neutron energy. New measurements on stable isotopes are thus called for.

5. The n_TOF contribution

The study of neutron-induced reactions has always been of great importance for a large variety of fields, from Nuclear Astrophysics, where neutron capture processes are at the bases of Stellar Nucleosynthesis of heavy elements, to fundamental Nuclear Physics for studies on, for example, neutron-induced symmetry breaking and nuclear level densities. Finally, neutron cross-section data are fundamental for applications to nuclear technology. In recent years, studies performed towards the development of advanced nuclear systems have shown that for many nuclides, current nuclear data are characterized by an insufficient accuracy [7]. The necessary update of the neutron cross-section libraries rely on the availability of new and accurate measurements that can only be performed at advanced neutron sources.

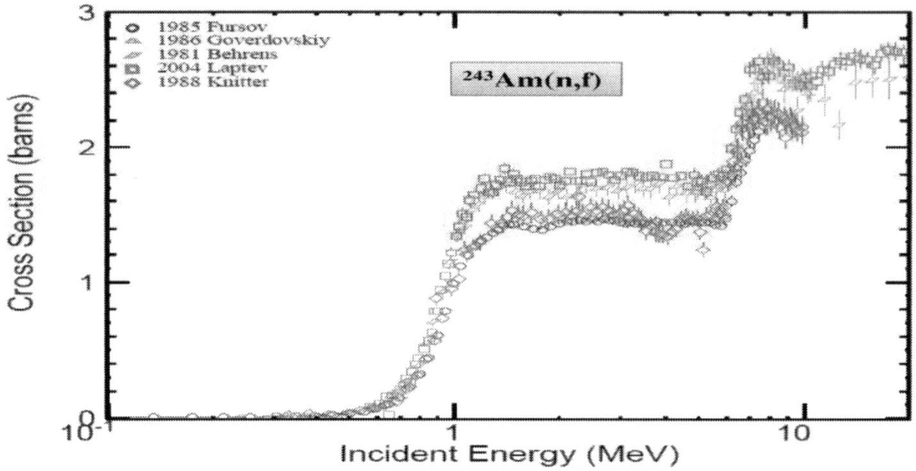

Figure 4. The neutron-induced fission cross-section of ^{243}Am. Data from the most recent measurements tend to cluster in two groups, with a discrepancy of 15 %.

With the aim of fulfilling some of the requests of nuclear data for advanced reactor systems, as well as for collecting fresh new data of interest for Nuclear Astrophysics, an innovative neutron facility was built at CERN: n_TOF [8]. Proposed by Rubbia, n_TOF is a time-of-flight installation based on a spallation neutron source. The neutron beam is produced by 20 GeV/c protons from the CERN Proton Syncrotron accelerator, impinging onto a large Pb block, surrounded by a water layer acting both as coolant and moderator of the neutron spectrum. A scheme of the facility is shown in figure 5. The main features of the neutron beam are the very high instantaneous neutron flux and the wide energy spectrum, covering over 9 orders of magnitude, from thermal to approximately 1 GeV neutron energy, as shown in figure 6. Another advantage over existing neutron facilities is the high resolution in neutron energy (reconstructed from the time-of-flight), thanks to the 200 m long flight path used. Several detection systems have been setup at n_TOF. Among them, a 4π Total Absorption Calorimeter made of 40 BaF_2 crystals was built for measuring capture reactions, while fission measurements are carried out with a Fast Ionization Chamber and a stack of Parallel Plate Avalanche Counter. Finally, an innovative Data Acquisition System, based on fast Flash ADC, has been adopted for the first time at a neutron facility, in order to handle the very high count-rate produced at n_TOF as a consequence of the intense neutron beam and of the natural radioactivity of the samples.

5.1. Recent results

Together with the innovative features of the n_TOF facility, the high-performance experimental setups and state-of-the-art acquisition system allow to collect, in some cases for the first time, accurate neutron cross-section data even on highly radioactive samples or on isotopes available in small quantity. In the first experimental campaign, data have been collected on several reactions of interest for the Th/U fuel cycle, for Generation IV fast reactors, and for Accelerator Driven Systems aimed mostly at nuclear waste transmutation. A list of the cross-sections measured at n_TOF can be found in [9]. In most cases, the accuracy of 3 to 5 %, required for the

Journal of Physics: Conference Series 168 (2009) 012024 doi:10.1088/1742-6596/168/1/012024

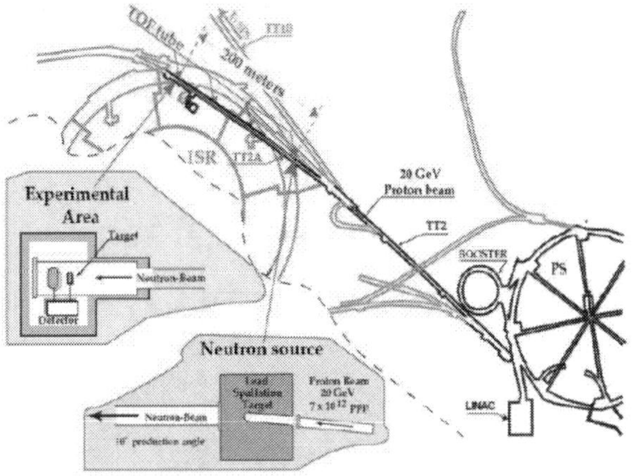

Figure 5. A schematic map of the n_TOF facility at CERN. The proton beam from the PS accelerator complex impinges onto a massive Lead block, located in the TT2 tunnel. A water layer acts surrounding the block acts both as coolant and moderator. Neutrons produced in the spallation process propagate in a vacuum tube. Several collimators, shielding walls and a magnet are position along the flight path. At the end of the 200 m line, the experimental area hosts different experimental apparata.

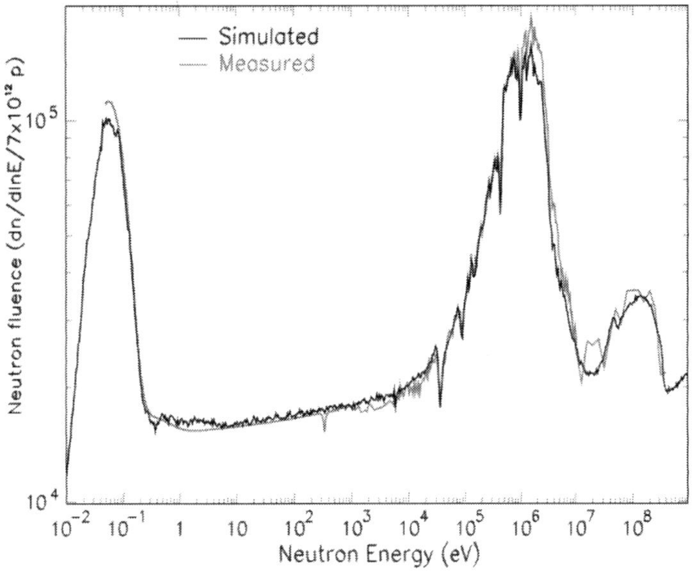

Figure 6. Simulated and measured neutron fluence per proton pulse, available for measurements in the experimental area of the n_TOF facility at CERN.

applications to emerging nuclear technologies, has been reached in the n_TOF data. Accurate capture cross-sections have been obtained in particular for the ^{232}Th(n,γ) [10], as shown in

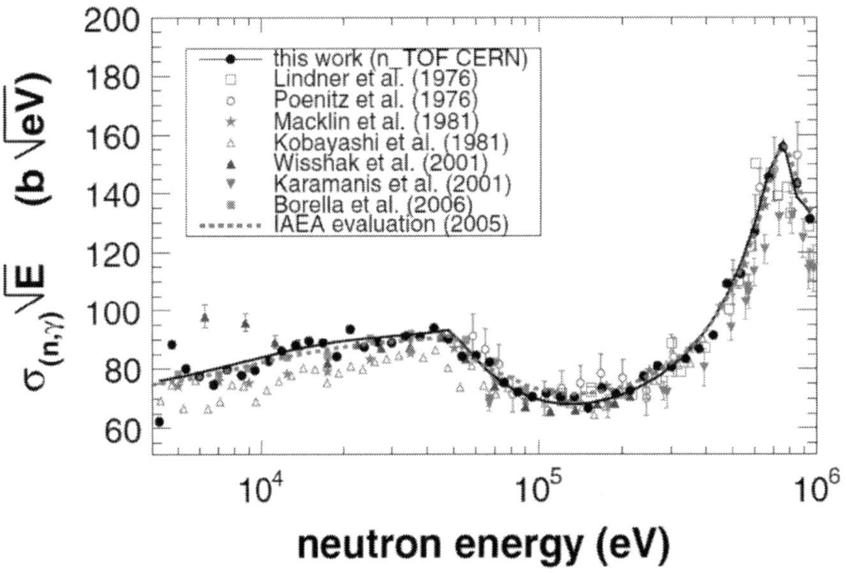

Figure 7. Simulated and measured neutron fluence per proton pulse, available for measurements in the experimental area of the n_TOF facility at CERN.

figure 7, and ^{233}U(n,f) [11], the two most important reactions involved in the Th/U fuel. For the same application, the fission cross-section of ^{232}Th and ^{234}U have been also measured and are now in the process of being released.

Similarly, high accuracy data have been collected on the capture cross-sections for some of the most important transuranic isotopes involved in transmutation processes, in particular ^{237}Np, ^{240}Pu and ^{243}Am [12]. Finally, accurate neutron-induced fission cross-section measurements have been performed for the minor actinides ^{241}Am, ^{243}Am and ^{245}Cm [13, 14]. As shown earlier in this publication, the current accuracy on these isotopes is much poorer than requested for the reliable design of Generation IV nuclear energy systems. The n_TOF results, now in the process of being published, have shown severe shortcomings in current databases of evaluated neutron cross-section, and have demonstrated their inadequacy for the new nuclear science applications. In most cases, the n_TOF data, combined with advances in nuclear theories for fission, will provide the basis for new, more reliable evaluation.

Most of the results obtained by the n_TOF Collaboration in the first experimental campaign have been published and uploaded onto the experimental database EXFOR, making them available to the nuclear data community, and for the needs of new applications. A second experimental campaign, with measurements of interest for applied nuclear science as well as for Nuclear Astrophysics and fundamental Nuclear Physics, is now starting and will hopefully produce more interesting results on studies of neutron-induced reactions.

6. The surrogate reaction method

The n_TOF results represent an example of advances in the knowledge of basic nuclear data for applications, that can be obtained by improvements in the neutron facilities, experimental setups and techniques. In particular, the high instantaneous neutron flux of the n_TOF neutron beam has allowed in the past, and may lead in the future, to measure highly radioactive samples, even with activity of the order of hundreds of MBq. However, some measurements cannot at present be performed neither at n_TOF nor in other existing facilities. This is, for example, the case of neutron-induced reactions on ^{233}Pa, an isotope of relevance for the Th/U fuel cycle. Due to the relatively short lifetime of 27 days, it is almost impossible to produce and handle samples of this isotope. Furthermore, in the cases of minor actinides, the natural radioactivity of the sample would completely mask the signals coming from the neutron-induced reaction.

A new experimental technique is being employed, since a few years, to collect nuclear data that would otherwise be impossible or very difficult to obtain. The technique allows to determine the cross-section for neutron-induced reactions through an indirect method, also called surrogate method, proposed in the 70's [15]. The method exploits alternative reactions (for example few-nucleon transfer reaction) that lead to a compound nucleus with the same A and Z of the one formed in the neutron-induced reaction. The decay probability of the CN (for example in the fission or radiative de-excitation channel) measured in the surrogate reaction is then combined with the compound nucleus cross-section for the neutron-induced reaction, obtained from optical model calculations. The product of the two represents the cross-section for the desired neutron-induced reactions, provided that the decay widths is dominated by the statistical level density, and the angular momentum of the CN formed in the surrogate reactions is not much larger than the spin-cutoff parameter of the level density distribution.

In the last few years different measurements of surrogate reactions have been performed, allowing to obtain information on neutron cross-sections for difficult cases. In the work of Petit *et al* [16], the mentioned ^{233}Pa(n,f) cross-section, very difficult to measure directly or reliably estimate from theoretical calculations, was determined from the surrogate reaction ^{232}Th(^{3}He,p)^{234}Pa. Interesting results were also obtained for the fission cross-sections of minor actinides ^{242,243}Cm and ^{241}Am [17].

7. Conclusions

The fast growth in the global energy demand, consequence of the increase in the world's population as well as by improvements in the life standards in many previously underdeveloped countries, will pose severe problems in the future. The current situations of energy production, which relies heavily on the use of fossil fuels, cannot be sustained for long, due to the predicted decline in the availability of this energy source and, most importantly, due to the severe environmental impact of fossil fuels, in particular related to the production of CO_2 and associated climatic changes. Therefore, safe, clean and cost-effective energy sources have to be developed, in a timely fashion. While energy savings and developments in the use of renewable sources may help reducing the dependence on fossil fuels, in particular in the short and medium term, there is little doubt that an important role will have to be played by the nuclear energy, even in the field of transportation, by large-scale production of hydrogen. The public acceptance of nuclear energy, however, is not very high, at present, mostly due to the safety, proliferation and, especially, waste management issues associated with current reactors. A greater importance of the nuclear energy source requires the development of new generation systems addressing present concerns. In this respect, a new class of reactors, the so-called Generation IV systems, are now being investigated, with the aim of making them available by 2030, when most of current nuclear reactors will have to be replaced. The main concept of the Generation IV reactors is the possibility to recycle a large fraction of the actinides produced during reactor operation, which would allow a much more efficient use of the uranium resources and, more importantly, a large reduction of the volume of long-lived nuclear waste to be stored in geological repositories.

The development of Generation IV reactors requires R&D on several field. Among them, advances in the knowledge of basic nuclear data are fundamentally needed. In particular, the design of fast reactors able to burn Pu isotopes and minor actinides (Np, Am and Cm), requires accurate neutron capture and fission cross-section data on these isotopes, as well as other information such as fission fragment distribution, neutron multiplicity, delayed neutron emission, etc. A large effort is therefore being devoted, by both the experimental and theoretical Nuclear Physics community, to update the current knowledge of basic nuclear data, and make it more suitable for the application to the field of emerging nuclear technology. In this respect, an important contribution is being provided by the n_TOF Collaboration, which is performing high-accuracy measurements at the innovative neutron beam at CERN. Some measurements with highly-radioactive isotopes will require improvements in the neutron facilities, experimental technique and instrumentation. For some very difficult isotopes, the surrogate method represents a valid alternative to direct measurements, and is now being widely employed in various laboratories around the world. A large effort is also required on the side of theoretician and evaluators, to interpret and combine the large amount of experimental data, so to produce more reliable databases.

In conclusion, in the next few years, it will be necessary, for part of the Nuclear Physics community, to dedicate large efforts to collect nuclear data with high accuracy, needed for the development of new generation energy systems. Although a huge task, it has to be done, as an important contribution to solving one of the main problems of the society, i.e the continuous quest for energy. Recalling the motto of the Generation IV Forum, it will be our contribution in preparing today for tomorrow's energy needs.

References

[1] Data taken from *Key World Energy Statistics, 2008. International Energy Agency (IAE)*, 2008 Edition
[2] See http://gif.inel.gov/roadmap and http://nuclear.gov
[3] http://www.nea.fr/html/dbdata/hprl/index.html
[4] McLane M 2005 *EXFOR Basics, Technical report, International Atomic Energy Agency*
[5] See http://www.nndc.bnl.gov or http://www-nds.iaea.org
[6] Aliberti M *et al* 2006 *Ann. Nucl. En.* **33** 700
[7] Aliberti M *et al* 2008 *NEMEA 4, Neutron Measurements, Evaluations and Application* ed A Plompen, EUR 23235 EN 99 and 127
[8] Abbondanno U *et al* 2002 *n_TOF Performance Report* CERN/INTC-0-011 INTC-2002-037
[9] http://www.cern.ch/n_tof
[10] Aerts G *et al* 2006 *Phys. Rev.* C **73** 054610
[11] Calviani M *et al* 2009 submitted to *Phys. Rev.* C
[12] Cano-Ott D *et al* 2007 *International Conference on Nuclear Data for Science and Technology, Nice, France* (EDP Science) p 627
[13] Calviani M *et al* 2007 *International Conference on Nuclear Data for Science and Technology, Nice, France* (EDP Science) p 335
[14] Calviani M *et al* 2008 *Neutron Measurements, Evaluations and Applications* ed A Plompen, EUR 23235 EN p 65
[15] Cramer J D and Britt H C 1970 *Phys. Rev.* C **2** 2350
[16] Petit M *et al* 2004 *Nucl. Phys.* A **735** 345
[17] Jurado B *et al* 2007 *International Conference on Nuclear Data for Science and Technology, Nice, France* (EDP Science) p 331

AUTHOR INDEX

Institute of Physics
Dirac House, Temple Back
Bristol BS1 6BE UK

CURRAN ASSOCIATES INC
proceedings
.com